Thomas Hobbes:

Thomas White's *De Mundo* Examined

Thomas Hobbes:
THOMAS WHITE'S
DE MUNDO EXAMINED

The Latin translated by
Harold Whitmore Jones

Bradford University Press
in association with
Crosby Lockwood Staples   London

Granada Publishing Limited
First published in Great Britain 1976 by Bradford University
Press in Association with Crosby Lockwood Staples
Frogmore St Albans Herts and 3 Upper James Street
London W1R 4BP

The original Latin text, edited by Jean Jacquot and Harold
Whitmore Jones, published in France 1973 by Librairie
Philosophique J. Vrin, Paris

This translation copyright © 1976 by Harold Whitmore Jones

All rights reserved. No part of this publication may be
reproduced, stored in a retrieval system, or transmitted,
in any form, or by any means, electronic, mechanical,
photocopying, recording or otherwise, without the prior
permission of the publishers.

ISBN 0 258 97001 4

Made and printed in Great Britain by
William Clowes & Sons, Limited
London, Beccles and Colchester

ACKNOWLEDGMENTS

Thanks are expressed to the Bibliothèque nationale, Paris, for leave to issue a translation of Manuscript fonds latin 6566A, and to the Librairie Philosophique J. Vrin for endorsing its publication, especially as regards the reproduction of the geometrical diagrams from the 1973 Paris edition.

I am grateful especially to Dr Jean Jacquot and Mme Freda Jacquot for their help and encouragement over fifteen years. Transcription of the Latin text had proceeded some way when I became aware of, and was invited to co-operate in, the work leading to the publishing of the Paris edition; I look back with pleasure on what has been a long collaboration. In addition to assisting in the proof-reading of the present volume, Dr Jacquot read and commented at length on the early versions of the translation; many turns of phrase in the present text, and a number of the notes, are his. I am also indebted to Mrs Stephana Babbage, who worked through the greater part of my draft, amended several errors, and threw light on places that had been dark to me. The following I thank for their advice in particular areas where I needed help: Dr Eric G. Forbes of the University of Edinburgh, Miss Alice Browne, and Miss Catherine Coventry. For such mistakes and shortcomings as remain I alone must accept responsibility.

H.W.J.

CONTENTS

The source manuscript carries no contents list. As explained at page 12, some chapter titles of the original are here modified.

INTRODUCTION
1

CHAPTER I
Our author denies that philosophy should be treated logically (fol. 5 ff) 23

CHAPTER II
Is the world of finite size? (fol. 7v ff) 28

CHAPTER III
Are there several worlds? Does a vacuum exist? What do rarity and density consist in? (fol. 14 ff) 39

CHAPTER IV
On the immobility of place. Where the spirits are found (fol. 18v ff)[1] 51

CHAPTER V
Is there fluid matter beyond the moon? Are there living creatures there? Is quantity the same as indivisibility? (fol. 28 ff) 55

[1] The foliation is irregular in this chapter and in Chapters 29 and 30.

CHAPTER VI
On Aristotle's demonstration that the heavens are imperishable (fol. 33 ff) 65

CHAPTER VII
Whether in the heavens there are found bodies similar to our own (fol. 45 ff) 78

CHAPTER VIII
On the generation and nature of comets (fol. 53v ff) 87

CHAPTER IX
Whether the universe contains self-illuminating heavenly bodies (fol. 65 ff) 99

CHAPTER X
The telescope incapable of further improvement? (fol. 81 ff) 114

CHAPTER XI
The earth is not a magnet? Constitution of its effluvia (fol. 97 ff) 128

CHAPTER XII
Can the indivisible naturally re-form? Whether changing a single thing means creating a different universe (fol. 107v ff) 137

CHAPTER XIII
Is there need for a world that contains every created thing other than itself? (fol. 117 ff) 147

CHAPTER XIV
Meaning of the term 'motion' (fol. 127 ff) 157

CHAPTER XV
Is the cause of a wind blowing within the torrid zone the sun? (fol. 134 ff) 164

CHAPTER XVI
On the tides (fol. 140v ff) 170

CONTENTS

CHAPTER XVII
The tides are not due to the motion of the earth?
(fol. 153 *ff*) 179

CHAPTER XVIII
That according to White the earth has no resistance
to motion (fol. 170 *ff*) 193

CHAPTER XIX
Why the earth keeps its axis parallel. The cause of the
precession of the equinoxes (fol. 190 *ff*) 212

CHAPTER XX
Does the earth's motion carry the earth's atmosphere
round with it? (fol. 202 *ff*) 222

CHAPTER XXI
That a body moves of itself or *per accidens* (fol. 214 *ff*) 233

CHAPTER XXII
The force of circular motion does not hurl off bodies
resting on a radius (fol. 232v *ff*) 249

CHAPTER XXIII
The angle of contingency is not less than every right angle
(fol. 242 *ff*) 255

CHAPTER XXIV
The sun does not directly cause the moon to move
(fol. 262 *ff*) 276

CHAPTER XXV
Astronomy is subject to error; mistakes concerning circular
motion; how the planets' motions may be assessed
(fol. 282 *ff*) 299

CHAPTER XXVI
On the Introduction to White's Third Dialogue
(fol. 286 *ff*) 304

CHAPTER XXVII
The motion of the universe is due to an external principle
(fol. 290 *ff*) 309

CHAPTER XXVIII
The world has not existed for ever (fol. 312 ff) 337

CHAPTER XXIX
Something has always been in existence (fol. 322 ff) 348

CHAPTER XXX
That *ens a se* is unique and the cause of everything else (fol. 351 ff) 361

CHAPTER XXXI
The existing world is the best of those creatable (fol. 358 ff) 390

CHAPTER XXXII
How His own goodness moved God to create the world (fol. 365 ff) 400

CHAPTER XXXIII
That God created a world which is free (fol. 370 ff) 406

CHAPTER XXXIV
Being able to create a different world does not make God changeable (fol. 376 ff) 411

CHAPTER XXXV
The basis of the Stoics' 'fate'. God is never the cause of evil (fol. 384 ff) 421

CHAPTER XXXVI
The uselessness of basing judgments upon astrology (fol. 398 ff) 435

CHAPTER XXXVII
The earth was founded for the uses of man; man enjoys free-will; free-will does not conflict with providence (fol. 406 ff) 443

CHAPTER XXXVIII
Why do human affairs seem to be ruled by chance? (fol. 422 ff) 459

CONTENTS

CHAPTER XXXIX
Why most men fall from happiness (fol. 444 *ff*) 481

CHAPTER XL
Must the world come to an end? (fol. 453 *ff*) 491

APPENDIX I
Geometrical figures in the text 499

APPENDIX II
From Kepler's *Epitome astronomiae* (relevant to folio 266) 517

INTRODUCTION

I

Manuscript fonds latin 6566A in the Bibliothèque nationale at Paris is devoted to a single subject: Thomas Hobbes's criticism of the printed book *De mundo dialogi tres* (1642) of Thomas White. Of the events leading to the composition, and to the retrieval, of Hobbes's study a detailed account will be found, together with textual data, bibliographical apparatus, and an index of names and topics, in Jean Jacquot's and the present translator's edition of the Latin text. Their *Thomas Hobbes: Critique du* De mundo *de Thomas White* (Paris: Vrin, et Centre national de la recherche scientifique, 1973) also includes supporting manuscript material not previously printed. In addition, a commentary volume is contemplated that will take in parallels between passages in the Paris manuscript and the work of thinkers other than Hobbes. The immediate task, however, is to familiarise the reader with the background against which Hobbes produced his reflections upon White, and this can be done selectively under four heads: Thomas White the man; the authenticity of the Paris manuscript; the annotations entered in it by Marin Mersenne; and the *De mundo* itself. The remainder of this narrative (Sections VI–IX) is devoted to the literary problems that arise in the translating of the text.

Thomas White (1593–1676) has so far received scant attention from modern historians of philosophy, though he numbered among his friends some of the best-known thinkers of the mid seventeenth century.[1] He was an English Catholic priest and the intimate of Sir Kenelm Digby, and the greater part of his life before the Restoration was passed abroad. He produced in all nearly forty books, the *De mundo* being third in the order of composition of those of which he was sole author. Like Hobbes, in fact (whose translation of Thucydides, his first publication, appeared in 1629), White was approaching middle age when he began to issue works on his own account. The contents of the *De mundo* we shall consider later, but a word must first be said on the place of the book in White's publications at large. These were mainly theological—we shall not here examine his career as an ecclesiastic—but some, such as the *Institutiones Peripateticae* (1646) and *Euclides metaphysicus* (1658), and also parts of others, return to topics occurring in the *De mundo*. Of the personal links between him and Hobbes, however, little is known; yet despite the sharpness of some of the latter's comments in the present work it is not improbable that the men were on good terms. Indeed, their familiarity endured into their old age, but they remained opponents in philosophy.

II

That the text under examination is Hobbes's there can be no question, even though the manuscript is a copy made by two amanuenses—both good workmen but apparently accustomed to dealing with theological material rather than with natural science. The volume specifies neither White's name nor the title of his book nor, except on its spine, its own authorship. Conceivably, though this is by no means certain, these details appeared on a title-page or other leaf which has been lost, probably when or before the manuscript was bound. (Several pages are assembled in the wrong order; notes to the present translation indicate where.) But to its recipient—if it was specifically addressed to one individual—or to those among whom the treatise circulated, perhaps unbound, it would have mattered little whether the identifications were present or not, White's book (or at least its existence) and the authorship of the present text being known to Hobbes's circle. Detailed evidence of the work's authenticity, much of it self-corroborative, appears in

---

[1] An article contributed by the present writer to *Notes and queries*, New series vol. 20, no. 10 October 1973, pp. 381–388, adds to the accounts of White in the *Dictionary of national biography* and elsewhere.

the Paris edition, but three facts may be recalled here. First, the manuscript bears the name 'Hobs' on the spine. Second, a great many of its ideas appear, some in near-identical phrasing and many at once recognisable, in Hobbes's printed books. And third, correspondence of the time shows us over what period (late 1642 and early 1643) Hobbes was engaged upon the project.[2]

Discussion of the precise period of composition of the dissertation appears in the Paris edition, but a few words may be added here. By comparison with that of the printed *De corpore* and other of Hobbes's published Latin works (with which Hobbes sought, and employed, assistance in the translating) the style of the treatise contains inelegances–particularly in sentence construction–which suggest that he did not, or could not, take the opportunity of polishing and thoroughly revising his narrative before having the scribal copy made. One or two examples of sentences that would benefit from recasting are pointed to in the notes, for example at folio 100v, and the question is taken up in Sections VII and VIII below. As we shall see with respect to White, some obscurities met with in him are also the result of hasty composition–we have already noticed his large output. With Hobbes, however, the haste is no doubt explained by his wish–or others' encouragement–to reply to White while the *De mundo* was new from the press and hence topical.

Next, the question must be asked: Why was the present manuscript not published in its own day? As we have it, the text is in near-finished form (but, as we have just seen, not incapable of improvement) and the pages neatly enough written to be acceptable copy for press. In the latter aspect, a parallel situation arises with Hobbes's *A minute or first draugt* [sic] *of the optiques* (British Library: Harleian MS. 3360), which is also carefully penned by a scribe, perhaps for presentation purposes. Neither work was published, however, until Molesworth printed parts of the second. Again, the present treatise cannot have been thought to be a personal, private document, for not only may it have circulated, as noted above, but also the censure of White is less severe than the vituperation that

---

[2] E.g. Descartes to Mersenne, 26 April 1643 (*Œuvres de Descartes*, ed. Adam et Tannery, III (1899), pp. 657–658). Descartes, who quotes from fol. 142 below, does not mention Hobbes by name, but he writes in the context of his recent reading of Hobbes's *De cive*.

Hobbes was to level at Wallis; and, particularly later, in the period between Bentley and Warburton, what scholars have been loath to publish personal polemic, much of it of limited appeal? We must therefore seek an explanation elsewhere, and the most likely one is that Hobbes wished to retain most of the material for inclusion in the large philosophical system he was then pondering. This would explain in particular why the document was not printed soon after its composition: publication being delayed, there is little doubt–though the matters at issue between Hobbes and White had not grown cold–that the intention to print would recede, the material being thought, by Hobbes and others, to be subsumed in the *De corpore*. But perhaps the explanation is more mundane: Mersenne, or someone to whom he lent the treatise, mislaid or forgot about the volume, and it passed from sight until almost our own day. At all events, Hobbes himself does not seem to have pressed for the work's publication.

III

Let us now turn to the matter of the annotations inserted in the Paris manuscript. To establish these as Mersenne's we need have no recourse to the evidence of the handwriting, for not only does Mersenne's scientific correspondence inform us about Hobbes's activity in 1642–1643 but also Mersenne draws upon the present treatise in his own *Cogitata physico-mathematica* (1644). The circumstances under which he became acquainted with the material he there takes from Hobbes are, however, unclear. Did Mersenne first meet it *in situ* on reading the whole treatise, and then decide to adopt it? Or had Hobbes shown him a separate draft earlier? We shall return to this point when we have given details of the indebtedness itself. In the ballistics section of the *Cogitata*, in addition to borrowing from folios 307 and 338 below, Mersenne mentions Hobbes twice: first ('Ballistica', pp. 81–82) merely as a 'vir subtilis' in connection with a theory about the motion of the sun and the earth, and second ('Ballistica', Preface, *sig.* a) by name, as the inspirer of Mersenne's own Proposition XXIV (pp. 74*ff*). Here one instance additional to those given in the *Critique* may be singled out. At p. 75 of 'Ballistica' Mersenne discusses illumination from a point-source of light. On folio 68v of our text he inserts an explanatory phrase of his own, which comes in exactly the same place in the corresponding passage in the *Cogitata*. Whether Mersenne added this phrase to an already existing draft of that work, or to his copy for press, or even in printer's proof, or whether he incorporated the words in the course of composing the *Cogitata* I do not know. But it is very probable that in borrowing Hobbes's material Mersenne was working directly from our manuscript and not from a copy or from Hobbes's foul papers. There may have existed, of course, a transcript or transcripts of the entire manuscript additional to the present one, or Mersenne may have requested that fair copies of parts of it be prepared for him. But it seems far more likely that he read the treatise as we have it, annotating as he went, and this before he published his *Cogitata*.

The annotations themselves, however, need not be discussed in detail here, though they are informative about their author and about the thought-processes of his contemporaries. Many are emendations of Hobbes's Latin, but collectively they are not a commentary on the present work. In the notes to the body of the text, therefore, I refer to them only occasionally, and usually this is when they throw light on Hobbes's meaning. Mersenne's corrections of obvious errors made by Hobbes or by his copyists are not separately indicated, nor are the places where he clarifies the expression while retaining the sense. In sum, Mersenne misses little; one or two instances where he allows words such as *frena* (folio 235) to pass are recorded for reference. At least once (442) he verifies a point by consulting a copy of the *De mundo* itself.

IV

The *De mundo* is modelled on Galileo's *Dialogo...sopra i due massimi sistemi del mondo* (1632), and its dialogues likewise have three protagonists, cast in much the same roles as their counterparts in Galileo. White's book has never been translated, but John Hall of Durham (1627–1656; he died young), we are told by John Davies in his *Life* of Hall prefixed to Hall's *Hierocles upon the* Golden verses *of Pythagoras* (1657), entertained 'intentions of putting the Dialogues *de Mundo* into English, which very designe, in part, argues a settlednesse, as to the acknowledgments of God and religion'. The *De mundo* has a list of chapter headings at the start, and at the end an index of the topics covered, but (as is common in similar books) the index is not comprehensive. The chapters carry unnumbered subheadings, but the chapter numbers and the article numbers in Hobbes's commentary do not match the corresponding places in White.

White's narrative is distributed among the three speakers Ereunius, Andabata and Asphalius. Whether the setting is authentic or contrived I cannot say: Andabata meets the others (all are common acquaintances) when the latter are journeying to Reims. We are not told, either, exactly where the meeting takes place. Again, White seems to see his role as that of listener rather than participant.[3] None of the three characters, therefore, expresses White's own views directly; indeed, in the *Critique* Hobbes names a particular speaker very rarely, contenting himself with transcribing selected extracts. In this connection Hobbes's cross-references (to which we shall recur) within his own treatise may be mentioned. Rather than taking the form of specific allusions to chapter and article, though one or two do so, the references are given, for example at folio 338v, in general terms: 'as was said earlier', 'as we shall see', and the like. This shows that the treatise was planned as a coherent and consequential unit; that the chapters were written in the order in which they now stand, even though they often draw upon ideas previously existing in Hobbes's philosophical drafts; and that the whole is not a series of *ad hoc* references to a source text. (Confirmation of this fact appears in a note to folio 81.) Hobbes's achievement, then, consists in his producing a harmony by reconciling a previously existing (though constantly modified) draft of a philosophical scheme with a consideration of the specific issues raised in the *De mundo*.

V

To demonstrate White's general philosophical position as shown in his book it is not necessary to outline his attitude towards every problem he raises for discussion; this task Hobbes does for us directly, or by implication, taking his material in the order in which White presents it. To give an idea of the *De mundo* as a whole, therefore, only two things need be done: first, to complement the topics chosen for scrutiny by Hobbes by listing those he does *not* refer to, and second,

---

[3] 'Andabata,...utroque vestrûm [*sc.* Ereunius and Asphalius] comitante, in aream progressus'–*De mundo*, p. 2.

to indicate briefly some of the Galilean questions examined by White. To list in their original sequence and contexts the points on which Hobbes is silent would have the advantage of showing how White moves from one question to another, but in summary form such a proceeding becomes fragmentary. In what immediately follows I have chosen, instead, to group the points under a few broad headings, believing that White's advance from topic to topic is sufficiently illuminated by the items chosen by Hobbes. When the latter does not do so, I have tried to make clear, in the main text, which views are his and which those of White: the differences are not always apparent, however, for on some issues the two men are largely agreed.

First, then, White has not shaken himself clear any more than has Galileo from occult qualities, and he sometimes allies them with metaphysics. He begins with a consideration of quantity, place, substance, and space. Rarity and density he sees (*De mundo*, p. 31) as 'complexions of quantity', and the question of what constitutes space includes the consideration of motion and rest within space (pp. 32–33); hence 'spirit cannot be "in a place"' (p. 35). In the same context arises his consideration of the relationship between quantity and divisibility. Next, after considering some astronomical matters–the largest section in the book and one which leads him to theological questions– White turns again to kinetics. He recalls Galileo's experiments to illustrate 'common motion': the shooting of an arrow from a moving vehicle; the recoil of guns; and the firing at objects in flight (p. 223). His belief that incorporeal things, though themselves unmoved, can yet move other things (p. 278)[4] leads him, after he has postulated that bodies can be in imperceptible motion or in a state of change (pp. 280–283), again to theology. Indeed, theological considerations are among those for which he has said, at the beginning of his book, that the universe is finite (p. 9*ff*); and he returns to this theme by querying whether causation is independent of an external agent that itself is independent of other agents (pp. 220–224). No body has existed *ab aeterno* (p. 301); all bodies have been created. The material with which the book closes includes (p. 441) the assertion: 'Of motions that will occur in the future, none can be infinite'.

Among the arguments not examined by Hobbes are some

---

[4] This example, paralleled in Hobbes, is referred to in Section VII below, in a grammatical context.

which he presumably considers uncontroversial and some he thinks so absurd as not to need refutation. But of the following not all can be firmly assigned to one or the other class. White recalls that we, the supporters of the New Science, are better informed about the heavens than are the Aristotelians, because they only hypothesise, whereas we demonstrate by observation and experiment (pp. 44–47). Circular motion, he says, is not 'proper' to the heavens (p. 50), because motion is not necessarily 'proper' to elements, and the heavens are elemental. Exposing the 'impostures' of the Chaldean astronomers (p. 67), he concludes that there is evidence of corruptibility within the sun (pp. 66–67). After considering Pliny's views on comets, he speculates that there may be self-illuminating celestial bodies situated beyond those visible to us (pp. 84–88). The reason White gives why the stars do not crash down upon the earth is interesting: he believes (p. 100) that they must be attracted by some force counter to, and stronger than, that exerted by the earth. Can this force be attractive–as with rubbed amber–or magnetic? Whatever it is, it is not the Aristotelian heaviness/lightness or what Aristotle calls *genesis* (a process that White ascribes to a *generans*). Next, some (such as Hippocrates) consider the moon humid. This hypothesis White accepts, because (p. 158) the moon's light is of a fiery nature, and lunar vapours are aroused by the sun, though they may be imperceptible to us; these the moon lacks the power 'to reduce to a mean (*modulus*)', and they remain, creating dampness. Again White asserts that, in explaining phenomena such as this, he does not have recourse to occult qualities.

Hobbes deals fully with White's speculations concerning the cause of the tides, but he omits White's observation (*De mundo*, p. 245) that in the Red Sea the tides do not follow the monthly pattern recorded elsewhere. Hobbes does, however, examine–taking the topic where it occurs (*DM*, p. 155) and as he reaches it–the descent of bodies in free fall, his views being those expressed in Chapters VI and X below.

Questions concerning human liberty and ethics illustrate well Hobbes's using the *De mundo* as a whetstone for sharpening his own ideas. 'The exercise of liberty is motion', says White (pp. 399–400), and he believes that we gradually feel our way from doubt to certainty. The will (*voluntas*) does not, he declares, move itself directly, but is moved by a passion that it excites round the heart; this passion sends spirits to the brain, and these arouse certain images (*species*) which cause the *facultas aestimativa* to vibrate, such vibration imparting motion to the will and effecting decision-making. The question of liberty, particularly in a religious or political context, is investigated by White elsewhere, especially in *The grounds of obedience and government* (?1649; 1655), *An apology for Rushworth's Dialogues* (1640), and *Quaestio theologica...[de] humani arbitrii liberta [te]* (1652).

Little need be said about the second means by which we can study White's position in philosophy: through pointing to parallels between him and Galileo. Most of these are taken up by Hobbes, so mention of three will suffice. The first topic is the sphericity and magnetisim of the earth; the second, the nature of motion; and the third, the concept of force.

VI

So much for the background. I turn now to show what literary problems have here to be faced; what aims are to be borne in mind; and what procedures have been adopted. In what follows–the tone may be sometimes thought querulous rather than apologetic or justificatory–words such as 'usually', 'normally', or their equivalents appear quite frequently. This indicates that we shall be dealing, not with rules to be rigidly adhered to, but rather with principles to be generally followed.

The present-day translator of a seventeenth century Latin text encounters a difficulty at the outset: in comparison with the vocabulary, the grammar, and the styles of classical writers those of neo-Latin authors have been little studied; precedents and exemplars relative to later manuscripts and to printed books are therefore correspondingly few. Hobbes's vocabulary and style will be considered below. To begin with, in the present undertaking one of the difficulties lies not so much in translating Hobbes (and, as will be shown, this is testing enough) but rather in translating White, whom Hobbes himself correctly described as often being crabbed and obscure, both in his matter and in his manner. Two characteristics of White's style are worth mentioning in this context: in constructions using reported speech (*oratio obliqua*) White tends to delay his verb, producing a sequence of accusatives not always readily or reliably distributable; and he could have paid more attention to his word-order, to enable us to isolate his sentences' constituent phrases more easily. Among the aims of the present translation, therefore, is that of making the two authors intelligible to readers without, or with little, knowledge of later Latin and (a matter considered below) without a knowledge of its specialist terms and their implications: those assumptions, accepted procedures and modes of argument since abandoned. That is one reason why it is difficult, without misleading the reader, to make Hobbes and White speak the language of the twentieth century. Indeed, like politics, translation is the art of the possible.

I shall next describe the general principles I have adopted. First, at the risks of over-translating and being prolix, I have tried to make Hobbes's meaning unequivocal or, if doubts exist, I have drawn attention to them. (This latter point will be expanded in a moment.) I try to distinguish—but such differentiation is not always possible—between interpretation of the narrative as affected by textual matters such as punctuation, conjectural emendation, and so on (a number of cruces remain unresolved), and interpretation of the narrative as it stands translated: 'What does Hobbes *mean* here?' With the latter issue, comment has been kept brief, and is sometimes tentative only; indeed, a few of the notes ask questions rather than answer them—a legitimate procedure, since no translation is final? Some of the difficulties arise from the role played by the punctuation of the original, as with a classical text (whether or not the latter's earliest state has any). Like its spelling and capitalisation, the punctuation of our manuscript is wayward, even by seventeenth century conventions; and it includes some inconsistencies in its underlining of words and phrases (it does not employ quotation-marks). One should notice particularly the tendency of seventeenth century writers, or copyists, to place commas both before and after an adjectival clause. The reader therefore cannot always tell whether these clauses are restrictive or continuative: one of several such instances occurs on folio 6 ('deeds...importance'). Detailed discussion of this problem is not possible in the present volume; but the question occurs with an *adverbial* clause at folio 409, line 8.

Again, I place alternative renderings, where these seem worth consideration, in the notes, giving the original Latin only if essential; and words not occurring in the Latin text I put within square brackets, such insertions acting also as a small commentary inside the narrative. These glosses, though one hopes they are not distracting to the eye, themselves pose a problem.

They are not by any means all removable, leaving the text to run uninterruptedly without them: sometimes, in addition to enclosing words supplied by me, they replace pronouns in the original that do not relate to specific antecedents; sometimes they are pronouns added where the Latin requires no pronoun; sometimes they contain the Latin words—usually single ones—being translated; and sometimes they repair or expand some ellipsis in the original. (This list is not comprehensive, and the procedure is applied selectively: consistency is unattainable.) In the last instance particularly, to remove the brackets leaves jagged edges; but the procedure adopted, perhaps finical to some, seems better than the only other courses open: either silence, or the inclusion of a complicated *apparatus criticus* and sigla. In this connection the reader may feel that the alternative versions are given too freely; but surely it is a translator's duty to show where there exist interpretations valid in their several contexts but different from the ones he has chosen?

Some further points on procedure may be mentioned. I have occasionally italicised words or phrases, added parenthesis-marks or round brackets, and placed expressions within inverted commas—these minor changes without separate notice. One or two cross-references (in addition to Hobbes's own) are supplied, but for fuller details the Paris edition should be consulted. Where, for clarity, I have inserted numeration I have not thought it necessary to adhere to a single system. The geometric figures in the manuscript narrative are gathered up in an appendix; in this way, especially in Chapter XXIII, they may be identified more quickly and accurately than if they were placed, as in the original, somewhere near the appropriate contexts. When Hobbes's Latin refers to White as 'author' (never 'auctor' and never by name) I render by 'our author' or by 'White' so as to avoid giving the impression that Hobbes is speaking of himself ('the author'). Similarly I say 'his' (i.e. White's) 'page' where Hobbes's narrative lacks the possessive; but detailed page references (given in the Paris edition) to the *De mundo* are not supplied, many occurring within our text itself. When Hobbes claims that he cannot understand a word or expression used by White I normally print the Latin. In translating proper names I give the well known in their familiar form, 'Descartes' for example; the lesser known I normally modernise where possible, as with 'Fromond' and 'Rothomann'. As for my English spellings, strict adherence to rules is not possible, especially where hyphenisation is the issue. I have settled each time for the form that best suits the immediate context, as with 'cease-to-be', 'free-will', and 're-create' (112v, etc.), (but 'reassemble'), rather than attempting absolute consistency.

Chapter and article headings I retain as they stand in the original–at the heads of chapters–rather than moving each article title to the beginning of the section to which it belongs. The latter course, though at first seeming advantageous, would rob readers of the ability to see the substance of each chapter at a glance. (This detail is not repeated in my list of contents, the items of which, for the reader's convenience, are adaptations from those that head the chapters in the text.) Indeed, some individual article titles are comprehensible only if read in conjunction with those immediately preceding them. Moreover, even as regards the opening chapter I consider it unnecessary to detail White's position on every point raised by Hobbes. As argued above with reference to the manuscript as a whole, it is necessary to give only as much of White's argument as will set Hobbes's work in context: the present is an edition of Hobbes, not of White. Lastly, anything resembling a close literal translation the reader would find intolerable, as will now be explained.

VII

I have broken up several of Hobbes's longer sentences, though some divide reluctantly (I return to this question below); I have rendered many passives by changing the construction to the active; I have simplified a number of impersonal constructions; where Hobbes prefers two negatives, or a litotes, to a positive, I often follow him–even though this can in my opinion lead to awkwardness–on the grounds that to make alterations can sometimes impair the sense; reported speech has often–to advantage, in my view–been rendered by direct; I have occasionally taken some liberty with the tenses, Hobbes's own practice being sometimes lax here;[5] and I have inserted additional paragraph divisions without separate notice. (As to the last, a

[5] His tense sequences are sometimes difficult to render in English, and occasionally (e.g. fol. 215) he is given to using the pluperfect where the context would seem to call for the perfect (see also fol. 74v). In *oratio obliqua* constructions, too, the question arises as to the tense of the infinitive.

few instances where justification of this step, or of my ignoring the original divisions, seems called for are pointed to in the notes.)

Hobbes's use of pronouns, mentioned above, requires comment. They occur frequently and are rarely ambiguous in the Latin; but in the English it is sometimes necessary to repeat the relevant nouns. And, also regarding pronouns, one particular difficulty arises with the verbs 'to move' (*movere*) and 'to change'. Hobbes does not invariably make clear whether he wishes to be understood as speaking transitively or whether intransitively (for instance at folios 325v and 410): that is, whether something moves itself or whether it moves other things. In some contexts this leads to obscurity: a *se* inserted where required would have made all clear. (Unfortunately, in a different context noticed below, provision of this pronoun has the reverse effect. We can understand, say, 'God conceals Himself', but what of 'God wills Himself'–*Deum Se velle*, folio 382–used absolutely?) Under 'parts of speech' attention may also be drawn to Hobbes's use, acceptable in the Latin, of the adjectival *ille*, 'that', in passages where it is more natural in English to say 'this'. If he does not reproduce this predilection of Hobbes's the translator stands charged with inaccuracy; if he does, he may be accused of discourtesy towards his own language. I have opted to retain the word 'that' wherever such procedure seems desirable. I have sometimes had to take decisions, too, on whether a context calls for the English definite article or whether the indefinite, or, indeed, whether to use either would be incorrect. Such instances identify themselves, e.g. Chapter XXXIII, Article 6.

Four points are to be dealt with next: Hobbes's Latin vocabulary, his grammar, his use of technical terms, and his Latin style. Unfortunately, even though many ideas present in our manuscript occur also in the English versions of Hobbes's Latin writings that were first printed in his lifetime, these counterparts are not as helpful to a modern translator (or retranslator) of Hobbes's Latin as may be thought. It will be remembered that Hobbes's English writings were produced during a time when the vocabulary of English was expanding rapidly. Instances of neologisms, some of which did, and others did not, establish themselves, include 'conamen,' 'conation', and 'conatus'. (The last, to which we shall return, is found in Glanvill also.) Furthermore, the meaning of some words, especially of those derived from Latin, was fluid. Many words that have altered in some of their senses during or after Hobbes's day are familiar, such as 'predicament' (Chapter VII below), 'complication' (folio 114), and 'consist'; less obvious instances are the word 'probable' and the innocent-looking 'particle'. Again, by the seventeenth century the word 'series' was beginning to assume the general meaning 'collection' rather than 'a group characterised by some factor common to all its constituents'.

If the significance of some English words (and phrases such as 'by and by') was changing, then as now, we should not forget that many of the Latin words themselves commanded a number of meanings. Among these I am several times uncertain whether I have chosen correctly. *Dicto* (folios 359v and 426) may mean 'say repeatedly' or 'say categorically'; and with *tollo* the context does not always indicate whether Hobbes means 'sustain, lift' or whether he intends 'remove'. *Concipio* and *percipio* are similar instances; they are hardly distinguishable one from the other. It is not always obvious, either, in what sense Hobbes uses words such as *focus* (which, from the relevant context, I give as 'combustion-point'); *eventus* ('result' or 'occurrence'?); *perfectissimus* (the pleonasms 'most perfect' or 'completely finished'?); *simplus* ('single' or, in specialised usages, 'simple'?); *comprehensibilis* ('able to be taken hold of' or 'visualisable'?); *defectus* (395v, 'lack' or 'fault'?); *potentia* ('potential' or 'power'?); and *appetitus* ('searching after' or 'yearning for'?). There are of course others. This, then, is a further sense in which the narrative is sometimes more difficult to render than a classical text; and the immediate contexts of such words are of little service. In addition, Hobbes borrows words from classical Latin and gives them a new dress, as did Kepler with the word *inertia*; for example, Hobbes uses *cymba* (folios 153v, 160, 164v) to mean 'a vessel or container' and 'a ship'.

Since some of Mersenne's corrections concern grammatical issues alone, a word on Hobbes's grammar is in place here. The translator's difficulty lies in deciding with what deference, near-absolute or only partial, Hobbes and his contemporaries treat the rules and conventions of classical Latin and, indeed, the extent to which individual practices varied; furthermore, authors (Mersenne included) were not always self-consistent. For all the above reasons, and sometimes from mere lack of care–as with White–the Latin narrative can be ambiguous or unclear. Readers perusing the English, therefore, may not realise that their difficulties (some of which a translator cannot resolve) can well be of Hobbes's own making.

VIII

The question of technical terms is important. These fall into two broad classes (though the latter sometimes overlap): terms of scholastic philosophy and terms of pre-Newtonian physics. It should be noted that the former class includes adverbs such as 'determinately' and *simpliciter* as well as nouns. Understanding of the terms, and of the concepts behind them, met with in Aristotelianism and its medieval accretions Hobbes can take for granted in his contemporary reader (one wonders what factual *données* he assumes the latter to be possessed of); but nowadays even the professional philosopher may need assistance. At this stage we may remind ourselves that, even if he condemns the scholastic logic as taught him in his student days, Hobbes's normal procedure is to adapt these traditional methods very little; he may change the emphases and may redefine some of the terms, but he follows the old ways of proceeding in philosophical disputation. Such is the case in his present exercise. It will be recalled, too, that some seventeenth-century thinkers use such terms (as White sometimes does) as a means of *approaching* a problem–a practice castigated by some exponents of the New Science.

The list of philosophical terms is a formidable one; the following occur within the present text alone: *ens, esse, mediatè, praesens, numerus, modus, sensibilis, demonstratio, denominatio, intrinsecus, externus, operari, essentia, adequatus, spiritus, actio, passio, generans* (as noun), *species, intellectus, adminiculum,* and *prima materia.* I have usually chosen not to translate the like by words such as *agent, patient, movent, mobile, generation, corruption, mode, accident, passion*: the English 'equivalents' are rarely accurate, and a group of such words can, so translated, read as pure jargon. So I have supplied rough counterparts where possible, or have left the words untranslated, as, for example, the word *ens* at folio 109v, but 'patient' at folios 385vff. (The former term can in particular give trouble, for the English noun 'being' cannot distinguish between *ens* and *esse*, and it masks the old distinction between 'existence' and 'that which exists'.) I do not think this procedure is inconsistent with that which I adopt for Hobbes's *own* technical terms, which I examine below. In particular, some of the philosophical language, if turned into the vernacular, now has associations that did not exist before. For example, *verè et realiter* given as 'really and truly' (folios 217–217v) suggests street-corner gossip rather than the precise use of a specialised vocabulary, and versions such as 'indisputably and in actual fact', 'properly and correctly', and 'correctly and faithfully' have a certain lameness. Contrariwise, in one situation the task of the translator is easier than that of the Latin author. This is where Hobbes and White, in coining words and phrases, have to find equivalents for *ad hoc* nouns which the English (as at folio 16v) can quite happily form by adding the suffix 'ness' to certain adjectives, adverbs, or prepositions, or by some other device. Presumably the reason for which Hobbes would attack those medieval thinkers who (folio 330v, note; and compare 316v) create unusual substantives is not that the procedure is inadmissible but rather that the results are, in his view, grotesque.

The difficulties are no fewer when one has to deal with the technical words of the earlier natural science, including *conatus* (already noticed), which I render by the words *effort, force, impulse, tending*, and the like (Hobbes himself translates *conatus* as *endeavour*). Not only do such terms as *nisus, impressed force*, or *momentum* (as used before Newton), dealing with concepts now abandoned, lack precise modern equivalents, but also their meanings were not universally fixed. The list of technical words is extensive: *gravitas* ('weight' or 'gravity'?; for 'weight' Hobbes also has *pondus*); *moles* ('mass', 'bulk', 'volume'?); *principium* ('principle' or 'beginning'?); *pulsus, pulsio, impulsio; actus, actio; individuitas, indivisibilis, indivisus; affectio* (217), *affectus; finis, terminus, scopus, meta* (432)–such terms as are similar one to another are not necessarily interchangeable, and the preceding lists are by no means exhaustive. The four occurrences of the word *umbra* on folios 35 and 35v seem to be in three different senses; and *gravitas* may have different meanings on folios 40 and 40v. Technical words, particularly in astronomy and geometry, whose meaning *was* fixed, however, I define if Hobbes himself does not; but this step has been taken with diffidence, for the specialist will know them anyway, and the general reader, if he wishes to grasp the relevant arguments, will need a fuller commentary and explanations.

INTRODUCTION

There are yet further difficulties. First, the supersession of old terms by modern and the use of the former in a new sense mask traps for the unwary, unless the context warns. When, for example, Hobbes speaks (folio 201v) of the sun's being in Cancer or Capricorn he refers not to the tropics but to the zodiac. Second, as do other philosophers, Hobbes takes existing words, Latin and English, and, as we have seen with Kepler, re-defines them or gives them new meanings. One word which occurs quite frequently in the manuscript sums up this problem. It is the verb *determinare* and its cognates. In most contexts I take it as meaning 'to set bounds on', 'to limit', rather than 'to decide' or 'to settle', and I usually render *determinatio* as 'propensity'. Hobbes normally translates the verb as 'to determine', even in contexts that do not always make clear the sense he intends; and even when he uses the English word *determine* his meaning is not always apparent: 'The will determineth itself, and...external things...work upon it not naturally but morally'.[6] The problem is particularly difficult to resolve when *determinare* is applied, in *oratio obliqua* constructions, to God, as in *Deum Se determinare*–a different problem from that noticed earlier: *Deum Se velle*.

Particular comment is called for on Hobbes's own technical terms, which are frequently met with, especially in his discussion of the mechanics of perception, and the meanings of which probably constitute the greatest poser in the present undertaking. Generally Hobbes takes pains to define such words (as with 'names' and 'signs'), but with four he does not directly do so here. These last are *phantasma*, *imaginatio*, *sensio*, and *imago*. In favour of keeping these terms and their like unchanged are the facts that each is employed in a specialist sense and that the reader may wish to know which is being used in a given context. Further, to render each by a single unvarying English word could produce an awkward and stiff-jointed narrative, even (as we shall see) were it possible to devise exact equivalents. To the retention of such terms, on the other hand, there are at least three objections; and even if readers accept all or some of these last they may not agree on their order of importance.

---

[6] *The questions concerning liberty, necessity and chance*, in *English works* (ed. Molesworth, 1839–1845, hereinafter referred to as *EW*), v. 451. The Latin works in the same edition are given as *LW*. Reference is also made to the British Library copy of the *De mundo*, which has contemporary marginalia.

First, inclusion of the terms as they stand in the Latin looks unnatural on the English page on which several are encountered. Again, their admission would suppose that Hobbes is consistent in his use of them; but a glance at, say, folio 308 shows that this is by no means certain. And third, in the English versions of his *De corpore* and other works Hobbes himself accepts anglicisations. My decision is to attempt English translations of these specialised terms; and the reader anxious to learn the actual Latin words in use in the more important passages I refer to the notes immediately following, from which the words can readily be found (folio numbers are sufficient identification).

Phantasma I render either by 'impression' or by 'fancy',[7] but in a few places I think it best to use words such as 'conceit' and 'fantasy' (this last, of course, needs discreet handling in modern English).[8] *Imaginatio* (also used by White) appears as 'mind-picture' or as 'imagination',[9] but occasionally it emerges more naturally as 'concept', 'thought', and 'awareness'.[10] *Imaginatio* and its adjective *imaginarius* must also be considered in conjunction with the word *sensio*. For this latter I usually prefer 'perception'[11] to 'sense', since Hobbes seems normally to reserve *sensio* for an act of, rather than the faculty of, sensation; the latter is *sensus* (e.g. folios 84v and 427). Likewise *imaginatio* is sometimes the image-making faculty, sometimes the image or mind-picture (in Hobbes's specialised use). The second is his meaning in the well-known 'Imagination is nothing but *decaying sense*' of *Leviathan* I.ii, in which 'decaying sense' matches *languescens...sensio* of the corresponding account in Chapter 25 of *De corpore*.[12] To *imaginatio* Hobbes devotes Chapter 2 of his *De homine*;[13] *sensus* as differentiated from *sensio* is studied in its Chapter 1. Next, *imago* is perhaps best kept as 'image',[14] but some instances seem to require 'picture' or 'reflection'. The word 'image' is not always satisfactory, for another reason: it alone can render *species* in the optical sense.

[7] Fols 307v, 339–341, 343v, 344v, 345v, 348, 353, 430 (the adjective), 434.
[8] Fols 14v, 19, 28v.
[9] Fols 8v, 15, 19v, 308, 310, 312v, 314, 329v, 340, 341, 341v, 344, 347v, 349, 352v, 365, 410v, 427.
[10] Fols 342v, 344v, 429v, 431v.
[11] E.g. fols 307, 308, 308v, 339v, 340, 349.
[12] *LW* I.323; cf. *EW* I.396. Compare *sensibilis/sensitivus*.
[13] *LW* III.
[14] Fols 15, 18v, 308, 313–314, 340, 344v.

(No problem arises with *species* as opposed to *genus*.) In Hobbes's usage the word *notio* seems sufficiently elastic to command, among the words by which, for variety, I render it, translations such as 'idea' (folios 352, 353), even though the latter is found in its own right in some passages of the Latin, e.g. at 361. All in all, then, the procedure adopted seems to fit the circumstances best.

There remains for notice Hobbes's own Latin style as here seen. It will be remembered that he assisted Francis Bacon with the translating of the latter's *Essays* into Latin, but neither has Hobbes absorbed much of the Baconian philosophy at large[15] nor do the Latin styles of the two resemble one another. Usually terse and nervous (but not laconic) in the phrase, Hobbes is often garrulous and repetitive in the organisation of the present work–despite what is said at the end of Section IV above–and sometimes within his sentences themselves. In this dissertation he shows a penchant for the syntactically loose sentence that proceeds by cumulation and that sometimes, through the use of copulatives, runs to inordinate length. Though often acceptable in English, this practice is far from being universally so; hence the order in which the data are presented within the individual sentence has occasionally been altered in the present translation without, it is hoped, impairing the sense. Conversely, on occasion Hobbes can be over-concise. His vocabulary has its limits, and in order to give the translation interest and variety I have tried to make its range of English words as wide as possible. In particular, opportunities of introducing figurative language to enliven the English version are few; the latter is therefore less attractive than it could be were such chances available, but this deficiency is sometimes offset by Hobbes's flashes of irony (usually at White's expense) and his occasional *obiter dictum*.

IX

A little must be said, in conclusion, on two matters: the importance of the manuscript for the student of Hobbes's

---

[15] Aubrey gave Hobbes a copy of Bacon's *The elements of the common laws of England*, which perhaps stimulated Hobbes's interest in this side of Bacon's work and was one of the influences on his writing the posthumously published *A dialogue between a philosopher and a student, of the common laws of England*, if the latter is indeed Hobbes's. See fol. 38.

philosophy, and the choice of a title for the present translation. The occurrence here of some of Hobbes's ideas before they reached print is self-evident; what is less obvious is the different degrees of scrutiny to which Hobbes subjects White's opinions: outright dismissal, rejection after a consideration of the balance of possibilities, or acceptance after due weighing. Such was the background against which Hobbes's own *idées données* became clarified and developed, finally to reach the form in which they now stand. As regards choosing a title for the translation, the question is not as easy as may be imagined. I have decided against the word *critique*, which in English suggests a preoccupation more with a work's manner than with its matter, and against *polemic*, which suggests unrelieved censure. Of the remaining possibilities none, for one reason or another, seems unexceptionable. Several, for example, imply that the debate between Hobbes and White continued after the penning of the present work. (As we infer from an observation of Anthony à Wood in the *Athenae Oxonienses*, under the entry 'Glanvill', this indeed happened, though the argument proceeded only verbally.) Again, while striving to secure brevity, I wish to avoid the use of a subtitle, and to show that Hobbes was attacking only certain of White's views. Lastly, *Thomas White on the world: his case evaluated*, or the like, suggests that White wrote in English. Such, then, are some of the criteria I bear in mind when making my choice: *Thomas White's De mundo examined.*

H. W. Jones

Errata
The printing of the Latin text and its translating have been an undertaking of some magnitude and complexity. It is hoped that such errors as have escaped correction will, with the assistance of readers, be rectified in the projected commentary volume.

CHAPTER ONE

On Problem I [of the First Dialogue], in which our author denies that philosophy should be treated logically

Definitions: (1) Philosophy, metaphysics, physics, ethics, politics, mathematics, and (2) logic, history, rhetoric, poetry. 3. That philosophy must be treated logically. [(4) *The manuscript has no article heading.*]

1. If we are to judge whether philosophy should or should not be treated logically[2] it is useful to know what the following are: philosophy, logic, and the other arts by which we expound, and discourse upon, any kind of subject. Now, philosophy is the science of general theorems, or of all the universals[3] (the truth of which can be demonstrated by natural reason) to do with material of any kind. The first part of philosophy, and the basis of all the other parts, is the science where theorems concerning the attributes of being at large are demonstrated, and this science is called *Philosophia prima*. It therefore deals with essence, matter, form, quantity, the finite, the infinite, quality, cause, effect, motion, space, time, place, vacuum, unity [*unum*], number, and all the other notions which Aristotle discusses, partly in the eight books of his *Lectures on physics* and partly

---

[1] Fols 1–4v are blank.
[2] After some preliminaries in the *De mundo* White proposes as the first main topic for examination *Philosophiam non esse logicè tractandam*. Devoting three articles to prolegomena, and without identifying the context of the phrase, Hobbes plunges into his consideration of it.
[3] Perhaps, '...or of all the universal theorems...' Or, 'all' may precede 'general'.

in those other books which were subsequently called *Tôn metà tà physiká* (it is these latter that gave *Philosophia prima* its present name, Metaphysics). The other part [of philosophy] concerns beings, distinctly and separately from one another; it demonstrates the reasons for natural effects in natural bodies taken individually, and on that account it is called physics or natural philosophy as exemplified in Aristotle's *Of the heavens, Of coming-to-be and passing away*, his *Meteorology, Of the soul,* and others of his works. One part [of philosophy at large] concerns the passions, the manners and the aims or purposes of men, and is called ethics or moral philosophy. Another concerns human society and discusses civil laws, justice and all the other virtues; it is called politics or civil philosophy. Another part considers the relations of space to space, time to time, figure to figure, number to number; this part constitutes geometry and arithmetic, which are usually combined under the name of mathematics. This last title came into use because the writers on geometry were acknowledged to have shown–that is, to have made evident and beyond doubt to their pupils and readers–that what they taught was true. On that account the pupils were said not only to have heard, but also for certain to have learnt something; likewise from *manthanein*, that is, 'to learn', geometry and arithmetic were termed the mathematical sciences. [It seems that] of those who professed the sciences other [than mathematics], some have spoken more plausibly than others have; but none of them has taught anything that was not open to question.[4] From this it appears that all the sciences would have been mathematical had not their authors asserted more than they were able to prove; indeed, it is because of the temerity and the ignorance of writers on physics and ethics that geometry and arithmetic are the only mathematical ones.

Some parts of philosophy, such as astronomy, mechanics, optics, music, are mathematical; others, still untouched,[5] deal with quantity and number, not merely in theory but with

---

[4] The verb on which this sentence depends is missing; perhaps Hobbes intends: 'Everyone agrees that...' or the like.

[5] This seems a curious remark. Surely Hobbes does not mean 'which others or I have not yet dealt with'? The words also form the first instance where the meaning is ambiguous because the MS.'s punctuation is not consistent in its handling of relative clauses, it not making clear whether they are restrictive or continuative.

reference to the [real] movement of celestial bodies or that of heavy ones, or to the action of shining or sound-producing bodies; [these parts] must therefore be counted among the mathematical sciences. So much, then, for philosophy.

2. In order to express what they wish [to say], writers are in the habit of using [certain] devices; there can be as many of these as there are different purposes in communication. Now, there are four legitimate ends of speech. Either [i] we want to teach, i.e. to demonstrate the truth of some assertion universal in character. We do this, first, by explaining the definitions of names in order to eliminate ambiguity (this is termed 'to define'), and second, by deducing necessary consequences from the definitions, as mathematicians do. Alternatively, [ii] we wish to narrate something, or again [iii], our aim is to move our hearer's mind towards performing something;[6] or [iv], we want to glorify [certain] deeds and, by celebrating them, to hand them down to posterity. The art by which the first is accomplished is logic; the second, history; the third, rhetoric; and the fourth, poetry. Logic is a simple form of speech, without tropes or figure; for every metaphor has by its very nature a double significance and is ambiguous. Metaphor is therefore opposed to the aim of those who proceed from definitions, these last being employed deliberately in order to eliminate equivocation and ambiguity. Historical style admits of metaphors, but [only] of such as excite neither sympathy nor hatred; for its end is not to move the mind but to shape it. It must not be sententious, either; for an aphorism is nothing but an ethical theorem or a universal assertion about manners, and the end of history is to relate deeds, which are always of importance.[7] Aphorisms and metaphors are appropriate in rhetorical style, however, as both are of service in moving the mind. Lastly, the poetical style, although it admits the kind of metaphor used as ornament, does not (so long as the poet speaks *in propria persona*) use aphorisms. That part of the poem where the poet speaks *in propria persona* is the relation of a particular fact, whereas every maxim is universal; but when he speaks in

FOL. 6v

---

[6] Cf. Horace: *Ars poetica*, 100.
[7] A further instance where it is doubtful (because of the punctuation of the original) whether the adjectival clause is continuative or restrictive (cf. fols 17 and 46). Second, Hobbes may mean 'singular', to contrast with 'universal' at the beginning of the article; cf. the close of Article 3.

the character of another person the poet's style must be appropriate to that person. Besides, the aim of the poet is to hand down illustrious deeds to future generations; poetry must therefore be metrical, so that it can be remembered for its beauty, its measure and its harmony.[8]

3. Now philosophy, i.e. every science,[9] must be treated in such a way that we shall know from necessary deductions that the conclusions we draw are true. Philosophy should therefore be treated logically, for the aim of its students is not to impress [others], but to know with certainty. So philosophy is not concerned with rhetoric. Again, its students seek to know the necessity of consequences[10] and the truth of universal propositions: therefore philosophy is not to do with history, and much less with poetry, for the latter relates deeds of great moment,[11] and it deliberately sets aside truth.

FOL. 7

4. Yet our author says (his page 7): 'All sciences are not to be banished to such mathematical islands, but are to be called forth from narrow places like those', and he adds in the margin, in less ornate language: 'Philosophy must not be treated logically'. Why? 'Do you think', he asks, 'we can have any metaphysical, natural or moral science when the contest being waged on all sides in prolix speeches is a rhetorical, not a philosophical one?' Then he says that the reason why philosophy must not be treated logically is that truth and sound knowledge cannot be acquired through a contest in rhetoric; which is tantamount to saying 'Philosophy must not be treated logically because *rhetoric* teaches nothing certain'. This, in fact, is a good reason why philosophy should not be treated rhetorically; but when he says: 'Philosophy is not treated logically, and in the same way as the mathematical sciences are', he is saying: 'Philosophy must not be taught, nor must its principles be clearly[12] and firmly proved'. He also says that, by perusing[13] these very *Dialogues*, the reader must not hope to see anything proved true–an ill-omened statement to make at the start [of

[8] Cf. fol 313v.
[9] Either, 'philosophy includes all other disciplines', or 'every study must be treated philosophically whether or not it is a part of philosophy'.
[10] I.e. if consequences must necessarily follow from premises.
[11] Perhaps, as above, 'individual deeds', especially the important ones.
[12] Or, 'nor are its principles clearly...'
[13] Perhaps, 'on perusing' or 'after perusing'.

his work]. He might have stated that metaphysics, physics, ethics and many other parts of philosophy *had not yet been treated* logically; but that they *should not be so treated* was the last thing he ought to have said. [Finally,] the reason why authors often fail to keep their promises is that, first seeking glory, and esteeming truth only afterwards, they are competing for renown and for the applause of their hearers. This intention has produced, in fact, an art like that of the Sophists, which the ancients called *eristic*, or the art of controversy, but one in which there is little honour; perhaps our author thought this art was logic.

CHAPTER TWO

❧

On Problem 2 [of the First Dialogue]:

That the world is not infinite in

magnitude

Explanations of (1) to divide, to bring together, one, the whole, number, part, end, beginning, and (2) finite, infinite. 3. That it is impossible to know by natural reason whether the world is infinite or not. 4. An examination of our author's argument by which he tries to prove that the world is finite. 5. Three arguments proposed by him to prove that an infinite [world] is possible.[1] 6. 'Universal', 'Particular', what they are. An examination of the answer to (7) the second argument, which is based on the power of God;[2] and to (8) the third argument. This reply [of mine] gives definitions, as accepted by mathematicians, of the point, the line, the surface, and the solid.

1. If, when we state: 'The world is finite', what we say is true, then the world *is* finite;[3] if the same applies with [the word] 'infinite', then the world is infinite. But we cannot know if either word is correctly applied to the world unless we know the meaning [of 'finite' and 'infinite']. Suppose the word 'world' also implies 'finite', that is, from something's being called 'world' the same thing is necessarily called 'finite'. Then we know that the world is finite; just as we know that

---

[1] On the term *possibilis* compare fols 13v, 358, 363.
[2] For 'power', perhaps 'potential'. It is not always clear in which sense Hobbes uses the word *potentia*.
[3] Cf. fol. 320v.

[the term] 'animal' can be applied to man because someone who says 'man' understands 'a certain animal'. If, however, the word 'world' also signifies 'infinite', then we know that the world is infinite. [Again], if 'world' signifies neither ['finite' nor 'infinite'], we do not know whether the world is finite or not; this is discoverable only by defining our terms. We must therefore ask: What are the things we call 'finite', 'infinite' and 'world'? But we must first ask what is meant by 'to divide' and 'to bring together', and what is meant by 'one', 'the whole', 'number', 'part', and 'end'. We are said to divide something when we consider first the thing and then something smaller contained in it.[4] For example, when I am considering a man as a rational animal I am not 'dividing' a man; but after I have considered the same man as being composed of a head, shoulders, arms, etc., then I am said to have divided him. Insofar, therefore, as I have considered a man in the first way, that is, as undivided, he is said to be 'one'; hence the general definition of 'one' as 'that which is undivided'.[5] [Suppose I consider him] in the second way. Then any one of those things which have been considered in man, if it is smaller than man himself, is called a part, such as the head, the shoulder, etc. [Again,] if this part be re-united with [the remainder of] that which has been divided, [the product] is immediately termed 'the whole'. From this it follows that 'a part' is what, being the smaller, is taken from, or considered in, that which is the greater. 'The whole', however, means 'all the parts together'; and what, before division, we called 'one' we call 'the whole' [even] after division.

FOL. 8

There are people who understand 'division' only as 'the separation and dispersal of a continuum', in which case a knife, a sword, or some other instrument is necessary to the nature of division (among these persons seems to be our author, who says somewhere that air is more divisible than water, and water than earth)—as if half of a marble column is not a part of it unless the column is broken, or the half can be 'a part' only if division takes place, or again (they say), as if a quantity, than which a smaller can always be given, were not divisible into such parts as may be further divided, simply because it cannot,

---

[4] Cf. fols. 32v, 107ff, 357v.
[5] For 'way', perhaps 'philosophical mode'. 'General': as accepted in Hobbes's day (not his own definition)?

FOL. 8v

on account of its smallness, forever be split and part torn from part. There are also others who, though not believing it necessary to division that the thing to be divided be broken, still think its parts must be conceived in the mind: in other words, that we must conjure up a vivid image of a part. Yet if this were true it would be impossible for a quantity to be divided into parts infinitely divisible, since by perpetual division we should finally arrive at a quantity so small that no image of it could be formed. We cannot, in fact, imagine a quantity smaller than what can be seen, because no image can exist except of a visible thing. Nevertheless, time, whose parts can neither be separated from one another by force nor conceived of as images, can yet be divided. There is therefore no need, if division is to be performed, that the thing to be divided shall be broken or even that it shall be represented by a sign or conceived in the mind.

In order to know, by correct reasoning, what smaller quantity, such as the half, the third, or the quarter, is contained in every [given] quantity, it is sufficient to consider these quantities by their names and appellations, i.e. to refer them to arithmetic. That any quantity, however small–so small even that it eludes every sense and imagination[6]–contains another quantity smaller than itself, and this in every proportion, clearly appears from the following. Say that, given the said quantity, however small, two other, separate quantities are added to it in a stated proportion. If the total, which, because it has magnitude, can be divided, is divided in the same ratio, it follows necessarily that the small quantity is also divided in the same proportion.[7] In the same way, to achieve bringing-together [compositio] it is not necessary to unite things physically, or even in the imagination, but only to do so by reflection, or by introducing the names of things into ratios or calculation.

To compound [componere] is merely [to add] unity to unity, and so subsequently to reconsider all together every number you

---

[6] The word *imaginatio*, which is later translated as 'mind-picture' (fols 308, 312v, 329v, 340, 344, 349), Hobbes uses in the plural also.

[7] Hobbes's meaning seems to be as follows: Say x is the original quantity, however small. Quantities y and z are added, each being proportional to the other and to x. 'The total is divided in the same ratio' presumably means: 'In whatever ratio you divide the whole, x is divided in the same ratio', and hence the proportion of x in sample a is to the proportion of x in sample b as a is to b.

have considered separately.[8] Number takes its origin from bringing-together as much as from dividing. From the former, in fact, [arise] whole numbers, as 2 is made by 1 plus 1, 3 from $1+1+1$, etc. From the latter we make the numbers called fractions: for if a unit be divided into two equal parts each becomes a half, and similarly if [a unit] be divided so that one part of it is to the total in any other proportion, [then] a fractional number is made that expresses the proportion of the division; for instance, if the part cut off is to the whole as 2 to 3, the fractional number $\frac{2}{3}$, or two-thirds, is made.

We all know that we can, as we please, add number to number and subtract the smaller from the greater. Now, one part situated between two parts and touching on either side another part of the same whole is called the middle part; but another, not situated between two parts but touching only one part of the same whole, is called an extremity. As there are two extremities, that from which we commence reckoning is called the beginning; the other, the end. We can begin with either, as we please. For instance, let there be a whole, AE, and parts AB, BC, CD and DE.[9] Since BC touches AB on one side, part CD on the other, it is called the mid-part; and for the same reason CD is [also] a mid-part; but AB, which touches BC only, is an extremity, and so for the same reason is DE. If we start counting the parts from AB, then AB will be the first, or the beginning, and DE will be the end; but if we start from DE, DE will be the beginning, and AB the last part or the end.

FOL. 9

2. That is finite, therefore, in which a part can be taken such that no further part can be taken of the same whole. That is infinite from which no part can be taken unless a further part can be. When we have understood this it is evident that no part of any whole can be infinite; for to any given part another part can be added, otherwise the whole will not be obtained. No number whatever can be infinite, because we can increase any number by one unit in order to make it greater. The common statement that number 'is infinite' does not mean that a certain number is infinite but that, by the addition of one unit or several units to the number, [another] number is always

FOL. 9v

---

[8] Presumably: 'Numbers considered individually disappear as entities when absorbed into totals themselves forming entities'.
[9] Figure 1.

produced. Hence we see, not that a thing is infinite, but that the term 'number' is indefinite.[10] In the same way, when we seek to know whether a body is infinite, we are seeking only this: 'Does the extent to which we can go by compounding (i.e. by mentally adding bodies to other bodies) match that to which that body is enlarged and exists in reality?' And because by 'world' is understood 'a body extended in all directions', to ask: 'Is the world infinite?' is the same as asking 'Is body infinite?'

3. Now, the very definition of the word ['number'] tells us that, in this sense,[11] number is infinite. For, since number is made up of units, that is, of one added to one as often as we please; and since it is in our power to add units, a task for our intellect, then a number which we make in this way will exist. But the existence of no body depends on our mind or thought; therefore even if we conceive of bodies unendingly added to bodies, it does not follow that these bodies exist in the things of nature. Neither can the definition of 'world', that is, 'the aggregate of all the bodies which exist',[12] tell us whether or not another body can be added to the world. That knowledge is therefore to be derived from a knowledge of God's will, namely by asking: 'Was it His will to create the world finite or to make it infinite?' 'The world is finite' is therefore a dogma, not of science but of faith, and neither physical nor metaphysical reasons can determine it.

FOL. 10

4. Our author, whose footprints we are following in these our meditations, thinks he has demonstrated that the world is finite, and this, as he himself says, by reasoning no less solid than the demonstrations of Euclid and Archimedes; also that he has overthrown the arguments [produced by] himself [for refutation], having pointed to the impossibility of the world's being infinite. His reasons must therefore be briefly examined. To prove that the world is finite he puts forward only one argu-

---

[10] Perhaps *indefinitus* is to be taken as 'without limit' rather than as 'vague'. See the Introduction, pp. 16–17, on the verbs '*definire*' and '*determinare*', and fols 377, 419.
[11] As defined in Article 2.
[12] From fol. 453 and elsewhere we learn that this was an accepted definition in Hobbes's day.

ment, and this can be reduced even to a single syllogism as follows:

If the world is infinite, then some part of it is distant from us by an infinite number of paces (or other measures);
But no part of the world can be distant from us by an infinite number of paces;
Therefore the world is not infinite.

The minor proposition, [the truth of] which nobody could doubt, he has confirmed with ample proof; but he has remained silent about the major, and this latter is false. What, indeed, does the inference mean: 'The world is infinite; therefore a part of it is infinite'? When he says that number is infinite, because any number can have another added to it, is the consequence that any number is infinite? Or, if you consider any part of the world or any number of paces within the world, will the number of paces under examination be infinite because there is still another pace [you can take]? The world can[13] be greater than any number of paces whatever, but not so as to suggest that a certain number of paces is infinite. He seems to have thought, 'The nature of "infinite" consists not in infinity's being endless but in the fact that its end is infinitely removed from us'. That a part of the world is infinite, however, follows not from the world's being without limit but from this limit's being infinitely distant from us.

5. Of the arguments proving the possibility of an infinite world the first is as follows: That is possible which, being proposed, is followed by nothing inconsistent with it; but from the proposition: 'The world is infinite' nothing inconsistent follows; therefore [that the world is infinite is possible].[14]

The second argument is: What God had the power to do is possible; God could create an infinite world; therefore it is possible that the world is infinite.

The third argument is: What is intelligible is not impossible; but it is intelligible that the world is infinite; therefore that the world is infinite is not impossible.

The minor [premise of the third proposition] is proved as follows: God, or even a creature, can conceive of every sort of

FOL. 10v

---

[13] Or, 'may'.
[14] For the last bracketed words the MS. has merely 'etc.'

magnitude, and of course the magnitude of one pace, two paces, etc.; these magnitudes are certainly infinite in number; therefore [God understands] one magnitude made up of all these, which is an infinite magnitude.

6. Coming to answer these arguments, [our author] says (his page 19) he has shown that the assertion 'The infinite is possible' is inconsistent with the well-known principle of reasoning: 'The particular is included in the universal'. But he has not shown it, nor could he do so, for the two statements are not contradictory. To make this more fully understood we must explain what 'universal' is.

It is obvious that any individual thing is one, and singular. Like Peter and John, each man is an individual, and because there exists no man who is not one of a number of individuals, it follows that no man is universal.[15] In the same way we prove that no stone, no tree and, in sum, no thing is universal. What then, is 'universal' if a thing is not universal? Every word should be considered [as being] equivocal. For example, 'number' in the statement 'Two, or three, is a number' is the name of a thing; but when we say '"Number" means "two" or is the name of "two"', here 'number' means the word 'number' itself, and is the name of a name. Thus when we say 'Peter is an animal' then 'animal' is the name of a thing–that is, of Peter himself; but when we say '"Animal" is a genus' here 'animal' is only the name of a word. Again, as I have shown before, no thing can be universal, but a name can.[16] A thing [called] 'animal' is always some one animal out of the number of animals which exist, but it is not more than one (for an animal is not *any* animal, but a certain one); therefore it is always singular. The word 'animal' is certainly a single word, yet it is the name, not of any one animal only, but of any animal. When we have grasped this it is immediately clear that '*Universal' is a word that is the name of several things, but a word that is the name of a single thing is singular.* Yet a single existing name may be used to signify several things, and for this reason it is

FOL. 11

---

[15] Or, 'a universal', and so in what follows. See also the next note.
[16] Both the use of the word 'before' rather than 'above', and the words 'but a name can' (it is unclear whether Hobbes is claiming that this was shown above or whether it is a new assertion) suggest that the reference is not necessarily to the closing words of folio 10v but may relate to some earlier work. The topic 'names' is dealt with later. Cf. fols 5*ff*, 19.

called universal. A thing is properly said to be 'one' and not 'universal' or 'singular', because these are only the appellations of names. As man is one thing, so 'Socrates' is one thing; but the name 'man' is universal and the name 'Socrates' is singular. As a rule, 'universal' is correctly defined as 'What is said, or declared, about many'; 'singular' as 'What is said of one only'. 'To be said or declared', however, pertains not to things but to appellations.

Again, some universal names signify more things than other universal names do. A name which signifies more things embraces more things; but when [a name] indicates not only single things but a universal thing also, it then covers fewer things. So logicians usually say that the name (be it single or universal) included in a more universal name is a *part* of the name which contains it. That which includes it they call 'the *whole*', but not such that a 'whole' of this sort is taken as 'an integral whole'; for 'Socrates' is not part of 'animal' as a hand or limb is part of a man. And in the same way that a proposition is true when the universal is said of the particular–for instance, 'Socrates is a man'[17]–so when the whole is said of the part, as in 'The hand of a man is a man' or 'A part of the infinite is infinite', the proposition is not true.

FOL. 11v

Therefore that very well-known principle of reasoning: 'The individual is included within the universal', is what logicians call a declaration concerning the whole; i.e. that a name describing the universal also describes the singular. Hence if 'Man is an animal' is true, 'Socrates is an animal' is also true. Yet the assertion, 'The infinite is possible' is, says [our author], incompatible with this principle, because, [he believes,] if the infinite is attributed to the whole it will also be attributed to the part. He is deceived here by the equivocation of 'part' and 'whole'. 'Infinite' is not a more 'universal' name than 'part', and the arguing from the whole to the part is the inclusion not of the singular within the universal but of the lesser within the greater, because 'Every man is an animal; therefore this man is an animal' is sound argument; but if someone reasoned like this: 'A thousand paces make a mile; therefore one pace is a mile', or like this: 'Paces which exceed all number are infinite; therefore paces which do not exceed all number are

---

[17] I.e. when the universal takes in the particular, as in '"Mankind" includes "Socrates"'.

infinite'? could anyone fail to see how absurd such reasoning would be?

FOL. 12   To the first argument in favour of infinity's being possible–which is deduced from the fact that nothing irrational follows from the assumption–he makes no reply. Nor do I see what reply there could be [i.e. from him or from anyone], since [the case] is absolutely established by formal reasoning.

7. He answers the second argument by denying (at his page 20) the antecedent, which was: 'God can bring it about that the measure of what is in His power is also the measure of what can exist in itself', which he says is not true. That is, he asserts that, whatever God can perform, not everything that He can do can be done, i.e. can be brought into existence (which is what 'to be done' means). This is manifestly absurd, because to say 'God can make or do something' is the same as saying: 'Something can be made or done by God'; and 'can be made' is the same as 'can actually exist'. Indeed, not only is a potential useless unless it can be brought to fruition in an act; it should not even be called 'potential'. Therefore when he says: 'God cannot equate the measure of quantity in His own potential with what is in the nature of things', he is saying that something is possible but also is impossible. Besides, he calls this measure infinite, whereas every measure of quantity is finite; for (to settle the matter) in things 'measure' is the same as 'distinct limits', but to the infinite there are no limits.

8. The third argument he answers on his page 21, conceding that 'What is in fact impossible is also intellectually impossible'.[18] But 'what is infinite is specifically in some intellect', this he denies. What does this 'specifically' mean? Is it any different from 'conceiving [one thing] distinctly' from other things? And what is 'distinctly' if not 'determinately', i.e. 'finitely'? He denies, therefore, only this: that the infinite can be conceived finitely. But how, [one asks,] can the infinite be conceived 'non-specifically'? 'In [our] understanding of

FOL. 12v   abstract substances there are', he says, 'infinite species of difficulties.[19] These last are not distinct individually but are numerous in a wonderful unity, and indeed are infinite in number at

---

[18] Perhaps, 'What it is impossible to do it is also impossible to understand'.
[19] 'Difficulties': in the sense of 'barriers, objections'?

the same time.' This confusion of unity with number and with infinity is not only 'wonderful' but absurd and impossible; for 'There are many species which are also one species' is a contradiction. And even if [the words cited] were not a contradiction, whence can he know how the species of things are contained 'in [our] understanding of abstract substances'? Further, as he cannot know this, why did he think it should be asserted here in a context which concerns the philosopher, not the theologian? That it may be so, however, he shows on his page 22 by the example of bodies and mathematical surfaces: 'Suppose we concede that a surface encloses infinite lines and that a body exceeds countless surfaces, and in fact in such a way that there is not a surface, not a line which does not actually exist. How much more, then (to put matters simply), must we allow that things spiritual contain the corporeal, strong though the latter be?' In this example I see only one clear fact: that undoubtedly he understands equally well what line, surface and body and what abstract substances are; but that he has seen how we say 'A line is in a surface', or 'A surface is in a body', no more than he has seen how incorporeal substance contains corporeal substances and surpasses them. Everyone familiar with geometry knows that each part of a body is body, every part of a surface is surface, and, similarly, that every part of a line is line. From this it is obvious that a line is not *in* a surface, nor a surface *in* a body in the way that a smaller magnitude is in a greater or an infinite magnitude.

Now in measuring every body a threefold procedure has had to be followed, the first step concerning length; the second, breadth; and the third, thickness. [a] Length relates to the path described by the motion of any body whose size is not being considered. That is why geometers have designated such a body by a small mark, which the Greeks call *stigmé* and *kentron*, the Latins *punctum*—not because a point can be so small that it has no quantity but because they wanted as small an object as possible to be seen where size was not to count. The length described by the motion of such a point they called a line.[20] [b] The second way of measuring is that designated by the movement of a body when only its length is considered, and not its breadth. That is why geometers have been accustomed to represent such a body by a thin line which the Greeks call *grammé*.

FOL. 13

---

[20] Cf. fols 34v, 248.

It is not, in fact, without breadth, but is considered as being so. The path described by the motion of such a body they called 'surface'. [c] The third way concerns what is designated by the movement of a body when its surface is considered, but not its thickness. This movement produces a solid, or what is called a mathematical body. Hence lines and surfaces are not something contained in a body that is being measured, as parts of it, but are as it were the various tools of those doing the measuring. That is why they are called mathematical, i.e. instructive, because these words were used by geometers for teaching-purposes only. In defining 'line' and 'surface' they wished to avoid two things: first, the taking into account, when measuring surfaces, of the thin surfaces present in the material lines of the figures (which, for their demonstrations, it was necessary to draw materially); and second, the inclusion, among the parts of the actual solids, of the thin envelopes of the bodies whose surfaces they were measuring. When we understand this it is easily conceded that the line is not included in the surface and the surface in the body, and that the length of roads should be calculated without regard to their width, as indeed is generally done, and both these things[21] without regard to thickness; for so do people who sell land, without taking account of its depth.

Now if bodies are, [as White says,] in spiritual things in the same way as lines are in surfaces, or surfaces in solids, then we should consider small spiritual things as though they were not spiritual, just as we consider a small surface–by calling it 'a line'–as if it were not a surface.

Those who attempt to explain the Divine Intelligence in terms of our own, thinking that supernatural dogmas can be demonstrated by the reasoning used in philosophy, must necessarily fall into such absurdities. So the arguments for the possibility of infinity remain unshaken; nor can they be easily refuted except if, at the same time, those adduceable for eternity shall collapse also, because eternity is to time as 'infinite in magnitude' is to quantity.

---

[21] Also the things mentioned at the beginning of the folio?

CHAPTER THREE

※

On Problem 3 [of the First Dialogue]:
Can there be several worlds? Does a
vacuum exist? And in what do density
and rarity consist?

Explanations of (1) 'imaginary space', (2) 'real space', (3) 'place', 'vacuum', 'distance',[1] 'contiguous things', 'here', 'there', 'quantity'. An examination of (4) [our author's] first argument proving that there is one world only; (5) of the second; and (6) of the third. 7. Contiguous things are not 'one thing'. 8. An examination of the argument against the possibility of a vacuum. 9. Of the nature of rarity and density, and of the thermoscope. 10. Of the expansion of gunpowder when ignited, and why it occupies more space fired than when not fired. 11. Why, in pneumatic guns, the air rushes forth. 12. A scrutiny of our author's opinion about the nature of rarity and density.

1. It is from the notions of space and situation that our author derives an argument to prove that there is one world only; so we shall start by defining what is properly to be understood by these and other words concerning this subject. It is obvious that, by its action on the organs of sight, any visible object produces a certain image or conceit of itself when we register some object we have seen.[2] For example, the sun acting upon the eye makes it see a shining circle which we believe to be that

---

[1] Its being followed by 'contiguous things' would seem to require 'distant things'; yet the context below demands 'distance'. Cf. fol. 403.
[2] Or, for 'when...seen', 'as is the general opinion (on this matter)'.

FOL. 14v

of the sun. If we view the sun directly, the circle appears to be in the sky, but if we see the sun reflected by water it seems to be beneath the water. That is, the image always appears in that straight line at one end of which is the eye; so the image hardly ever appears in the place where the actual object is. In reflected or refracted [as opposed to direct] vision the object and the image are situated in different straight lines;[3] in direct vision, even, the sun and the heavenly bodies, and every distant object, although in the same straight line with the eye, are yet much more remote from the eye than their images are. That is to say, the objects are further away than they seem.

Now, the images so produced consist of colour and shape; but shape is finite space. Where the object lies, however, we find neither the image nor even its constituent shape. The apparent area of the sun or of any other object is therefore not inherent in the object itself but is merely imaginary. For how can the circle of the sun, which seems so small and so near, be a quantity inherent in, and equal in magnitude to [adequatus], an object as remote and as vast as that of the body of the sun? Moreover, the objects we view fade, and take with them their dimensions, though there remain in the mind their images, i.e. [their] shapes, or the spaces in which they [first] became visible. If this were not so, the images of absent objects would not be retained, nor would any memory or dream of things possibly exist.

Again, just as one can remember the life-appearance of a man who died some time ago,[4] so even if the whole world were destroyed except for one man, nothing would prevent this man from having an image of a world which he had once seen, that is, from visualising a space extending from him in all directions as far as he wished. Therefore 'imaginary space' is nothing else but 'the image or conceit of body'– I say simply 'of body', for the image of a white body is not only a space but a space

FOL. 15

that looks white. Alternatively, we shall say that 'space is the image of a body inasfar as it is body; a white space, the image of a white body inasfar as it is white; a finite space, the image of a finite body inasfar as it is finite, and a square space, the image of a square body'. Hence it is clear that the existence of space

---

[3] Cf. Chapter X.
[4] The text is here ambiguous, and the point of the parallel not clear. See fol. 456v.

depends not on the existence of body but on that of the imaginative faculty. If in a given space there is air, the space does not disappear if the air does, but it remains the same and can be filled either by water, or by another body, or by no body at all; in the same way, if the world were to vanish, then unmoved space, provided that it did not disappear at the same time, would still exist.

2. A mind-picture derives from the action of some agent we suppose to exist, or to have existed, outside the mind of the person who imagines [something]. Granted that this agent exist, we usually call it body, or matter. It follows, therefore, that bodies would exist even if there were no mind-picture at all. Next, it is impossible to admit the existence of any certain body without at the same time realising that it possesses its own dimensions, or spaces. So this space which, when inherent in a body, as the accident in its subject,[5] can be called 'real', would certainly exist even if there were no being to imagine it. I define 'real space', therefore, as 'corporeity itself, or the essence of body taken *simpliciter*, inasfar as it is body'. Hence a body is to 'imaginary space' as a thing to the knowledge of that thing, because our entire knowledge of existing things consists in that mind-picture produced by the action of these things on our senses. 'Imaginary space', therefore, which is the mind-picture of body, is the same as our knowledge of existing body.

But suppose that no mind-picture exists, yet some finite body does, which we shall describe as follows: 'that in which there is not, but yet could be, other bodies'. Suppose, for example, that the world is finite, and that no creature endowed with imagination exists. Shall we, or shall we not, say that there is space outside the world? On the supposition of a finite world, of course there is no real space outside it; but there is no imaginary space either, because we have excluded the possibility of a being possessed of imagination. So we must say that there is no space at all [either inside or outside the world], and yet there is absence of body. This, however, is enough to make the existence of several bodies possible; for the absence of body is understood through the mind-picture we have of bodies, and is therefore the same as what we call 'imaginary space'. This last

FOL. 15v

---

[5] By this last phrase Hobbes probably means: 'If an object *happens to have* spaces within it...' Cf. fols 123, 221v.

is, in fact, absence [of body]. Inasmuch as it is outside us it is a mere fiction and is not *ens*; but the fictions themselves do exist within us, being the movements of the mind, which are called mind-pictures.

3. Every time the real space of any body shall coincide with some part of imaginary space we call that part with which it coincides the place of that body. But if the real space of no body coincides with a certain imaginary space, then we call the latter 'a vacuum'. The 'distance' between two bodies, or even between two vacant spaces, is the shortest space lying between them. So if some intermediate part is removed from a continuous body, the distance between its extreme parts is not thereby removed.

From the foregoing it is easy to define which bodies are called 'contiguous,' i.e. those between which there can be no body. For contiguity it is not sufficient that no body be actually interposed; it should not be possible to interpose any. The uneducated believe that the entire space between their heads and heaven is empty. But if some philosopher inferred from this that they touched the sky with their heads, then undoubtedly they would say that philosophy were madness; but the same uneducated persons believe that they are [actually] in contact with all the things between which and their own selves no body whatever can be interposed.

[The words] 'here' and 'there' do not signify a place, nor do we speak of them unless we point in a certain direction with the finger or with a stick, or unless we mention some known neighbouring object, so that the person addressed can himself look with his own very eyes for the place of the thing sought. Either [i] 'place' is equivalent to the location of an object, as when we say 'here' or 'there' or we answer the question 'Where?' or we point to a certain place loosely, as 'in a field', 'in a house', 'in a box', which are rather vague descriptions for the place of the thing sought; or [ii] the eye of the questioner is directed by the pointing of a finger towards the thing to be reached or seized.[6]

The 'quantity' of a body is the same as what geometers call 'ratio'. If you ask how large a field is, the answer is given as a measure, say 'a thousand paces, i.e. that which is to one pace

---

[6] The positioning of [i] and [ii] is uncertain.

in the ratio of a thousand to one'. Quantity may be defined thus: 'It is the determination of a dimension, by comparison with another dimension [already] known'.

4. Having understood this, let us examine by what reasoning our author demonstrates that there is one world only. If we assume the world to be finite, then to infer from an understanding of the term 'this world' that there cannot be several worlds is no less difficult than to infer from an understanding of the term 'this sun' or 'this animal' that there cannot be several suns or several animals. He takes as the start of his demonstration [what is called] an identical proposition, i.e. this: 'What is nothing is certainly nothing', for 'nothing' is just as much as 'certainly nothing'.[7] But an identical proposition cannot commence a demonstration, because we can infer from it only what is assumed in the premises. For instance, let there be any identical proposition, such as 'A is A' and, in order to form a syllogism, let us take any other proposition of which one term is A, such as 'A is B'. From these premises 'A is A' and 'A is B' can be inferred only that A is B, [and that B is A], namely what was assumed without demonstration. Next, [our author] assumes that 'Non-being has no effect, no knowledge and no parts'. This is indeed true as regards the effect and the knowledge; but that non-being has no parts is false, for whatever can be divided will have parts; and we have shown in the previous chapter that imaginary space can be divided. To anyone who considers the matter it is evident that, just as the nose is a part of his face, so the mirrored image of the nose is a part of the image of the face that appears. The apparent face or nose is not *ens*, being, but is an appearance, a conceit, and a figment born of the motions of the organs of sight. But having assumed this false basis, he proceeds to demonstrate in the following way that imaginary space has no parts (his pages 28 and 29):

FOL. 16v

'If there are two worlds, either they will be contiguous or they will be separated from one another by a certain interval;

'Suppose first they are separated. Then one will be "here" and the other "there". But the "hereness" and "thereness" signify "towards this or that part". There are therefore parts

---

[7] The words from 'for' to the end may form part of the illustrative proposition itself.

in the space which we imagine to be outside the world. Hence there are parts in non-being, contrary to what was assumed.'

This reasoning is worthless, because what has been assumed is false. Besides, when we direct a questioner to the right or left, or upwards or downwards, by saying 'here' or 'there', we are not on that account saying that there are parts in imaginary space, nor are we dividing[8] it, but we are dividing real space; it is in real space that we are directing him. So apart from assuming false premises our author uses the word 'parts' ambiguously.

5. Next he reasons in the following way:

FOL. 17

'Simple and ignorant men, when questioned about the placing of something, answer according to one concept of "place" derived from the senses and from nature, namely that it is something that surrounds us and that, as Aristotle says, "it is a sort of immobile vessel". Now if outside the world, where there is nothing, were placed another world, the latter would be situated in nothingness, and therefore neither world would have a place; so they could not be separated by place.' Though what simple and ignorant men feel is not of great consequence to philosophy, I shall not concede to him that such men think place to be something immobile that surrounds us like a vessel. That is a more scholarly error than would occur to an ignorant man; to philosophers alone is it permitted to be learnedly insane.

But if it were true that 'what is situated in nothingness has no place', then it would follow that not even a single world was in a place; and if indeed what is not in a place does not exist, then not one world will exist. But if one world can exist, though it is not in a place, several worlds will also be able to exist, though they are not in a place [either]. It will follow that several worlds can exist, even if they cannot be separated by place. The Aristotelian definition of place, namely 'the surface of a surrounding body', is not true; if it were, the consequence would be that wine sealed in a cask could be transported from here to the Indies and yet not be moved, for what is always in the same place is not moved, and the wine, being surrounded by the same surface of the cask, according to the above defini-

---

[8] Or, 'distinguishing'.

tion [of 'place'],⁹ remains constantly in the same place and is not moved. Besides, a place is the equivalent of a thing put in it; but how can the thing placed (which is a solid, that is, a body having three dimensions) be the equivalent of a surface? The idea that 'Place (if it is the surface of a surrounding body) is a certain immobile vessel' is against the teaching both of Aristotle himself and of all the philosophers, who stipulate that every body is movable.¹⁰

6. Finally he employs this reasoning: FOL. 17v

Things between which no entity is interposed are not distant from one another;
But between two worlds no entity is interposed;
Therefore two worlds cannot be distant one from the other.

Here the major premise is false; for (as has been shown above [at folio 15v]) those things between which some entity can be interposed are distant from each other even if the entity is not actually interposed. Given three contiguous bodies, A, B, C; the distance between the outer ones A and C depends not on the existence of the intermediate body B but on the separate existences of A and C, for their distance apart is the relation of one to the other, not [their respective relations] to some third one, and therefore not to the middle one.¹¹ Yet he affirms that A and C are distant [one from the other] because B is an entity placed between them. That is, were God to reduce B to nothing, then, [says our author,] A and C ought to touch one another straight away, and this without having to move or needing time in which to reach one another. But, [I observe,] if for them to come into contact motion is necessary, then, to reason accurately, during the time that they are being moved they will be distant from each other even if there is no body between them.¹²

7. Judging that he has sufficiently demonstrated that there cannot be several worlds distant from one another, he then supposes them to be contiguous (his page 29). But this very

---

⁹ Cf. also fols 218, 268.
¹⁰ That the closing adjectival clause is continuative rather than restrictive (see fol. 6), i.e. that the comma must precede, is made clear by the manuscript's having a semi-colon in order to stress the point.
¹¹ Figure 2.
¹² Perhaps, '...even if no medium intervenes'.

phrase 'to be contiguous' he takes as the equivalent of 'to be one'. Therefore he immediately changes the word 'contiguous' into the word 'joined', which seems to savour more of unity. But I do not see why two worlds should not always remain two even if contiguous, just as two eggs are two. If we call [two worlds] 'joined' instead of 'contiguous', that act of joining them does not prevent them from being either two [separate] worlds or a [single] world made up of two worlds. One world, however, could not be composed of several worlds unless there could be several worlds. If therefore the world *is* finite, philosophical reasons still cannot determine whether there are several worlds; or whether one only; this is therefore a dogma of faith, in which we must rest content, not with the subtleties of philosophers, but with the authority of the Church.

FOL. 23[13]

8. This discussion is linked with the question: 'Can there be a vacuum?' which he treats briefly, on his page 30, in the following way:

'If a vacuum exists, there exists a place without body, i.e. a concave body without anything to fill the cavity. So the sides of this concave body will close up because there is no entity [*ens*] between them. But if the sides are closed up, this shuts out a vacuum; therefore as soon as we suppose vacuum we must reject its existence.' But it is false that the sides between which no *ens* is actually interposed are joined, because *ens* can still be interposed. So this whole argument falls to the ground.

9. The common people do not use the words 'rarity and density' in the way that philosophers do. Non-specialists call something 'rare' when its parts do not cling to one another; in this sense Virgil spoke of 'a few scattered swimmers sighted in the wide expanse of sea', and in this sense [too] we would say that a heap of dust is more dense than a little cloud composed of scattered grains of the same dust, or that one cloud is less rare than another, though they are both made up of tiny drops of the same water. But men of science made the observation that certain bodies, but air principally, could be compressed into a smaller space and could again occupy their former space.[14] And still more air could spread into a larger area, but in such a

---

[13] The leaves are misbound; hence the foliation is not here consecutive.
[14] Perhaps, 'regain their former volume'.

CHAPTER III: 9

way as not to appear greater [in volume] at one time than at another. The philosophers were at a loss when they came to give an explanation of this. Some, indeed, denying [the existence of] vacuum, said that it happened because the same body, remaining the same in quantity, was sometimes more rare and sometimes more dense. For instance, a grain of gunpowder does not occupy much space before it is fired, but when it is exploded we see it fill a space a hundred times greater; in the same way, compressed air occupies a smaller space than dilated air. Therefore these philosophers call 'more rare' that which is contained in a larger place; 'more dense' that which is contained in a smaller, the quantity being constant. But how, since place coincides with what is put in it, the same place could sometimes contain more, sometimes less matter is incomprehensible. Other philosophers in fact, admitting [the existence of] vacuum, hold that bodies are unified because they are permeated with empty spaces, just as the non-specialist considers the rarity of visible bodies to consist in the interposition of invisible air. But they cannot thus account for all the phenomena which lead us to believe that bodies are sometimes contracted, sometimes dilated. For the cause of the dilatation and compression of the air in a thermoscope can be sufficiently explained even if we deny [the existence of] vacuum.

FOL. 23v

The name 'thermoscope' is given to a glass flask with a large bulb [A] and a long but narrow neck, AB, some lower part, B, of which is immersed in water contained in another vessel, CB.[15] If the air in the flask is heated before the latter's extremity is immersed in water, so that the air enclosed in it is hot, the water rises slowly as the air gradually cools. Suppose, then, that the water rises to [a line] DE. Let us see how this can be explained if we admit the existence of vacuum.[16]

We concede that the parts of the heated air have been separated by the agitation produced by the heat, and that many empty spaces have been admitted [into AB] as a result of this separation of the parts. Hence the whole hollow space AB is occupied by air and the interposed vacant spaces. We also concede that, as cooling takes place, all these parts of the air which formerly were separated are re-united within the space ADE,

FOL. 24

---

[15] Figure 3.
[16] Pleading similar to that which follows occurs in a letter from Hobbes printed by F. Tönnies among 'Siebzehn Briefe...an Samuel Sorbière' in *Archiv für Geschichte der Philosophie*, III Berlin, 1890, pp. 211–212.

47

and therefore that all the vacant spaces are congregated into one space, DEB; but I see no necessity for the water to ascend in order to fill that void. Since nature has admitted a vacuum, i.e. one equal in quantity to the space DEB (dispersed into particles), why will she not also admit the same amount of vacuum [if it is] united in the space DEB, which is equal to all these vacant spaces? But even if she does admit this vacuum, there is no reason why the water should, contrary to the nature of heavy bodies, ascend at all. I see one way out of this difficulty, if the same philosophers accept this [alternative] hypothesis: the quantity of vacuum in the whole world is fixed and limited, so that it cannot become greater unless the boundaries of the world are pushed outward, or smaller unless they are contracted. Then, the air having been heated in AB, and more vacuum being admitted than there was before, there will necessarily be less vacuum in the rest of the world than previously. In consequence, other bodies situated outside the glass flask are compressed more than usual, as long as the vessel remains hot. Therefore as the flask cools they occupy their normal space again, and compel the water to fill the vacuum in DEB. This is indeed difficult [to admit], but I cannot see why nature should be so opposed to a great quantity of vacuum, unless because other bodies receive thence a compression greater than suits their own motions and natures. As regards [the reason for] rarity and density, I myself affirm nothing, for I prefer ignorance to error. But if anyone explains it clearly, I think he will be indeed laying bare those inner mysteries of physics.

10. As for the expansion of gunpowder when fired, those who admit the existence of those little vacant spaces will easily account for it. They will say that the innumerable particles of each grain [of powder] are scattered about when set fire to, and that what appears to be fiery is not any one continuum, but is many little sparks separated by empty spaces. The reason why they appear all together as one continuum is that which makes bodies, when lit and shaken rapidly, seem elongated, even if they are round, and circular when they are moved rapidly in a circle. Each particle in a fired grain of gunpowder is moved at an incredible speed, and therefore the particles seem longer than they are. Hence the sight of them fills not only the space they themselves occupy but also the one occupied by the adjacent vacuum. The necessary consequence is that an

appearance is produced as though of one continuum, though actually the producing-agent itself is not all of a piece.

11. To admit [the existence of] vacuum in the way I have suggested can despatch the difficulty of [explaining] the compression of air in the guns newly invented, which are called pneumatic, and in children's pop-guns made of elder-wood. Air is added by being forced in, until the little empty spaces are filled, and as a result the particles of the enclosed air are deprived of their natural movement, and strive to recover their former space (as inflated bladders, if compressed [and then released], spring back). It follows that, as soon as they are given an outlet, instantly and swiftly they burst forth. Such, in my opinion, is what should be said about rarity and density. Let us now consider what our author says about their nature (his page 31).

12. 'Such things', he says, 'are rare and dense according to the proportion of quantity to the subject.' What he means by those words 'the proportion of quantity to the subject' I do not know. Everybody agrees that proportion exists only between things of the same kind. Consequently there is no proportion between quantity, which is accident, and a subject, which is substance.[17] There can be proportion between quantity and quantity, and between two *quanta*, but not between quantity and *quantum*.[18] If anybody says, incorrectly, that there is proportion 'between quantity and a *quantum*', when he should have said: 'between *quantum* and *quantum*', the only proportion there will be will be that of equality; for quantity can be neither greater nor less than the subject in which it is inherent. But let us hear the reason for [his] statement. 'Nothing', he says, 'is clearer than that air is more divisible than water, and water than earth. So undoubtedly the proportion of divisibility, or of the quantity, of air to air, is greater than the proportion of divisibility of water to water, and likewise that of

FOL. 18v

---

[17] Hobbes seems to mean: '...no proportion between dimensions, which are an attribute, and the Aristotelian *hypokeimenon* (material), which is substance'.

[18] The distinction Hobbes is drawing here seems to be between *quantity* as used in formal logic: 'the connotation or range of a term', and *quantum* in mathematics: 'an amount that may be measured or expressed in symbols'. The general drift, however, is clear: Like can be compared only with like.

water to earth.' In saying that air is more divisible than water he obviously calls 'softness' divisibility; for soft substances are more easily cut with a knife than are hard substances, and more readily allow their consistent material to be separated. But in speaking of divisibility or quantity he shows that in his opinion divisibility of that sort (i.e. softness) is the same as quantity. Therefore the harder bodies are, the smaller they are, as they have a lesser divisibility, that is, their quantity is less.[19] So the water in a jug, since it is as divisible as the water in the ocean, will be equal in quantity to the ocean! If this is not raving, what is? But say we accept, in some way or other, what he says, how can the consequent be true: 'If air is more divisible than water, then the divisibility of air has the greater proportion to its subject, i.e. water'? Just as the divisibility of air is greater than that of water, so air is more divisible than water. Thus one draws one's comparison between the greater and the greater and between the lesser and the lesser, and therefore the proportion can be the same on each side. But enough of all this; what 'to divide' is, and what 'quantity' is we have explained at length above.

FOL. 18v

---

[19] Hobbes is ironically quoting the consequence of White's reasoning.

CHAPTER FOUR

❧

On Problem 4 [of the First Dialogue]:
On the immobility of place;
On the place of spirits[1]

1. What the intellect is, and how things are said to be 'in the intellect'. 2. That no immobile surface exists except in the intellect, i.e. in the imagination. 3. That incorporeal substances are beyond the grasp of the intellect; that they are believed in through faith, and are not known through philosophical reasoning.

1. So as to prove that several worlds cannot exist, our author has adopted the Aristotelian definition of *place*, namely: 'Place is the surface of a containing body, but is immobile'.[2] Now, he has realised [i] that all bodies are movable, and [ii] that every real surface is the surface of some body. Therefore [a] he had either to admit that place is imaginary space (he had denied this), or [b] he had to explain how the real surface of a movable body is not itself also movable. In order to prove the second to be the case he has used an argument based on the nature of the intellect. We must therefore say something about the latter.

You must realise that, when you look at anything, a certain image of it is left behind. This image (caused by the action of the thing you have seen, even though you have ceased looking at it) is called the idea or picture of the object viewed. Like-

---

[1] Under the second theme (discussed in Article 3) Hobbes deals with the question of where incorporeal substances *are*.
[2] Cf. fol. 17.

wise as a result of the action of objects on the rest of the senses something remains to mark this action. For example, as regards our hearing, a sound is the fantasy of the motion by which an object acts on the organs of hearing; and as regards our sense of smell, a scent is the fantasy or manifestation of the motion of the object that acts on our sense of smell. Most students of philosophy call these images of this type *things*, inasmuch as the images exist in the mind, considering them as being the things themselves; so instead of saying 'the image of a thing' they say 'a thing in the imagination'. Now, since the images of things are aroused when we hear words, or appellations, or names–all these, by common consent, are applied to things–we are said to 'perceive' those things. This is wrong, however; correctly speaking, 'to perceive' is 'to have that image of a thing for the arousing of which a name had been imposed', and this is the same as to be able to define it. Thus 'perception' is to do not with things themselves but with the words and terms by which we express our judgment about things. So it is said to be of universals. Things are not universals, but names are, as was said above in Chapter II [at folio 11]. Now–to define the term–'to understand a name' is to remember, by means of a name, those things which have to be considered in some matter, and for which a name has been imposed. Thus the name 'man' is understood when this word brings to mind not only a human shape but also its reasoning-capacity.[3] That is why brute beings lack an intellect: because they have assigned no names to things, they cannot possess an intellect, which is concerned with names; and because they lack reasoning-power, for there is no reckoning without a syllogism, nor is there a syllogism without a proposition, nor a proposition without terms. When, therefore, we say that 'some thing is in the intellect' this is the same as saying that this thing's image is in the mind, but conjured up by its own name. Consequently a thing 'in the intellect' and an imaginary thing are the same; likewise with space 'in the mind' and imaginary space.[4] If we allow these conditions we shall see our author establish the immobility of place from the fact that place is imaginary

---

[3] Owing to textual criteria to do with the positioning of a caret, the rendering of these two sentences is tentative only.
[4] This interpretation of 'imaginary' ('in, or relating to, the imagination') holds in what follows. See also 313v.

space,⁵ a thing he had denied earlier. But we must run through his argument.

2. 'Words', he says on his page 32, 'reflect the mind.' (We note, by the way, that this is utterly ridiculous, for what resemblance can there be, pray, between a word and the mind? And how is it that, if 'words reflect the mind', the languages of all nations are not alike, as their minds are?) 'Now, the mind contains virtually no one definite conception, but by comparing several sensations it holds, as it were, composite notions. And', he says, 'they err who seek afterwards in reality the same things that they find through the intellect.' Again, 'So it is that, when they hear that "place is immobile surface", they immediately seek, among actual things, some surface that by its nature is immobile; and when they do not find one, they exclaim that "place" has been badly defined.' By these words he means: 'The image of a surface, e.g. of running water, is not the one and only image of that water but is made up of the surfaces of the waters that follow one another in a perpetual flow, and also of the surfaces of all the bodies that could take the place of water'. This cannot be so, as the surface of any body is carried away with the passing of that body. He ought to have said that [i] the imaginary surface caused by the sight of the running water was an image of the water, not as water, but as body, and that [ii] the image therefore represents not that particular water but any water, or air, or body of the same size and shape.⁶ Hence it is the surface which was in the mind, i.e. the imaginary surface, that is immobile, not the surface associated with the water by chance. He immediately concedes this by saying: 'They err who seek in reality what they find in their mind'. This is the same as saying that an immobile surface exists not in things but only in the mind, i.e. not as a real surface but as an imaginary one. Rightly, then, do people exclaim that, if surface be considered real, 'place' has been ill-defined as 'immobile surface'.

Next he says on page 33 that 'a flowing surface can be motionless if in the intellect of the average man some reason associates it with something immobile'. But what is 'for a mobile surface to be linked by the intellect with something

FOL. 27⁷

---

⁵ The original itself is ambiguous: 'from the fact...space' goes with 'shall see' or with 'establish'.
⁶ Cf. fol. 111.   ⁷ Further misbinding of the original leaves.

motionless' other than 'to picture an unmoved surface'? For when an object is moved its real surface cannot be retained by any bonds except those of the imagination; nor is 'to be linked together by the intellect' anything but 'to be understood'. So the matter again comes to this: place is imaginary space. This contradicts what, in the previous problem, he had assumed in order to prove the unity of the world.

3. On his page 33 he had put the question: How are there[8] incorporeal substances in space, and how, unless they are in space, is it true (as every old woman and boy says) that 'what *is* nowhere does not exist'? Agreed, the question is a very hard one, for it is impossible to conceive of matter other than by the criterion of its magnitude. Now, what is conceived as having size is universally called corporeal and material. All nations have designated substances in this way: indeed, the ethnics,[9] noticing that, not only when they slept but also when they were awake, certain images appeared before their eyes, did not know that these were apparitions aroused by some violent mental passion. They were able to consider them as external things, possessing dimensions; but because these images were seen to fade easily, they could not be considered bodies. Wishing to give them a name for insubstantial bodies, they have called them spirits, for, being ignorant of nature, they recognise bodies not by their dimensions but by their opacity, i.e. whatever the sight can penetrate they consider to be a vacuum. This could explain how people came to believe in the existence of innumerable daemons, both good and bad ones, and of other incorporeal substances.

FOL. 27v

But since it cannot be known by natural reason whether any substances are incorporeal, what has been revealed supernaturally by God must be accepted as true. This is the way, therefore, in which Christians, bowing to the authority of Holy Scripture, not to the reasonings of philosophers, have classed substances; but to do so constitutes a tenet of faith, not of knowledge. Those who bring the matter forward for discussion cannot explain how substances can exist in this fashion and yet not be situated anywhere; so such people do not confirm the Christian faith, but weaken it. Indeed, it is natural for many to consider as false what someone tries to prove true, but cannot.

[8] Perhaps this word should be omitted.   [9] Cf. fol. 300, note.

CHAPTER FIVE

On Problem 5 [of the First Dialogue]:
Whether there is a liquid substance
beyond the moon.[1] Whether material
subject to decay is found in the heavens.
In what sense 'quantity' is the
same as 'indivisibility'

1. Definitions: Motion; change; to act; to be acted upon. 2. 'Categories' and 'related things'. 3. Birth; decay;[2] generation; corruption; death. An inspection of (4), the argument for [the existence of] liquid; of (5), the argument against the motions of the solid spheres; and of (6), the argument by which he proves that quantity is liquidity.

1. Here we must define motion; change; to act; to be acted on; 'related things' (or, as Aristotle calls them, 'what things facing what things'); birth; death; 'to be generated'; 'to be destroyed'; and other terms similar to the last. Our author has reasoned from all these, but has not rightly understood them. He believes, however, that he has shown from them that a kind of liquid material does exist beyond the moon; that the heavenly spheres, were they solid (as the astronomers of mid antiquity

---

[1] The first of a number of topics, constituting chapter headings or titles of articles, in which much of the wording refers to White rather than to Hobbes.
[2] Perhaps, 'destruction'. Both in the title and in the narrative Hobbes argues round the words *interitus* and *mortalis*.

thought),³ cannot be moved one within another; and, lastly, that there is in the heavens some matter subject to decay.

Postulate: Whatever is said *to have been moved* seems to us to be not at the same distance from, or not in the same position relative to, any point we take as fixed, i.e. it is not seen to be in the same place. Now, to be moved (or motion) is the constant quitting of one place and the assumption of another.⁴ If that which is moved assumes, during the time when it is moved, other properties, just as if, while he walks, a man blushes, this is associated not with the motion I have defined here (for someone, standing in the same place, can both blush and go pale) but with change. So *change* is a movement of parts, or a motion which causes the whole to appear other than it was before. For we suppose that change in things arises from a change of the appearances, or of the fantasies excited in us by the things themselves acting on our senses. So if anything appears to us other than it formerly was, we at once say that it has been changed, whether it appears in the same place or whether in different [places]. For we cannot believe that a thing whose every part is exactly as previously [it was] nonetheless acts differently on us, the same living people, from the way in which the parts acted before.⁵ We shall therefore ascribe every change to motion–not, indeed, of a body as a whole, but of its internal, unseen parts. If indeed we were to see that motion of the parts we would call it, not a change in the whole, but the movement of its individual parts; and we would say that there were as many bodies as parts we saw severally moved. *Change*, therefore, is a movement, not of the whole, but of the parts; and *motion* is not a change in the parts but a change in the whole. In fact, a movement of parts is termed change only if it gives rise, in the mind of the viewer, to a different picture or image of a thing. Thus a sphere,⁶ which rotates on its axis in such a way that the whole remains unmoved, though each part is in motion, is considered unchanged [in

FOL. 28v

---

³ Cf. fol. 30.
⁴ The words 'in the same place' require complementing: 'as that occupied by the point' or 'as it (the mobile) was some given time-interval previously'. With 'to be moved...another' cf. fol. 454v.
⁵ In the manuscript there is a comma after 'a thing' and after 'previously'; but Hobbes can hardly mean that every thing must remain unchanged.
⁶ Retention of this comma supposes that what follows relates to any and every sphere. 'Rotates': here and now? Or does 'rotates' signify merely the ability to rotate?

such circumstances], because it presents the same image and appearance.[7] But if, because of the motion of the [bodily] parts, [someone] appears differently to the observer (as, without moving, a human being sometimes blushes and sometimes grows pale because of the movement of his blood) he is then indeed said to have been changed. Change, then, is this movement of parts by means of which something presents to the observer an appearance different from its former one. That which moves or changes something else is said to act: that which is moved or changed by something else is said to be *acted on*.

2. In the book he called *Categories*, i.e. 'appellations', Aristotle distinguished the names or appellations of things into ten types:

[*i*] Certain names are assigned to things because of the images or pictures that the things arouse in the mind. These names [quiddities] answer the question: 'What is it?' i.e. 'What is the thing whose image we have?' The category of beings [*ousia*], or essences, consists of these.

FOL. 20

[*ii*] Other names answer a question concerning a part of the image ('Page 9 [of White]: For parts of the image in the mind are its extent or size or shape, colour, and any other perceptible quality'),[8] e.g. the question: 'What we see or visualise, how big is it?' The answer is given in terms of measure, e.g. 'half a yard or a yard long'. Hence is drawn up the category of *quantity* from names of the units of measure.

[*iii*] The question: 'What is something *like* of which we have the image?' e.g. hot, cold, white or black; square or round; from these terms is formed the category of quality.

[*iv*] One can ask what, how big, and of what kind anything is *in relation to* something else, e.g. whether anyone is a son, a contemporary, or alike [relative to someone or something else].

These names are allocated to things. Because we compare those things with others, we have the category of relationships or 'what in relation to what'.

---

[7] The same appearance, whether or not the whole is in motion? Hobbes would not seem to mean, however, that the sphere rotates so fast as to appear motionless.

[8] In the manuscript this parenthesis is in the margin.

[*v*] Here he placed terms indicating motion in respect of a thing that moves others, i.e. *action*.

[*vi*] Names indicating movement as regards the thing moved or altered, i.e. the concept of being *acted upon*.

[*vii*] Names indicating place, e.g. in town, at home.

[*viii*] Names indicating *time*, e.g. in the time of the First or Second Olympiad.

[*ix*] Names of *postures*, e.g. standing, seated.

[*x*] Names of *conditions*, e.g. armed, clothed, ready, and the like.

I do not seek to show whether this way of distinguishing was sound or useful philosophy; here we shall consider only the nature of *related things*. These, both according to Aristotle, and as reflecting the truth of the matter, are the names laid down, not because of something in that thing [under consideration] but only insofar as the thing is compared with something else. It therefore follows that a name of this kind no longer fits a thing if either the thing itself or the other thing to which it was compared is changed. For instance, if two things are said to be similar because each is white, then if either is changed from white to black the term 'similar' no longer applies. In fact, *absolute names* will always fit the thing to which they are applied, so long as it is not altered. So what is said to be white will always be white–unless it is changed–no matter how other things are altered; and what is said to be a yard long will ever remain so–howsoever other things are transformed–so long as nothing is done to it. Hence something which, after being similar [to something else], becomes dissimilar is not necessarily changed thereby, nor is in any way acted upon. The white object remains white and is not acted on, even though it ceases to be the like of the other white thing which changes into a black.

3. Birth and death are sometimes taken in the [respective] senses of 'creation out of nothing' and 'destruction' or 'reduction to nothingness'. These processes,[9] however, can be carried out and understood by divine omnipotence alone, for we cannot in any way comprehend in our imagination how something can be created from nothing or how nothing can result from something. Sometimes, indeed frequently, the

---

[9] Singular in the text.

CHAPTER V: 4

terms are taken to mean 'generation' and 'corruption', i.e. change. This is not, however, 'any change you like' but only 'a change which makes us assign to, or remove from, a thing the name that answers the question: "What is the thing?"' Thus we do not say that man is produced because the material of which he is composed takes on a new colour or the characteristics of flesh or bones, or is subjected to another, minor change–we say it only when the material undergoes so great a change that, on the change's manifesting itself to us, it creates a picture such as leads us habitually to call any thing 'man'. In the same way the death or decay of a man is not *any* change, as when white becomes black or when a healthy man becomes ill, but rather a loss of feeling or of human shape. The 'generation' of a body, therefore, is the change because of which a new name, from those relating to the Category of Essence, will be appropriate to the body. So a human being who, once eighteen inches tall, becomes three foot tall has in fact a new name, from those in the second Category–that of Measures–but he still bears the name of man. If he be changed into a tree, we say that the tree is generated and that the man perishes. The death of a body is that change which will render inapplicable the name the body bore previously, [one taken] from those pertaining to the [first] Category: things in being, or Essence. The material in which consists the nature of body does not perish, however; just as a flask of wine poured into the sea does not cease to be body though the wine ceases to be wine.[10] Therefore only the decay of living and feeling creatures [as opposed to the inanimate] is termed death; and any living or feeling thing which exists and can die is termed mortal.

FOL. 30[11]

4. Let us now come to our author's arguments. First, he says that, as demonstrated on his page 37, because a vacuum cannot exist, some bodies are liquids. 'It is clear', he says, 'that the smallest bodies are moved and that, without the agency of some liquid medium, they cannot cohere into any mass. Obviously then, whenever they do so cohere, there is present [either] a vacuum or a liquid mixture, since their outlines do not always fill a [given] space. But a vacuum does not exist. Therefore there is some liquid among them.' This reasoning

---

[10] Or, '...ceases to exist'.
[11] Fols 21, 22, 25, 26 and 29 are blank. The foliation now becomes regular.

depends on a hypothesis which we have previously shown to be wrong[12]–that a vacuum cannot exist.

5. The ancients thought, and the moderns agree, that some liquid substances exist beyond the moon. The writers in middle antiquity,[13] however, denied it, yet were unable to explain the phenomena of the heavenly bodies otherwise. But after Copernicus had set astronomy to rights again,[14] people began to disbelieve in those solid, concentric (or eccentric) spheres and in epicycles; indeed, physical arguments began to be sought by which it could be shown that these were not even feasible. The main point was that that solidity or hardness which was supposed to characterise the heavenly spheres would prevent them from moving. This reasoning was based on the fact that, if an exterior cylinder tightly encloses an interior one, the inner one seems incapable of turning. But it is à propos to examine whether the cause of this may be that the cylinders exert pressure on one another because of the tightness of the fit. Further, even if they do not fit very tightly, there cannot be such a perfect contact between cylinders fashioned by hand, unless their surfaces are smooth and do not hinder the motion because of [their] intricate unevennesses.[15] An indication of this fact is that with the aid of grease, which fills in the rough parts, the cylinder inserted within the other one turns the more freely, even though the [inter-surface] cohesion is greater for greased things than for ungreased.

FOL. 30v

But against the hardness of the heavenly bodies our author employs a new kind of argument that he takes from the nature of 'related names'. 'Given', he says, 'two bodies, A and B. Don't you see what we say about them' (i.e. their *names*)– 'that some names are not altered unless either individual body is changed, but that others can be altered when the body we are speaking of is unchanged?' This means: some [names] are re-

---

[12] At fol. 23*ff*.
[13] Perhaps Hobbes means 'of the middle ages'?
[14] This concept seems to derive from Galileo's *Two chief world-systems*; Stillman Drake's version (1953, p. 341) translates similarly, but Hobbes may mean 'set astronomy on its feet again'.
[15] This is given as literally as possible, except that 'smooth' renders 'not rough' of the original. The drift is clearly that hand-made cylinders tend to be uneven and do not turn easily, being obstructed by their roughness; but if you oil them they turn more easily, even though the surface contact is now greater. The words 'do not hinder' imply that, if the surfaces are rough and ungreased, the motion is hindered.

tained unless the things that they are the names of are changed, but other [names] are not kept even if the things they are the names of are not changed. 'So that it will always be true that A is A, whatever change is wrought on B, if A remains unmoved', or that Plato remains Plato whatever may befall Socrates. 'Also (because A is similar or equal to B) when A is changed B can be changed'; or the statement 'Plato is Socrates's pupil' can be changed when Plato himself is not changed but only Socrates is. All these facts could have been clearly and concisely expressed thus: '*Absolute names* are always kept, unless the things be altered, but *relative names* are not always kept'. Now who isn't taken aback when he sees the principle attached, as here, to the liquidity of the heavenly spheres? Moreover, our author assumes another principle: What is the same remains the same, unless it has been changed. He has defended his proof as follows: 'Given A, a place, and B, something put there. If neither A nor B is subjected to any change, then A is always A, and B, B. And A in relation to B is always the same and affects it in the same way. If, therefore, on one occasion we had A and B, and A *did not* fix the position of B, then, as long as everything stays the same, A *cannot* possibly fix B. And if there has once been a place [for A], under the same conditions there must be a place for the same [B]'. And a little further on he appends the rest of the demonstration. 'Therefore, when nothing has been intrinsically changed, it cannot happen that the relationship of the one [A] with the other [B] is changed. If two spheres touching one another are so solid', i.e. so hard, 'that nothing of either can be worn away through motion, there is nothing intrinsic that can be changed in them through local motion.[16] If, therefore, you claim that the heavens are changed locally, you are claiming that they are changed extrinsically, i.e. that the change has occurred as regards place.'

FOL. 31

The question was whether, of two concentric spheres touching one another and so hard that nothing could be worn away from them through motion, the inner one could be turned [independently] on its own axis. But he comes, not to this conclusion, [i.e. that it could,] but to another, namely that if the heaven be locally moved it is moved extrinsically, i.e. according to its place. But perhaps he thinks that these last words contain the same point as the words in which the question was put.

---

[16] The discussion a little earlier was not about spheres but about cylinders.

Let us concede this, astonishing though it be, and see the force of the premises from which it is drawn.

'1. If neither sphere is changed', he says, 'then basically the relationship of the one to the other will remain the same.

'2. If the relationship remains the same, the outer sphere, which was the container of the other, will always be its "place".

'3. If one sphere wears nothing away from the other by movement, then there will be no intrinsic change in either.

'4. If the spheres are solid, one rubs nothing away from the other. Therefore if two spheres, an outer and an inner, are solid, the outer will always be the "place" of the inner.'

FOL. 31v Here it is hinted that the inner sphere can be turned on its own axis only if its 'place' should change. Unless the word 'change' suggests to him 'local motion', the first [premise] is false, because any two bodies (say two planets) change their positions relative to one another because of their different motions, since they are sometimes more, sometimes less, distant from one another. But if [by 'change'] he understands 'local motion' the third premise is false, for two unchanged bodies can be 'moved according to place', i.e. they can be 'moved locally', even though one rubs nothing off the other and one does not influence the other in any way. Hence the conclusion, 'the outer sphere is always the "place" of the inner' is not the one which should have been drawn, which is: 'the inner sphere cannot be turned on its own axis'. Nor does [the outer sphere] 'contain' the inner, for in order for a sphere to rotate there is no need for it to leave its situation; it needs only that its parts move and subsequently return to their place.

Next, from the fact that some substance existing beyond the moon is liquid he proves that there is some destructible material in the heavens. The way he reasons is as follows (his page 41): the first of his premises is: 'What can be divided can perish'. To this he has added a third: 'What can perish is mortal'; which produces this: 'What can be divided is mortal'. This conclusion, together with a fourth premise, 'What is liquid can be divided', produces: 'What is liquid is mortal'. But he has previously shown that there is a liquid substance in the heavens. So it emerges that there *is* something mortal in the heavens.

The term 'indivisible' is normally taken in a twofold sense: [i] for 'single' or 'one thing only', e.g. 'the sun', 'the moon', 'this star', and [ii] for 'that cannot be split'.[17] Of [ii] the atom is an example, because of its small size, or because it is infinitely hard or, as he believes (were there, indeed, any atoms), because of its toughness. Now in whichever of these senses we take the first premise, the second is either wrong or absurd, for if by 'undivided' we mean 'single, because of the change of the phases of the moon and of Venus', we may infer that those two heavenly bodies are subject to decay. But if 'undivided' is the same as 'indivisibility' the second premise: 'that which is divided changes that which is indivisible', is absurd, for what is thus indivisible cannot be divided. It is therefore not true that 'what is divisible can perish'. Certainly, in no way can a body perish, but it can be changed—and to the degree that it must [then] be called by another name. Again, the other premise, 'What can perish is mortal', is not a correct use of terms, for the death of any substance is not perishing but a certain prescribed change of percipient bodies into non-percipient. Nor is perishing, i.e. reduction to nothingness, what we term death. In the fourth premise he introduces something false, namely that what is not liquid is not divisible. Nevertheless he concludes that there is something 'mortal' in the heavens; but by the same reasoning he could have concluded that no material in the heavens is not mortal. For all body is divisible, the sun and stars being no less so than the liquid enclosing them.

FOL. 32

6. The weakness of all the previous reasonings arises from this: that our author has not realised what 'to be divided' is. [First], because he lays it down that 'to be divided' is the same as 'for part to be taken from part' the results are that liquids and soft substances are less divisible than hard and closely-knit ones; that the harder things are, the less easily are they divided; that [a period of] time, the continuity of which cannot be broken, is indivisible; and many other nonsensical statements. Second, since he decides that 'to be divided' is the same as 'to perish', it follows that no body can by its own nature avoid being destroyed, i.e. obliterated. The opposite of this is patently clear: no body can perish; for although any body can cease to be a man or a tree it cannot cease to be body, i.e. it cannot decay. Third, he had stipulated that 'to be divisible'

FOL. 32v

[17] For 'indivisible' Hobbes may here intend 'individual, *sui generis*, possessing its own *Ichheit* or identity'; compare the discussion in Chapter XII, fol. 107v*ff*.

(i.e. as he wants to interpret it, 'to be liquid or soft') is the same as '[to be a] quantum'; therefore hard bodies will lack quantity. All these ridiculous statements, and countless others that follow from them, show that the study of philosophy, except for the very few, is harmful rather than profitable for the knowledge of truth.

He proves that quantity is the same as divisibility, i.e. separability, thus: 'Everyone', he says, 'who answers the question "How big is it?" does so through the use of measures, e.g. yards, ells, pounds and the like, for the measure of anything is a part of it, but the only parts that exist are separable. If they are merely perceptible, they are imaginary, i.e. none exists'. Therefore quantity is separability. Since, however, it was shown in Chapter III that a mental division is the true one and that imaginary things are indeed divided, the whole of this argument breaks down. Furthermore, as he places pounds, i.e. weights, among the measures of quantity, does weight also consist in a separability of parts, because weight is quantity, and the more fluid something is, the more it weighs? So quicksilver, a fluid, will weigh less than steel, which is very hard. But to our philosopher this 'divisibility' is all things–liquidity, quantity, density, tenuity, sensitivity (for such things as are divided mentally are, he says, acted on by the mind, because 'are divided'[18] is a verb in the passive voice), and corruptibility.

FOL. 33

---

[18] Text has the infinitive.

CHAPTER SIX

❧

On Problem 7 [of the First Dialogue]: On Aristotle's demonstration that the heavens are everlasting. On 'simplicity' and 'compositeness'; whether objects thrown upwards pause before they start to descend; and whether something thrown upwards with very small force [momentarily] brings to rest something which, in its descent, meets [the thing] with great force

1. Definitions: Simple, multiple, mixed. 2. That all lines are simple. 3. Definition: A straight line, a curved line. 4. Simple and composite motion defined, and how straight motion and circular motion are simple. 5. Because their movements are simple and mixed, bodies are not, in consequence, simple and mixed. 6. On 'nature' and 'natural' and the principle of motion. 7. 'Contraries' defined. 8. It does not follow that, because their motions are opposed, moved bodies are 'mixed'. 9. That bodies thrown upward do not pause before they commence falling. 10. A paralogism of our author revealed. 11. A body rising with a lesser motion does not hold back a body it meets that is falling at a pace greater [than its own]; 'greater' and 'lesser' motion defined.

1. You see that we have skipped Problem 6, in which are found

the reasons why Aristotle and others have thought that the heavenly spheres are hard. [It contains] a conjectural account which, though plausible enough, is of insufficient use in [leading to] a knowledge of truth. We have therefore passed over it to Problem 7, which contains many topics relevant to motion and not unworthy of inspection.

It seemed to Aristotle that bodies are divided[1] into simple and mixed, and he thought that mixed bodies are indeed changed, but that simple bodies are not. He chose to say, therefore, that the heavenly bodies and the elements were 'simple', and that all the other sublunary things were 'mixed'. Again, he infers that opposition[2] of motions results in a 'mixture'. This opposition can, he declares, exist in straight motion but not in circular. He says that simple bodies are those whose motion is 'simple', and that 'simple' motion is 'that motion which is transmitted in simple lines'. He does not define 'a simple line', but merely 'a straight line' and 'a circular line'; nor does he define 'mixed' or 'composite' motion. We must therefore see what 'simple', 'composite', 'mixed', 'a simple line', 'a composite line', 'simple motion' and 'composite motion' are, and in what sense a body is called simple, composite or mixed.

FOL. 34

A 'simple' thing is 'one thing as compared with another of which it is a part'. We have defined a 'composite' in Chapter 2 above, namely, 'one thing resulting from the mental addition of one thing to another'. Therefore 'simple', as opposed to 'composite', is the same as 'one thing' compared with 'something else that contains it'. Further, 'simple' as opposed to 'multiple' is as one thing compared with another thing of which it is one of a number of parts. So 'multiple' equals 'composed of several parts'. Thus the shield of Ajax was said to be 'sevenfold' because it had seven layers of hide, but if it had possessed only one it would have been a 'simple' shield. 'Mixed' therefore means 'made up of heterogeneous parts' (i.e. parts which are dissimilar as regards their perceptible qualities) distributed at random, as when we mix powders of a different colour, or different liquids.

2. Yet how a line can be distinguished into 'simple' and

---

[1] Text: 'distinguished', in the philosophical sense of 'differentiated'. Cf. the start of Article 2, and folio 38v.
[2] Text: 'contrariety', which, despite its awkwardness in English, may be justified in that it here implies difference of kind as well as of direction.

'composite' is inconceivable, except insofar as one [line] can be added to, or subtracted from, another longitudinally. But as regards breadth, i.e. because lines *lack* breadth, one line cannot be 'added to' another. Therefore all lines are equally simple, both the curved and the straight; the hyperbolical and the elliptical, parabolical, and circular. So Aristotle was wrong when he distinguished lines into simple and composite, as is our author in chiding him and saying that a circular line is composite. Our author has been deceived in taking as composite the line in which something movable is moved by the thrust of two or more things moving at the same time [as one another]. On his page 48 he instructs us 'to reflect that the first point that flows because of the joint effort of two lines of direction has a more composite flow than [it would have] if it followed only one [line]'. This is false, for if a straight line AB is described by a moving body (at A) driven by a single wind blowing through CA, is [AB] less composite than the same straight line AB when described by a moving object (at A) driven by two side-winds EA, DA, blowing at the same time?[3]

FOL. 34v

But, he will say, the diagonal of a rectangle will be 'simple' because [i] it can be traced out by a moving object driven by a single mover, and [ii] only by chance is it traced by objects moving sideways. Every circular and curved line, [he says,] essentially demands that the moving object which traces it out be driven by several movers and on that account is, by its own nature, composite. But this, [I rejoin,] is a compositeness not of lines but of motions. Further, the words he adds, 'A curved line encloses breadth as well as length', are erroneous. We are correct and accurate if we state that a circle, or the three sides of a triangle, or the four sides of a quadrilateral do indeed bound the surface made by a long body moved through [another] long body, but that they do not enclose length and width. If, as he says, a curved line leaves the shadow of its own points, it does not, [I answer,] necessarily enclose length and width. A curved or [even] a broken line does indeed enclose the surface *on the inside of* itself. It does not, however, enclose the surface lying *within* itself, for that is not[4] in its

FOL. 35

---

[3] Figure 4. Though not so punctuated in the text, this sentence seems best interpreted as a (rhetorical) question.
[4] Text: '...itself, i.e. not' etc.

nature, a line being length only; and what is not in the very nature of the line, but only within the line's boundaries,[5] has nothing to do with the 'simplicity' or the 'compositeness' of a line. Therefore no line, straight or curved, in shadow or in sunlight,[6] is composite, and so this distinction of lines [into simple and composite] is false as much in Aristotle as in this our detractor of him.

3. Our author defines a curved line as 'one that goes outside the shadow of its ends'; so he seems to be relying on, and taking refuge in, [his] definition of a straight line. This is given as: a line whose middle parts stand [undeviatingly] between its limits. The definition stems from the fact that vision always travels in a straight line and hence that, if something opaque is placed in the straight line between an object and the eye, it obstructs the vision by forming a shadow.[7] But the definition is false, because even if vision did not exist at all, the consequence would not be that no line was straight. Nor is Euclid's definition better: 'That line is straight on which any point lies equally between its ends'.[8] For we can understand what is meant by 'to lie equally between two points' only if we say that 'to lie equally between' is 'to lie between, in a straight line'. But for the nature of a straight line to be well explained we should define it thus: 'A straight line is the only line of the same length that can be drawn between two points; and a curved line is that which can be traced out by more paths than one between the same points, the line's length remaining constant'.

4. Now if motion is to be distinguished into 'simple' and 'composite' this distinction must not be drawn from the fact that there can be one moving body, or more. For in that case no motion would be simple, because both circular and straight motion can be created by several moving bodies

---

[5] I.e. within that which the line encloses?
[6] I.e. whether the line 'leaves the shadow of its own points' (see above)? But there may be a reminiscence of Proclus's use of the analogy of the edge of a shadow against the latter's background in order to plead that a line has no breadth. See Euclid's *Elements* in Heath's edition (1908), I.159.
[7] Perhaps, 'by masking (the object)'.
[8] As worded by Hobbes, the definition is misleading: Euclid's words are 'A straight line is that which lies equally with itself'.

CHAPTER VI: 4

jointly producing it;[9] as straight motion across a diameter is compounded of the motions of the sides, and circular motion is compounded of [a] one motion through the semidiameter, this latter being a straight line, and [b] another motion that is always perpendicular to the semidiameter.

This division into simple and composite [motion] can be inferred from a difference of velocity, but not in the way that White says, which is: 'Simple motion is that whose velocity is everywhere equal' and composite motion, 'that whose velocity is unequal'. That first motion we do indeed term constant and uniform. We term the latter 'not constant', but not 'simple' and 'composite', for the motion of heavy bodies is activated by continued acceleration but is still said to be simple. Simple motion brought about by a difference in velocities must, I think, be defined thus: 'Simple motion is that by which any movable part [of a body] is moved as fast as any one of the remaining parts'. But composite motion is 'that by which one movable part is carried faster than another part', and in this way all straight motion will be simple.

FOL. 36

Given a moving object AC, one part of which is A and the other, C.[10] The body moves in straight motion to DB. Therefore at the same time as all of AC will advance towards BD, part A will move to B and C to D. Now AB and CD are equal straight lines. Therefore the velocities will be the same in either case. Likewise if AC were divided into any number of parts, all those velocities would be equal to one another. Circular motion, however, will be simple or composite according to the difference of the motion by which the circle is described. Given a moving body AB which describes as it moves the circle AD on centre C so that part B is carried to E and part A to D.[11] This must necessarily happen if that motion depends on the movement of the radius CBA; and if [the motions] are made with the aid of compasses it is clear that part B is moved more slowly than part A. So the motion of the moving body AB is slower at part B than at A. Hence the movement of all AB cannot be termed 'simple'. But if the mobile body A were to be carried round [some point] so that

FOL. 36v

---

[9] Hobbes is ambiguous. He may mean that the distinction into simple and composite motions depends on whether moving bodies are collectively, or only individually, in motion.
[10] Figure 5.
[11] Figure 6.

its axis AB were kept always parallel to itself,[12] [either] [i] in the way in which this happens when a circle is drawn, the hand alone seizing hold of the axis AB and leading it round [the centre],[13] or [ii] in the way in which Copernicus supposes the earth to be moved through the ecliptic,[14] any point on the mobile body AB will then describe a circle equal to that which would be described by any point on the rest [of AB]. The circular motion thus delineated would be simple. But whether any motion other than the straight and the circular can be 'simple' in the way that, I say, the single parts of a moved body are borne at speeds which are always equal to one another I do not know.

FOL. 37

5. A simple body, as opposed to a composite, is the same as the part with respect to the whole; as opposed to a multiple it is the same as [i] one of several parts with respect to the whole, or [ii] one thing with respect to several things taken as a group; as opposed to mixed it is the same as homogeneous with respect to heterogeneous. That a body can be said to be simple, composite, multiple or mixed, because of the diversity of its movements, is incomprehensible. Every object, be it composite or multiple or mixed, can still be borne along a straight line and round a circle, i.e. by simple motion; and every simple body can be borne through any curved line, i.e. by composite motion.

6. 'But', he says, 'nature is the principle of motion.' Therefore 'a simple nature' should be present in objects borne by simple motion. But what is 'the principle of motion' except 'motion from [one] place to another close by'? So where motion itself has not commenced there is no 'principle of motion'. Hence 'simple nature' means the same as 'simple motion'; and so he commits a *petitio principii*. Because every body, even the simplest one, e.g. *materia prima*—if such exist—or the most tenuous ether, or the body of any star, can be carried by motion that is

FOL. 37v

not simple, the nature of all bodies will be non-simple, and so every body will be simple and non-simple. But nature is not correctly defined as 'the principle of motion'. A stone that is

---

[12] See the right-hand diagram of Figure 6.
[13] Of dubious interpretation, these words are given as literally as possible.
[14] Hobbes pursues the subject of the parallelism of the earth's axis at fols 190*ff*.

thrown by the hand is carried upward–but it could not have been moved without [possessing] some principle; yet the nature of the stone is not [of itself] such as to initiate an upward movement. Suppose we amend the definition and say: 'Nature is the principle of natural motion'. It will be rendered faulty, because what 'natural motion' is we cannot understand before we know what 'nature' is. We shall say, therefore: The nature of any body is its potential to work or to act–a nature essential to it–i.e. included in, or to be inferred from, its definition. For example, the nature of an animal is the ability to feel, that of a man to reason, that of an isosceles triangle to have two right angles.[15] 'Natural motion' is therefore that to which a mobile body from its very definition may have a proved or inferred potential. Consequently to every body every possible movement is natural. Earth would not be earth, unless it could be moved even upwards,[16] because the nature of body is the ability to be moved in any direction. Thus earth would not be body unless it could be moved even upwards; indeed, if it were not body it would not be earth.

FOL. 38

Hence it is through its essence of 'earth-ness' that earth can be moved upwards, and its upward motion, brought about by something driving it on, will therefore be 'natural'.[17] In a word, 'that which can become' is natural; and there is a difference between 'natural' and 'vehement'[18] only because a motion whose cause we do not see we attribute to nature; but a thing whose cause we do see clearly we call 'vehement'. So the 'potential to motion' is not motion [itself] unless it is also the principle of motion, for the 'principle' of any single thing is its *pars prima*.[19] Here the thought suggests itself how, when youths are placed by their parents in the charge of philo-

---

[15] 'Isosceles triangle', Mersenne's filling of a lacuna, cannot of course be right. Read 'trapezium'? Is the presence of the gap due to Hobbes's momentarily forgetting the word (or example) he sought, or to the copyist's being unable to decipher Hobbes's word, whatever it was? *Dormitat Mersennus?*– even if the copyist read, and gave, Hobbes's *angulos aequales* (or a contraction; Hobbes surely did not write *angulos aequos?*) as *angulos rectos*.
[16] Rather than, as would be expected under the old physics, downwards (relative to the other three elements).
[17] It is not clear whether Hobbes intends the element earth or the planet, or both.
[18] I.e. 'forced' or 'unnatural'.
[19] Cf. fol. 118v.

sophers[20] in order to learn something new, they unlearn, many of them, what they clearly understood before. When busied in the classifications of grammar, boys know well (for example) exactly what the 'principle' of a thing is. They turn it into the French word 'commencement' and understand something by that word, for when you ask them what 'the principle' of a path is they will say–or at least imagine [that it means]–'the first step' or 'the first foot' or 'the first something'; the 'principle' of a chain they will say is its first link, and the first principle of any single thing is its 'first part'. But after they are elevated to [the rank of] philosophers and you ask them 'What is a principle?' they translate it into 'un principe', but what 'un principe' is they do not know; they know only that it is anything that cannot be denied. The same thing happens with other words, the sense of which they know well enough when boys, but they lose it afterwards when they feel that the terms are ineptly explained by philosophers. From an ignorant explanation comes that difficulty ascribed to the loftiness of metaphysics, and so youths are not made learned by terms wrongly used, but grow foolish with their masters.

FOL. 38v

But to return to 'nature', which our author distinguishes into universal nature and particular nature. I have shown above (Chapter II) that nothing is universal but names. Therefore if nature is a universal thing it does not exist. Neither the nature of any celestial body nor of any body beneath the heavens is universal, nor is the nature of the whole world universal, for the nature of any single body would [then] be the same [as that of any other], just as the nature of man is the nature of Socrates and of Plato, etc. So the nature of no object is 'universal'; and therefore there is no more 'a universal nature' than 'a universal thing', i.e. there is no 'universal nature' at all.

FOL. 39

7. Aristotle says that motion along a straight line is 'contrary' to motion on the same line [but] from the other end. He is right in respect of the struggle or effort [*pugna sive conatus*]

---

[20] Or, 'placed in the charge of seeming philosophers'. That the given translation *may* be the correct one is suggested by a passage in Hobbes's (if it indeed be his) *A dialogue between a philosopher and a student, of the common laws of England* (the comma of the title makes for ambiguity) that begins: 'This laudable custom of great, wealthy persons to have their children at any price to learn philosophy, suggested to many idle and needy fellows, an easie and compendious way of maintenance' (1681, p. 127; ed. Joseph Cropsey (Chicago, 1971), p. 124).

by which the motions repel one another but not in respect of the distance that separates the ends [of the straight line]. On the other hand, he supposes that 'contraries' are correctly defined as 'those that are the greatest distance apart'. 'And in straight movements whose ends (from which each motion approaches the other) are very far apart there is', he says, 'indeed contrariety. In circular motion, however, the same point is both the beginning and the end of a circle, and [the motion's] "ends" are not "apart" at all; so here there is no contrariety.' Nevertheless, consider two moving objects as they start to be moved from the two limits of a diameter and one object tends toward the other through the circumference of a semicircle. I do not see how these two motions are less 'contrary' than [they would be] were they straight, for [the bodies] are borne in opposite paths and from points of origin in the circle as far apart as can be. It is not 'distance apart', but mutual interaction, that creates opposition; for the definition of opposites, i.e. 'Opposites are those which, though of the same kind, are furthest apart', though Aristotelian and satisfying to our author, is not accurate enough. By 'distance' we here understand 'unlikeness'. Indeed, hot and cold are very dissimilar things, and, being actively opposed [one to the other], they are contrary, not because of their difference, but because of their opposition. White and black are very dissimilar but are not contrary, because they do not actively oppose one another. So 'straight motion' is the opposite of 'straight motion on the same line [but] from the other end'; and yet [the latter motion] is not dissimilar; nor, being of the same kind, is it the maximum distance away.[21] The definition: 'Contrary things are those whose motions are the most actively opposed', is true, and in this way not only straight motions but also circular ones may be contrary to one another.

FOL. 39v

8. Even if we allow that motions upon straight lines become 'contrary' but that circular motions do not, nothing may be inferred from this about a mixture in sublunar bodies that will not also be inferred relative to a mixture in the heavenly ones. For a mixture is caused by the breaking of the whole into particles; but it is lateral motion, not opposed motion in the

---

[21] Dissimilar to straight motion? 'Straight motion...other end', from however distant a source, is still a species of 'straight motion'?

same line, that conduces to the breakage. Things that meet in motions diametrically opposed repel, [but] do not shatter, one another. So the change which makes any bodies appear to us different and of a different appearance–the change which makes us say that some are fairly simple, and others mixed– is nothing other than a diverse motion in the invisible parts of the object that has been changed. This can be produced by almost every action except that resulting from direct opposition. Therefore in sublunary things the creation of a mixture is not due to the contrariety of motions, except insofar as the parts of the bodies that rush together are, by opposite motions, compelled to spring sideways. So the demonstration of Aristotle is neither true nor yet satisfactorily refuted by our author; and whether the same changes take place in the firmament as around the earth, or not, so far remains uncertain.

9. There follows the question: Do contrary motions presuppose rest at some point midway? This may be posed as two questions [in one]: first, does a body thrown upward by the hand pause before it starts to fall? And second, if any body of any mass and weight falls and meets any body that is rising, must one or both of them pause for a moment at the point of meeting?

As regards the first topic, the view of Aristotle is that the body does pause. This opinion our author impugns–but he holds it himself! He attacks it on his page 51, thus: 'If we concede that between two contrary motions there is rest, the body which is moved through the whole circumference of a circle will pause, within that period, four times; its motion will be interrupted this number of times; therefore we must seek four times [the reason] why it resumes'. But he does not say why the motion is interrupted four times rather than very many times, nor can anyone guess why it is, for such motion is equally weakened at every point, and therefore infinitely. Now, say he conducted his argument as follows: 'If a pause intervenes between the motion of ascent and that of descent, which are contraries, we must ask why [the body resumes] its descent after [the pause]. If it really paused, [its] cause of motion was missing; but gravity, or whatever force from whatever place it is that drives heavy bodies downwards towards the centre, was not lacking. So a reason must be sought for the descent other than that for which heavy bodies seek the centre, it being self-evident that bodies coming to rest cannot be moved except by the approach of a new cause'.

This seems to me to be a very weighty argument [of mine], yet it requires [acceptance of] an intermediate pause on the grounds that, in consequence, two contrary motions are compounded and continued into one motion. This White believes to be impossible. Let us imagine, then, that a body at A is moved towards B at a constantly and uniformly decreasing pace until [the movement is] very slow at B.[22] Therefore within the whole time that the body is on the straight line AB it is being moved. Hence if it were to move forward at increasing velocity towards C it would surely be moved during the whole time when it is on the straight line BC. Now it would be moved very slowly at the point B, but it would still be moved. So it would be moved for the whole time it was on the straight line ABC. Thus at no time would it pause. Why, then, will it not be able to be moved continuously in the same way over ABD, ABE, ABF or even ABA, for the same reasoning holds good everywhere? But possibly someone will say: 'If A rises through B in such a way that it begins to fall back, and [yet] it is postulated to be moved at B, it will be moved upwards or downwards. But it is postulated that the body is not moved further upwards and that, in position B, it is not yet moved downwards. Therefore it is not moved either up or down, i.e. it is at rest'. The falsity of this argument lies in the fact that we say that something at B is not yet moved, because we postulate, not that it has stood still at B, but that it has reached B, for anything moved is not in the same position even for a very short time. Anything that arrives at B can move away from B with continued motion towards A again. By the same reasoning rest may also be inferred [to exist] within a uniform motion directed always towards the same parts; for if a body moved forward from A towards C at a constant velocity we could likewise say that a body at B is not yet moved towards C or towards any other part, and is therefore not moved at all.[23] A most obvious refutation of the existence of a pause at the point of change-of-direction is the one I have given above, namely that if a body were at rest at B, a new cause would be required for it to be moved, and this of such a kind as would necessitate the body's moving in a determined path to A.

FOL. 41v

FOL. 42

---

[22] Figure 7.   [23] Hobbes's tense-sequence is here retained.

I add another proof: because a heavy body thrown by the hand rises, the force it acquires from the hand is stronger than the force it has from gravity; but because the force from gravity always acts on a projectile, it continually weakens the projectile's motion upwards. But the hand does not continue to operate: having cast the body, it abandons it, and the force due to gravity continuously seeks to secure an ascendancy over the force [initially] exerted by the hand. The result is that at length the forces on each side become at first equal, and then the power of the force from gravity becomes greater than the force from the hand. Hence this proportion of forces is continually being changed, so that for no [length of] time, however short, are the forces equal. But just as any point on a line is traversed in an instant, so also the equality in the progression of the proportions of the forces, [each to each,] disappears in an instant, and we find an equalisation of forces [only] when the body begins to fall back. Therefore the projectile does not remain there but arrives and starts back at the same instant [of the equalisation of forces].

FOL. 42v

10. Our author's own argument (page 53) aimed at proving that there is rest at the point of reversal-of-direction is unfolded thus: 'If time is needed for something to be moved, however small the movement, it will follow that a pause intervenes at the point of reversal-of-direction. Now just as any dense medium cannot be penetrated all at once, so a motion is necessarily slow at its commencement; therefore there will of necessity be a pause between the action of the source of motion and the cessation of the moving object'. This is like saying:[24] 'Given that any body shall have risen to B by the force of a mover at A. The body is unable to rise further by the same force. Let [another] mover, by which the body is to be sent downwards, be located at C. The body will be at rest at B for a time equal to that taken by [the second] mover in travelling from C to B to be brought to bear on the body there [at B]. Now for C to travel to B some time is needed; therefore the body will be at rest at B for a while'. Who does not see that the mover which drives heavy objects downwards[25] is constantly being applied? and that the mover is therefore not rightly supposed to be distant from B and to need time in which to move

FOL. 43

[24] Figure 8. [25] This mover? any such mover?

to B? So this reasoning is a paralogism, and our author has been deceived in that, when he heard that all movement is effected in [a period of] time, he took this to mean: 'A body needs time in order that it can *start* to be moved'. This is not so, although it is true that time is needed for anything to travel from place to place, i.e. to be moved.

11. In the following way he sets himself the resolution of an objection to his own conclusion. 'If indeed', he says, 'a rising body paused for a while at the point where it is turned backwards, it would follow that a ball thrown by a boy and meeting a meteorite falling from the moon ought to make the stone stop. This consequent is true, because when the ball is stationary the stone, which is [now] resting on it, must also be stationary.' This seems very true to me, but the reply that he brings forward is incredible, and does not even remove the difficulty in the objection. He says that the air in front of the stone drives back the ball before the stone reaches it! This is unbelievable. And even if we concede the point, the difficulty is still not removed. The ball is thought of as rising when it meets the stone. Now what if the ball were of lead and were shot from a wind-gun against the falling stone with such force that the impact made a slight hole (the size of the ball)? Would the stone, falling (as is supposed) from the moon, pause because of this?[26] We cannot say, then, that the ball has been held back by the air surrounding it before the collision took place. Not every motion, but only the one that is either *the greater* or equal, cancels out every contrary motion. This, however, is too obvious to require proof; because if the motion of the ball shot from the gun is less than that of the falling stone the latter will fall without interruption, though more slowly. I call the *greater motion* of both, not the velocity or the motion of the greater weight [*maioris ponderis*], but the motion made greater by the velocity's being multiplied into [*sic*] the weight. Suppose the velocity of the stone is less than that of the ball but the [stone's] weight is more. Then the velocity of the stone divided into its own weight produces a higher figure than does the velocity of the ball divided into *its* own weight. Then [if this is done] I call the motion of the stone *greater* than that of the ball.

FOL. 43v

FOL. 44

[26] Mersenne's alteration from: 'The stone would not therefore pause'. He retains the phrase 'falling...moon'.

FOL. 45[1]

CHAPTER SEVEN

❧❧

On Problem 8 [of the First Dialogue]:
That there are in the heavens bodies
resembling our own.

1. In what consists the variety of things. 2. There is only one material which is changed, and this is *materia prima*. 3. *Materia prima* defined. 4. That the variety of things is unknown to us. 5. That it is not determined by the distinctions drawn in the *predicament*[2] of substance. 6. That the division of things is not correctly set forth by our author. 7. That fire can exist even if smoke is not emitted and nothing consumed; and what solar fire is. 8. The new star in Cassiopoeia is no proof that stars are generated. 9. [The belief] that the mountains on the moon are larger than those on earth is not made plausible by any argument.

1. Do there exist in the heavens bodies analogous, i.e. (as our author here understands it) similar, to our own? This we cannot investigate by natural reason; for, of the things that we know are in being now, not one precludes the existence of any monstrous creation that the human fancy can conjure up. On the other hand, indeed, of the things we know to exist at present, or which we see have existed, none in any way makes necessary the existence of similar things in a place where these last are invisible. Indeed, we believe not only that there are in the heavens bodies like our own but also that, in their own time, these bodies of ours will themselves be found there. (This,

FOL. 45v

[1] 44v is blank.
[2] One of the ten Categories of Aristotle; see Ch. V above and fol. 49v.

however, is a dictate of faith, not a dogma of philosophy.) Yet it may be asked, with reference to the things in the world that differ one from another as to variety,[3] just as with reference to the world itself, whether this variety is necessarily finite or not. It seems, however, that there can be only one way of finding out, namely by beginning with the variety of fancies or images excited by the things themselves acting on our sensoria; for without these images a stone could enquire about anything equally as well as a man can.[4]

We term things 'unchanged' that appear in the same way as they have appeared previously, and 'changed' those that appear otherwise. So 'change' in things consists in the fact that, if one assumes the sensoria to be unchanged, the things do not [always] arouse the same fancy or image in the mind. In other words, change consists in some chance motion of the parts of an object. Now fancies, or images, in the mind, are really nothing but the motion excited in the brain by objects; therefore the cause of that motion must be a motion in the parts of the object, because motion, by its nature, can be created only from motion. (The fact that God, being unmoved, yet moves [others] is not natural but supernatural, and also above the human understanding: this, through faith, is to be allowed to the honour of God, to whose nature it is not fitting to attribute, except figuratively, anything we can understand.) So change in things consists in the movement of their parts, i.e. the invisible parts, and hence arises that change in things which is being enquired into. When, however, anything changes so extensively that a new name is called for because the thing takes on a new appearance, then we state that the thing has ceased-to-be ['corrupted'] which conjured up the earlier image, and that another thing has come-into-being ['generated'] which produces a new image.

FOL. 46

2. But say that things themselves do not perish through change, but that only their fantasies and images do. There must result one of two consequences: (*a*) there is only one thing whose

---

[3] Mersenne alters 'with reference to...as to variety' to 'with reference to the variety of things in the world that differ one from another'.
[4] I.e. 'Did not these images exist (in man), then a stone...'. 'Equally as well' is here preferred to the more natural 'just as well' (used elsewhere), which can mean 'equally efficiently' and 'additionally'.

FOL. 46v

species change;⁵ or (b) if there are more than one, these cannot act upon and change one another. They are such that all their parts are always at rest and always keep the same position with reference to one another, i.e. they cannot work on our senses and therefore we cannot perceive them; the result is that we cannot be aware or possess any knowledge of more than one thing subject to change. Recognising this, Aristotle introduced as the principle of things a *materia prima*, which could take on the form of everything perceptible and which could be changed in any fashion whatever, i.e. in an unlimited number of ways; hence the variety of things is infinite. This is accepted by those philosophers who believe that everything consists of atoms, for they make these atoms homogeneous, differing from one another in shape alone, as if to say that those atoms were that unique matter which Aristotle called *prima materia*.

FOL. 47

3. But so that we may the more easily understand what *materia prima* is, we must first take note that body and matter, or the corporeal and the material, i.e. whatever occupies the space that we picture to ourselves, are the same thing. Taken on its own [*simpliciter*], this thing is called body; but when compared⁶ with something made from it, it is called *material*. *Simpliciter*, wood is termed body, but this same wood, inasfar as a bench is made out of it, is [also] called the material of the bench. Next we must consider *materia* when subjected to a few changes, so that from these we can see what the material of everything is. Let us imagine, therefore, that we have seen only the following four kinds of body: water, spray, ice, and snow. We would say that water is the material both of the spray (into which it is converted by being violently shaken about) and of the ice and snow into which it is changed by the intensity [*vis*] of cold. But we could also say, by the same criterion, that the spray is the material of the water into which [the spray] is changed by dispersal, and that ice, or snow, is the material of the same water into which it is changed by heating; also that, in the same way, anything at all is, in turn, the material of anything [else].

---

⁵ This relative clause is treated as restrictive. As noted earlier, the text tends to preface every adjectival clause with a comma, and so here, and hence cannot distinguish between the distributive and the restrictive. See immediately above (close of Article 1).
⁶ Or, 'considered'.

CHAPTER VII: 4

If we wanted to know what the *materia prima* of all things is, we should find out either [i] what material existed first of them all—Aristotle, who wished [to show] that the world has existed for all time, did not admit this—or [ii] what the material is that is common to them all, since by *materia prima* Aristotle could understand no material other than the common. But the material of water, of ice, of snow and of spray is [itself] neither water nor snow nor spray, because nothing is the material of itself; in order, therefore, that there can be some material for them all, we must introduce a fifth substance. For the same reason, those who seek the material of which everything consists think this must be sought outside the nature of all existing things, because no existing thing can be the material of itself, any more than water is the material of water. Hence it is clear that *materia prima* is nothing but body in general, and that the former simply does not exist, any more than do the other universals which, as I said above in Chapter II, are words, not things themselves. This is why Aristotle correctly deprives *materia prima* of every quality and of every act, and ascribes to it only potential, i.e. mobility, and quantity, i.e. corporeity or material-ness. These are necessarily present in universal bodies, insomuch as the mind cannot discriminate among the latter. So if we take *materia prima* as being a thing, then any thing you choose is that material and thus exists as a thing; it ought not, however, be called the material of that thing, i.e. the material of itself, or *materia prima*, for this last, existing as something common [to other things], cannot be called single.

FOL. 47v

FOL. 48

4. Therefore body or *materia prima* can be changed, and its parts moved in innumerable ways; and by means of motions of this sort it can arouse innumerable fantasies in the minds of percipient creatures, i.e. numerous kinds of images. Granted that it is impossible to know what motions the separate particles of the whole world have, it follows that we cannot know how many varieties of things there are and hence whether or not there are in the heavens bodies like ours. It may be that there are; it may be that all the chimeras and monsters of the human imagination have their counterparts in the heavens; it also may be that there is not in the heavens any heavy or light object, any man or animal or tree—as being in fact the things we cannot know at all, because they do not work on our senses from so great a distance.

5. But our author points out to us a certain other path, by which he considers, by means of [a series of] continuous distinctions, all the varieties of things. From this it follows not only that the varieties are finite but also that there are not many of them. The first distinction he draws is between spirits and bodies. Next, body (which, he says, is the same as 'great')[7] he subdivides into rare and dense. Now, he previously defined the nature of these, [i.e. bodies,] as consisting in the ratio of the quantity of every body, [considered separately,] to body itself: hence, because of the make-up of the universe, the dense and the rare necessarily constitute [?respectively] the heavy and the light. Where there are heavy and light, he says, warmth, cold, [and] damp and dryness cannot be absent; where the latter are found there exist the elements that through their motions (which are contrary one to another), necessarily produce mixed bodies; and from these last originate plants and animals. But a rational soul, which he says cannot be created by these means, he infers is separate from all these and possesses an immortal nature. A little further on he adds these words, at his pages 59 and 60: 'I confidently assert that there is no other single kind of being in the universe except those we have enumerated. Every body is rare or dense' and has both degrees, or only one degree, or several degrees of rarity and density. Of such things, [he continues,] that which has several degrees is mixed, the other elemental. But of mixed bodies one is pure, another organic;[8] of these the latter is a plant. But of plants one is pure, another a mover of itself; of these things the latter is animal, etc. 'But since you see that separate things are confined, because of the necessity of immediate contradiction,[9] this clearly is beyond doubt: no other categories can remain hidden under the nature of body.'

This division is merely the Aristotelian category of substances, distorted; namely an instance of the dividing of names according as some are more, some less common, i.e. according as some signify more things, some fewer. This is equivalent to his saying that one name is common to all things, as 'being' is; another less common, as 'body'; another still less common, as 'animated body' or 'non-animated body'; another even more

---

[7] See below, fol. 50.
[8] I.e. 'serving as an instrument for something else'?
[9] 'Confined' (*coarctari* is the verb) may mean 'abridged'. White seems to mean: 'Because I must at once refute the validity of the classification...'

restricted, as 'percipient animated body', e.g. an animal, or 'non-percipient animated body', i.e. a plant; and so by subdividing infinitely until we reach some name that signifies one thing alone, e.g. Socrates, or 'this man'. Indeed, it is astonishing if, by arranging names in this way, we can deduce an argument for limiting the things we do not know and to which we have therefore not yet given names. But grant him the division he wants; say we go on for ever, as we must, through a contradiction (i.e. by asserting that every body is mixed or not mixed, and again, every mixed body is a plant or not a plant, and yet again, as he has it, that every plant is an animal or not an animal). It will follow that the variety of things is infinite, for every negative name has meanings without limit. But if we had to go on in this way to prove that the variety of things was limited, we would have no need of the whole [Aristotelian] *predicament of substance*; it would have sufficed to say that all body is mixed or not mixed; for in these two contradictory names all body, actual or hypothetical, is contained.

FOL. 49v

6. In the same division of things suggested by our author we find many points to censure. (1) When distinguishing by name 'spirit' from 'body' he has not shown how the mind can tell spirit from body. Substance cannot be understood by the human imagination unless it has dimensions or extension, i.e. size of some kind; he himself says that body is the same as 'to be large'. How, therefore, can it be that spirit–if indeed spirit may be thought of as substance–is not body? In my view he would have done better to say that certain substances are incomprehensible, and are therefore not recognised by the imagination but are believed in through faith, as are spirits; but that other substances are conceivable as bodies. (2) [I censure his saying that] rare and dense are contraries; for though this is so, yet he who previously took them as synonymous should not call them opposites. According to his teaching in Chapter 3 they are identical: so the same thing is both rarer and denser than the same thing! From what he teaches, it will follow that the same iron is rarer than air; on the other hand, if it *is* rarer, the consequence will be that it is denser! Say first that iron is denser than air. It therefore has greater quantity in the same mass and hence greater divisibility and, as a result, greater separability and greater liquidity, i.e. [iron] is rarer than air. Conversely, if we say that iron is rarer than air, it therefore has

FOL. 50

FOL. 50v

less quantity, i.e. less divisibility, i.e. less liquidity, i.e. less rarity, than air. Surely our author has philosophised very unsuccessfully on rare and dense? (3) He distinguishes 'mixed' into 'purely mixed' and 'not purely mixed'. But what is 'purely mixed' but 'mixedly mixed', and what is 'not purely mixed' but 'not mixedly mixed'? (4) He makes a plant a species of animal, as though an animal were a plant, or an animal and a plant were not two opposite species of living things. (5) Lastly, [I censure him] for not deducing the conclusion which he should have deduced, namely that bodies corresponding to our own exist in, and can be found in, the heavens; but he has left this to be inferred from conjectures to do with the moon; of these, several make it appear indeed probable that there is something on the moon like earth, but others are based on very doubtful experiments.

FOL. 51

7. The more the roughness of the moon's surface is examined by the better telescopes, the more it [is seen to] resemble our own mountains and valleys. From this we may surmise that the moon is of the nature of the earth.[10] If this be true, then it is not impossible that there may even be living creatures on the moon; but no experiments support the conjecture that (as our author does not doubt) human beings and our own native fauna are also to be found there; for some parts of the earth itself produce hardly any human creatures. Let us leave these matters and look at the experiments by which he considers that there is corruptibility in the heavenly bodies.

First, he says that the sun itself, unless it finds sustenance from without, is being consumed little by little. On the evidence that, with the aid of a telescope, Galileo saw sunspots, which White thinks to be vapours drifting out from its body, [White says that] the sun is fire. As a result, it will gradually be extinguished unless it is sustained by material that it chances

FOL. 51v

upon from somewhere else. But neither, [I reply,] is it credible that these spots are vapours, nor is it necessary for the sun's fire to emit smoke or to be fuelled. Smoke is nothing but particles of combustible matter, and disappears when the fire is fiercest, unless fresh fuel be thrown on. What material, I ask, that can emit smoke, can be found near the sun? [None.]

Furthermore, we can perform at any time an easy experiment

---

[10] Or, 'earth' (the element).

to show that there can be fire without smoke. If you place a burning-mirror opposite the sun, no smoke is seen before you move some dense matter to the combustion-point, and yet fire was present there before you do so.[11] Because, however, fire does not rise by agreement [*ex condicto*] every time fuel is thrown on, there was fire at the combustion-point before the fuel was added. Therefore fire can exist without smoke. Now, the fire at the combustion-point of the burning-mirror is of the same kind as the fire from which it is derived, namely the sun. If someone asks: 'What *is* this fire produced by the burning-mirror?' I answer that it is the motion of the air that reaches the combustion-point. This motion derives from the sun itself, which drives off the air touching it on all sides. This air, being so driven, also drives the air next to it; and so as the succession of pressures continues the whole surface of the mirror is [eventually] affected. Hence the motion reflected from, or transmitted by, the shape of the lens is all directed towards one spot, the very small combustion-point, whence its convergence becomes swifter, just as, for the same reason, a river flows more rapidly when its course is narrowed. Therefore the air entering the ignition-point at this speed is continually dispersed without smoke. Now, any wood we put at this combustion-point will be consumed in smoke, and this is what is called 'to be burnt'. Why, therefore, will the sun not be of fire, although nothing is actually consumed (just as[12] fire can yet exist at the combustion-point of the mirror even though no fuel is added which it can ignite)?

FOL. 52

8. Second, from the appearance of a new star, in 1572, in the constellation of Cassiopoeia, and of other stars elsewhere, he argues that there is corruptibility in the heavens. That star, he says, though short-lived, did not perish at once, but gradually wasted away. True enough: but we may not conclude that that star was not of the same kind as, and coeval with, all the other fixed stars. If the sky is liquid, as he believes, why could not the star, borne thither[13] by its own circular motion or by another motion, proceed in the same direction (where it was seen to

FOL. 52v

---

[11] Save at fol. 52, line 5, Hobbes has *speculum* where the context would appear to require *vitrum*, 'lens'. The foci of a conic section he calls (*EW* VII.317) 'burning points'. The final sentence retains his tenses.
[12] Cf. fol. 45v, note 4.
[13] I.e. towards the constellation of Cassiopoeia.

waste away) until it completely disappeared? This can be said of new stars other [than that under discussion], unless the star in Cassiopoeia were the same star that, [moving] through an orbit of its own, first appeared in Cygnus.

9. Third, it is likely enough that, as he states, there are mountains and valleys on the moon, and that the moon herself resembles the orb of earth. What should be inferred from this, however–that there are human beings on the moon–is too difficult a matter for him to broach. His opinion that the lunar mountains are huge by comparison with ours he supports by no argument, for the shape of the moon is absolutely circular to those who view it through a telescope. Whatever mountains there are are invisible to us not because (as he claims) they are covered in dense air, but because they are a long way from us. I do not know whether he or anyone else has ever seen during an eclipse of the sun those swellings on the moon's shape which he here adduces as proof of the size of the mountains of the moon. [The generation of] comets, the nature and origin of which our author will explain in the next Problem, forms the only argument in support of there being great changes in the heavens.

FOL. 53

CHAPTER EIGHT

※

On Problem 9 [of the First Dialogue]:
On the origin and nature of Comets

1. A short acount of the phenomena of the more recent comets. 2. Observers' opinions on their nature. 3. The formation of a comet explained and examined, according to our author. 4. A refutation of his opinion concerning the bending of its tail. 5. The explanation of a phenomenon observed by Libert Fromond. 6. A refutation of the two corollaries deduced from it by our author.

1. I myself saw that huge comet of 1618 in mid-December, when its beard [tail, or train] appeared, to my view, to be at its greatest length. I thought that its head was burning, for it seemed as though the illumination of the night air was projected from that fire. I pondered the fact that neither the comet itself nor its mane [*crines*] could at that time have fallen within the shadow of the earth unless the comet was indeed near to the earth (because the sun was about 20° in Sagittarius but the comet was more to the north than Arcturus was).[1] I did not know what to make of this, nor, when I had read other authors,[2] did I subsequently find anything but grounds for [further] doubt. Let me openly profess my ignorance on the formation and nature of comets–not only do I know nothing for sure, but also I do not put forward any conjecture worthy of consideration: all the men I have so far read declare, each in his own different way, that the natures and the phenomena of comets are

---

[1] There were at least three noted comets in 1618, as we are reminded by the *De tribus cometis 1618* of Orazio Grassi.
[2] Presumably, 'authors other than White'.

two different questions. Nor are they satisfied until, with regard to the question in hand, they have put forth ridiculous speculations on light. Let us, then, briefly recapitulate the comets' phenomena, which have been recorded by the ablest philosophers of the present century, and let us set down what has been thence inferred about their nature, for comets differ in the following respects: their manes, their periods, their situations, their paths, their movements, and their parallaxes.[3]

FOL. 54v

The five comets of the years 1531 to 1539 were all bearded, the beards being, according to the observation of Peter Apian, turned away from the sun.[4] The comet of 1556 was also bearded, the beard, according to the observation of Cornelius Gemma,[5] being likewise turned away from the sun. The comet of 1577 was bearded when Tycho Brahé saw it, and the beard was always turned away from Venus.[6] According to this last observer the same was the case with the comet of 1580. That of 1596, as viewed by Rothomann at Kassel in Hesse, was bearded; its beard was not directly opposite the sun, however, but it was turned a little towards the upper skies. The comet of 1618, as watched by Willebrord Snell, Thomas Feynes and Libert Fromond, was bearded; for a few days its beard pointed away from Venus, but afterwards the beard pointed directly opposite [contrarius] to the sun, as did the beards of its predecessors. The comet of 1618 described an arc of 50° with its beard or tail; but the comet of 1577, which Tycho observed,[7] described an arc of 22°.

FOL. 55

Now the comet of 1585, [as] observed by Rothomann at Kassel, had its mane dispersed spherically [in orbem]; the beards or trains of all these comets were sometimes brighter, and sometimes less bright, so that [in the latter instance] the fixed

---

[3] In the text these nouns are all singular.
[4] The text is followed in 'away from' (barbis soli aversis) but Petrus Apian is equivocal here: '...directly towards the part of the sun opposite (? the comet)': Cosmographia, per [Reinerum] Gemmam Phrysium [or Frisium]... restituta (1539 &c.). French translation of 1584, pp. 289–290: '...la queue ou cheuelure s'estend tousiours directement vers la partie opposite du Soleil'. In the work cited in the next note Gemma agrees with Apian: '...cometarum caudas semper in adversam projici solis partem' (I.117).
[5] C. Gemma: De naturae divinis characterismis (1575), II. 26–28, 135.
[6] Libert Fromond or Fromont (1587–1653): De cometa anni 1618 (1619), p. 116: 'Tycho deprehendit...in duobus anni 1577 & 1580 cometis caudas non a sole directe, sed a Venere fuisse aversas'.
[7] See, for instance, a collection of his letters written before 1592: Tychonis Brahi epistolarum...(1596), one section of which is devoted to this topic.

stars shone through them. But the head of this last comet was never transparent, though our author says it was. Indeed, [the comet] was not much larger than a star of the first magnitude.

The comet of 1556 first appeared during August; that of 1577 first on 13 November in Sagittarius, and continued for seventy-four days; that of 1585 appeared first about 8 October in Pisces, in longitude 23° and latitude 13°, and disappeared about 10 November; that of 1596 first appeared in July; that of 1607 in September, disappearing in October; that of 1618 first appeared in Scorpio in November, passing from view about 15 January 1619[/20].

The path of the 1577 comet was from 13° in Sagittarius to the star in the right shoulder of Pegasus, and formed part of a great circle; that of the 1585 comet was from longitude 23° in Pisces, latitude 13°, to longitude 16°, latitude 8° 0′ 0″, almost seven [degrees] of which were not outside the Zodiac. The path of the 1607 comet extended from Hydra through Leo, through the Coma Berenices, through Boötes and through Ophiucus. Lastly, the track of the 1618 comet was from Scorpio through Boötes and between the Ursae, where it wasted away. FOL. 55v

The motion of the 1577 comet was in the consequent[8] of the signs [of the Zodiac]; the proper motion of the 1585 comet was also in the consequent. That of the 1596 comet was in the consequent, but that of the great comet of 1618 was in the antecedent.[9]

The observers of these comets who also looked into the question of comets' parallaxes found either that there was none or that they were less than the parallax of the moon. Now, as regards [observing] parallax, the infallible rule is this: the time when you must compare the distances is [a] when there is no fault in the instruments by which the parallax is studied, and [b] when, in respect of the distances to be compared, the space between the stations is clearly marked.[10] FOL. 56

---

[8] I.e. motion from west to east.
[9] Motion from east to west; cf. fol. 64.
[10] By 'compare the distances' Hobbes seems to mean: 'When you note the angular distance between the two apparent positions of the star'; by 'in respect of the distances to be compared' he means 'the apparent distance apart of the two positions of the stars'; and by 'the space between the stations is clearly marked' he means 'the distance between the two places at which the star lies is clearly perceptible or large enough to measure (e.g. the night is clear)'. It is possible, of course, that he also envisages that the observer himself changes his position.

2. From this short survey of comets it is sufficiently clear that there is no phenomenon common to them all; for there are different types of mane, different positions, different paths, different movements, different places [in the heavens],[11] different durations and–except for the fact that the comets have a lesser parallax than the moon has–different parallaxes, so nothing can be universally laid down about comets, except that they are beyond the moon's orbit.

Whoever has observed the nature and origin of comets has based his conjecture mainly on a comet which he himself has seen. So Willebrord Snell of Leyden, who saw the comet of 1618 (which first appeared almost in the same sign as the sun), thought all comets to be red-hot or blazing masses cast forth from the body of the sun in the same way as blazing stones are hurled out from Vesuvius and Etna, and to be borne in a path determined by their very ejection. Through some occult quality, however, their trains, [Snell believed,] are mostly turned exactly opposite the sun; if it has happened otherwise, this is due to the aspect of some star, just as the declination of the magnetic needle is supposed by certain learned men to vary. Libert Fromond, following Apian,[12] considered that the same comet [1618] seemed to possess a beard, because the sun's rays are transmitted through a comet's head; he thought this to be the case because [its beard] was observed to be always opposite the sun. Because it anteceded the comet itself, Montl'héry, a Swiss, likened the same beard to the mass of ash turned up from beneath a fire, thence bursting into flame (this mass the French call *une fusée*).[13] Again, Rothomann, who observed the comet of 1585, i.e. the one with a mane, but this scattered spherically, thought that that appearance was due to the reflection of the sun's rays, in the way in which parhelia and paraselenae are thought to appear;[14] and therefore that the [comet's] material consisted of some cloud issuing, not from the earth, but from some planet.

---

[11] Cf. fol. 404v.
[12] I.e. Apian's work on the 1532 comet: see the next folio.
[13] Text, unaltered by Mersenne, 'Motherus Helvetius'. The reference is to a little known (and then unknown to Mersenne) work published at Yverdon in 1619: E. de Montl'héry: *De la trompette du ciel; c'est-à-dire du comète effrayable, qui l'an de Christ 1618 est apparu...*'Anteceded': presumably in space; but perhaps 'surpassed (in brightness)', which would explain the clause '*Because*...the comet itself'.
[14] Bright spots on the halos encircling the sun and moon respectively.

CHAPTER VIII: 3

Certain observers, especially the ancients, thought that comets were exhalations from the earth but kindled beneath the moon; others, the first[15] of whom was Peter Apian (the observer of the 1532 comet), considered the tails, or the beards,[16] of comets to be infinitely small bodies fluttering about through the upper air, these being illuminated by the sun's rays passing through the comet's head, which they think is a kind of cloud. This last view is also the one held by our author. He puts it forward, however, as if it were a new one when he says on his page 72 that the wise have hitherto considered comets to be of fire.

That the material of a comet is not a kindled exhalation from earth is first of all suggested [to us] by the comet's remoteness, for parallax presupposes a big distance. Nor can the earth account for [the existence of] an exhalation which, condensed into a cloud, would appear (at so enormous a distance) as large as the head of a comet together with its tail; the latter has been observed to subtend an arc of 50°. Nor, indeed, although the substance is celestial, is it likely to be a flaming mass, such as a scrap of the sun itself, as Snell thinks. This [hypothesis] is discounted by the fact that comets have never been seen close to the sun; also by the diaphaneity of a comet's tail, even near the head; and finally by the very novelty of the idea; for who will believe, unless on the surest grounds, that the sun spews forth its own entrails, as do sulphurous mountains [theirs]?

FOL. 57v

3. Our author supposes that the sun—either our sun or another sun that controls other planets—raises swarms of atoms from the rest of the planets in the same way as it does from earth, and that these can form clouds. He goes on: Given that such a cloud turns one side, and a very large one, towards us, and turns its other sides in different directions. If the sun were near the side laid open to our view and illuminated by the solar rays, then the cloud would appear shining white. But, if this cloud be raised to the required distance from us, it will appear as a small star, and, because the inequalities have been smoothed out by the distance, as a round one.[17] In this way, he

FOL. 58

---

[15] Perhaps, 'most important'.
[16] See fol. 58, note. Hobbes means that either the beards or the tails of comets are small bodies.
[17] Lack of punctuation in the manuscript makes interpretation dubious, but reference to *DM* 70 suggests that the above is what is intended. The

says—indeed, aptly enough—the head of a comet has been formed. Now, as that cloud, one side of which forms the head of the comet, can be penetrated by the sun's rays, the atoms which are further in will also be illuminated, [even those] as far distant as the opposite side. From this it must follow that we see a kind of long furrow of light, which we call either the comet's beard or its mane,[18] according to the extent to which the illuminated atoms can have different positions, each to each.

But if the beard of a comet is formed in this way, many quite unacceptable things follow. First, it is inconceivable that the beard of the 1618 comet was formed thus. Since one side of the cloud (the side constituting the comet's head) did not, on account of [the cloud's] distance, appear much greater than a star of the first magnitude (whose visible diameter subtends an arc not exceeding two scruples, yet the beard would seem to subtend full fifty degrees), the dimension of the comet cloud, over which the beard extended, was to the dimension of the side forming the comet's head in the ratio 1500:2, so that the whole cloud should have been of the same shape as [that of] the comet together with its beard.[19] The way in which sunbeams passing through a window into a bedroom illuminate the fine specks of dust that are dancing about depends on the shape of the window; hence they make what is illuminated in the air appear a parallelepiped if they enter by a rectangular window, and a cylinder if they come in through a circular one. In a similar way the same rays entering a comet's head, which appears circular, should make the image of the whole comet, together with its beard, appear as the image of a thin cylinder that is throughout of the same thickness as that head. Now if someone says that we are to attribute to refraction some phenomenon that seems other than it [really] is, he must show why also the rays entering the window are not refracted when they meet air which is, as it were, a cloud made up of specks of dust. Therefore a cloud should have been added to his[20] structure:

FOL. 58v

FOL. 59

'required distance' seems to be the distance to which the cloud must recede before it appears to be a star.

[18] This distinction could well have been drawn earlier: see fol. 57, note.

[19] It seems best to keep this breathless and obscure sentence (which in the text runs from the beginning of the present paragraph to the foot of fol. 58v) as far as possible as it stands.

[20] White's? or that of the hypothetical speaker just mentioned? What follows attempts to retain the original reading, taking 'structure' to mean

i.e. whether the side facing the sun and our view [of it]²¹ would be concave or convex. Hence the sun's rays entering the head of the comet would be dispersed and would cast a conical image, the top of which would lie in the comet itself; the image would not, indeed, be quite different from the image which in fact appeared, but it would not be similar either.

Second, if such were the reason for [the existence of] the manes of all comets, that comet with a mane should have seemed, when the mane was strewn spherically [*in orbem*], uniform, as it were a single, but a fairly large, star without a mane. Third, if the beards of comets are formed through the illumination of atoms in a mist, how comes it that the length of the mist is not, at some time or other, distributed in such a way that it markedly deviates from the straight line in which the sun and the head of the comet lie? Perhaps, if by 'the illumination of a comet-cloud' he understood, not the kind of illumination that appears on a wall, but such as appears in a mirror, it would be equally difficult to refute his opinion and to arrive at the truth.²² For in this way the head of a comet would be nothing but the sun's own image reflected from the concave or convex surface of a cloud–an image such as is seen in our convex and concave mirrors. Furthermore, the cloud which displays to us the comet's head would also have had to be not only concave or convex but also polished, e.g. of water or ice. In this way the rest of the [comet-] cloud²³ in the mirror would, if small and the mirror were of very small radius of [concave] curvature, have its mane quite widely diffused, but rather short, as with the comet of 1585.²⁴ But suppose the rest of the [comet-] cloud in

FOL. 59v

'fabric of the case being pleaded'; but here, instead of 'a cloud', one expects words such as 'a question' or 'a factor'. It may be best to amend the text, and read: 'Therefore the structure of his (i.e. our author's) cloud should have been added' (or taken into consideration). In either event the interpretation remains dubious.

[21] How can the same side of this so-called cloud face both the sun and the earth? Has an expression equivalent to 'turned away from' slipped out before '...our view would be...'?

[22] The word 'equally' seems to demand that the opening word of the sentence, 'Perhaps' (*fortasse*), be read as 'Even'.

[23] I.e. the part not producing the image of the comet's head. 'Rest', *reliquum*.

[24] Perhaps, '...more widely...shorter...[than it is now]'. 'As with...of 1585' is ambiguous: (i) The mane of this particular comet was (for whatever reason) short, or (ii) 'It was short for the reason proposed'. The same applies to the sentence next but one following.

the mirror is extremely deep and the head itself of the comet is like a concave mirror, the concavity of which is quite small, i.e. more nearly approaching the planar. Then the image would appear to have a more pendulous beard, with hairs less fanned out, as [was the case with] the comet of 1618. If one granted it not absurd to postulate [the existence of] mirrors of this kind between the sun and the [comet-] cloud, then by this hypothesis an explanation could be given for [the presence of] any shape at all among comets. But he advances no such theory, nor do I believe that one must be put forward; he stipulates only a mist of atoms, which, when lit up by the sun's rays, should expose to view a shape different from that of the cloud itself (unless the distance away of the shape happens to render it less uneven and crooked). This last is impossible and against the procedures of those skilled in optics.

FOL. 60

He achieves nothing by quoting some experiments (which he brings forward on his page 72) concerning the sun (or a candle) shining through a glass vessel with plane,[25] parallel sides and full of water, from the surface of which the sun (or the candle) is reflected to the eye, and in the body of which, if its other parts are made dark, a beam of light is certainly seen. But, [I say], whenever they are illuminated, atoms are always seen in the same position in which they [really] are, as occurs when the sun shines through the window. And in conic radiation through a convex lens that cone is luminous, a cone of illuminated atoms which would be formed, not by its own refraction, but by refraction through the glass. If, therefore, he indeed wishes [to say] that the solar rays penetrating the comet's head take on[26] a different shape from that of a cloud formed of atoms by refraction, he has to suppose that the head of a comet resembles a convex or concave lens through which only such refraction can take place as is similar to that permitted by the beard or the mane [*juba*] of a comet.

FOL. 60v

4. The tail of the 1618 comet sometimes appeared curved like an elephant's tusk. The reason for this our author assigns to refraction. 'Imagine', he says, 'that the tail consists of rays permeating a dense [but] diaphanous body. It is not at once

---

[25] Or, 'even, regular'.
[26] Or, 'If...he wants the solar rays...to take on...' (i.e. at present they do not).

clear that they are flung back at us in such a way that, obviously, they are not all perpendicular [to the body], and that necessarily the others are refracted. You now see that they have to bend. We must find out from the strict rules of optics why [this deflection is] in one direction or in another, and also the reasons for its extent and nature.'

In order to examine this question let us posit a dense [but] diaphanous body, i.e. his comet-cloud, at A, the sun at B and the eye, C.[27] Let us imagine, as he requires us, that AD is the tail of the comet. By what refraction, by what refracting agency, and how will AD be curved so that it appears to the eye at C to be curved like an elephant's tusk? Perhaps, if A is a dense enough body, refraction will take place in the direction of E, but AE will appear straight, although all BAE is not now one straight line, as BAD is. He who deals with matters the understanding of which demands a knowledge of the nature of vision is bound to fall into errors of this kind: yet our author thinks he can ignore the strictest rules of optics. For a man of science, error and the philosophical despising of any part [of his discipline] are alike disgraceful; but how more shameful it is to ally one's mistakes about vision (through a blatant ignorance) with a contempt for optics! But see what White says: 'Is it not at once clear that [the rays] are returned to us in such a way that obviously they are not all refracted perpendicularly (sic), and that necessarily the rest are refracted?'–as though a comet's beard were to appear to us, not by such reflection as that from an illuminated rough wall, but by refraction! Perhaps, if the comet's head lay between the sun and the eye, the sun could be seen by refraction; but how the comet's beard, were it nothing but illuminated atoms, may be seen by refraction passes comprehension. Indeed, teaching of this kind agrees well enough with the minds of those who speak of optics with contempt, and of those who glory that they have 'never handled a quadrant geometrically[28] and have never looked into the heavens mathematically'; that they recoil from the calculations of astro-

FOL. 61

FOL. 61v

---

[27] Figure 9.
[28] Later, White may well have handled a quadrant, for his friend Gerard van Gutschoven, Professor of Mathematics at Louvain (to whom he was to dedicate his *Exercitatio geometrica*, 1658), published (Brussels, 1674) *Usus quadrantis geometrici* on an instrument for measuring the altitude of any given arc. But perhaps the present reference is to the astronomical instrument alone.

nomy; and that they possess neither eyes, nor leisure, nor resources, nor a record of achievement, nor sufficiently strong a will [to undertake astronomical mathematics]–but only one that, from their early years on, has been accustomed to [dealing with] everything else [but optics].

FOL. 62

Lastly, those who admit that they have 'never looked into the heavens or understood the force of devices by which the firmament is made' nevertheless dare to philosophise about the skies 'by natural qualities' (as he puts it), as though without mathematics, without optics, and without looking into the heavens one could learn the 'natural qualities' of the firmament!

5. Libert Fromond noticed, on 11 December 1618 at five in the morning, that the head of the comet [he was observing] was near Arcturus, but that the beard stretched as far as to the left knee of the Bear. Next day, namely 12 December, 'I marvelled', he says in the first chapter of his *Liber de Cometa*, 'that, even though the whole of the Chariot was in view, not a hair of the comet's beard was apparent. At length, after the middle of prime when the sun had descended within the arc of the meridian, little by little [the comet] began to be delineated, its position being as on the day previous, and [itself being] in the same longitude[29] as, manifestly, the lights of dawn usually are. This is the time when the sun, the nearer it has climbed to the horizon, increases and directs all the more the rays it previously emitted'. He says afterwards, in Chapter 6, that that behaviour [of the comet; *apparitio*] is in no way [the basis of] the argument one must construct; for obviously the sun's rays had been forced through the comet's head by that beard.[30] Would that Fromond had chosen to explain (by suggesting some theory)[31] how, when the Chariot has risen but the comet's beard has not yet risen, this implies that the beard *is* only the solar rays piercing the comet's head! I see (I think) how it could have happened that the beard did not appear in the middle of the night but gradually rose shortly afterwards. Let the earth be at A, the earth's axis AB, the horizon AC, and the elevation

FOL. 62v

---

[29] Perhaps, 'of equal length'.
[30] This tortuous sentence seems to mean that one is unsafe in inferring much from the phenomenon described; but Fromond's printed text may be corrupt, even though Hobbes does not question it.
[31] For 'theory' (*schema*) perhaps, as at fol. 262, 'plan of the heavens'.

of the pole BAC, as seen from Louvain, where the observation was made.[32] Let there be a star at the knee of the Greater Bear near the horizon at C, and let it move in its daily arc to D. Because the sun was then about 20° in Sagittarius and the time was the middle of the night, the sun was sunk about 60° beneath the horizon, say at E. The comet will be in Boötes, i.e. at F: either more distant or less distant from the earth than the sun is, or equally distant. Lastly, let the comet's beard FG lie in a straight line opposite [to the sun] and wholly below the horizon: hence the beard will not appear in the middle of the night, but will appear in the morning, just as it appeared the day before; for if the sun and the comet move forward from E and F to H and I, the tail IK will therefore equal FG. Meanwhile if the star at the knee of the Bear moves from C to D in its daily arc the limit of the beard, K, will be seen at D, i.e. if the eye of the observer is at A. So we need wonder no further why the beard appeared gradually; I do not see, however, why this may not be explained in the following way: either the sun's rays passing through the head of the comet form the beard, or the beard is of some material that has been ignited, but is always opposite the sun.

FOL. 63

6. From this observation of Fromond our author draws two corollaries. First, as the comet was beneath the sun and [seen] in the middle of the night, its beard was in the shadow of the earth. Second, when the beard appeared like morning half-light above the horizon, the sun's rays passing through the head of the comet illuminated or came near to the earth. Each of these corollaries is impossible. As regards the first, (*a*), the sun, the earth and the earth's shadow make one straight line, and the sun, a comet and a comet's beard also make a straight line. So great are the distances, both of the sun and of the comet, from the earth that the last may be considered as virtually a point–at least as being not greater than the apparent magnitude of the sun. Hence those two straight lines (which have the sun as one common point and the earth's shadow, in which the beard is, as the other) must be coincident, and must form a single line in which lie the sun, the comet's head and the earth. This is manifestly impossible, because it was known that, at the time referred to, the sun was nearly 20° in Sagittarius and

FOL. 63v

[32] Figure 10.

the comet's head at Boötes's shoulder: these two places are at least a quarter of a circle apart. (*b*) Let the earth's shadow, in which the comet's beard is supposed to have lain hidden, be extended along the straight line AL, which is diametrically opposite the position of the sun. When the sun moves forward the comet does so also, and with the same motion, namely a daily one. It was, however, reasonable [for White] to hold that the beard lay in the earth's shadow, not only in the middle of the night but also for several hours afterwards, until it freed itself by its own motion [relative to that of earth]; for the comet's own motion was in the antecedent, i.e. towards the west. When, therefore, the sun's rays, which constituted the comet's tail, began to appear as they illuminated the earth, they ought to have come into view towards the west. [According to White, then,] the comet's beard has appeared as a dim morning light in the west, which is absurd. Moreover, on the night when the tail was first seen the sun's rays did not reach the earth itself; and what he adds on the finding of the comet's place is untrue; for the three sides of the triangle, the angles of which are formed by the centres of the sun, of the earth, and of the comet, could not be visualised in this manner.

CHAPTER NINE

On Problem 10 [of the First Dialogue]:
That there are many things in the
universe that shine of themselves

1. Motion is generated from all parts of a shining object at the same time. 2. Why vision is effected in an instant. 3. That there is alternate dilatation and contraction in the action of shining objects. 4. That the action of a shining object is weaker at a greater distance than at a lesser, and in what proportion. 5. By what things the shining of an illuminated body is increased or reduced. 6. What illumination is. 7. The reason for reflection at equal angles. 8. Why illumination at the perpendicular is stronger than illumination falling obliquely. 9. That luminous objects when combined shine more brightly than when dispersed. 10. That the author's argument against the illumination of the fixed stars by the sun is not valid. 11. That another argument of his against the same is unsound. 12. Galileo's demonstration that perpendicular illumination is stronger than oblique. 13. A proposed objection to Galileo's demonstration found invalid. 14. But this demonstration is not correctly refuted by our author. 15. He gives no reason why direct illumination is stronger than oblique. 16. What metaphysics is, and whence so termed.

1. The understanding of this and the following Problem depends upon a knowledge of the nature of light, and of the measures by which greater and lesser illumination is assessed. Our author had observed, as he himself says on his page 99,

---

[1] Fol. 64v is blank.

that catoptrics have not been dealt with carefully enough [by others].[2] He wished, therefore, to look into the subject more deeply; yet in the course of his study of the nature of refraction he says that his eyes have been dazzled by too much light and that they have warned him 'what it is sufficient to study, even in the sciences'. He means (as he wants these words to be understood) that even in the sciences a limit must be observed, or that there are defined limits on either side of which the truth cannot be found.[3] He is unwilling to make mistakes about the nature of light through knowing too much, and so he stops short before reaching an understanding of it, satisfied with the works of those he has accused of writing carelessly. The result is that when, in this following problem, he discusses the effects of light (what the nature of light is he does not know), he has made several mistakes, as will be shown after we have explained the true nature of light.

First of all, it is clear that all shining bodies are seen at the same moment, and on every side, by those around them. Next, it is also definite that vision is effected by the action of a shining body on the organs of sight. From these [axioms] it emerges that shining bodies, wherever they act from, act at the same time. Moreover, it is absolutely clear that every action is the local motion either of an agent itself or of the agent's parts, and that, when a shining object is viewed but does not change its position with its whole body, on all sides it acts on the organs of sight through the local motion of its parts.

2. It follows that the action of a shining thing is dispersed to any distance in an instant. Given any such body, e.g. the sun, whose centre is A and semidiameter AB.[4] Because its parts are moved from wherever they are and all at the same time (i.e.

---

[2] It was in *dioptrics* that White had secured a reputation for himself in France. So Mersenne to Haak, 12 February and 20 March 1640/41 (Samuel Hartlib papers deposited in the University of Sheffield, Set XVIII, Batch 2). The *Correspondance du P. Marin Mersenne* (Paris, 1945; in progress) in printing these two letters uses a transcript in the Birch MSS, British Library. (Courtesy of Lord Delamere and of Sheffield University Library.)

[3] For 'on either side of' perhaps, 'beyond (or within)'. It is dangerous, says the speaker Asphalius (*De mundo*, pp. 6–7), to profess concern for the glory of God and yet seek this beyond the evidence available within the confines of nature. Here Hobbes seems to mean that one should not delve too deeply into some matters.

[4] Figure 11.

at the same instant), its surface, [formerly] bounded by the circumference B, will now have wider limits (because of such a tumescence) in the circumference, say at C. This cannot occur without [the existence of] a vacuum, or of the little empty spaces that intervene between the [sun's] parts. But because it is possible to conceive of a vacuum, but not possible to prove that all space is filled with some kind of body,[5] there is no reason why the parts of the sun may not have a motion of this kind. Therefore, whatever corporeity there was in sphere B is extended by this motion as far as C; likewise whatever [corporeity] was first in the shell [*orbis*] bounded by the surfaces B and C (this shell I wish to suppose equal [in volume] to sphere B) must be driven further into another, outer shell, i.e. the one bounded by the surfaces C and D, and equal to this same sphere B. Again, that which was bounded by the surfaces C and D must be driven out into a shell still further out, one bounded by the surfaces D and E; and so on, infinitely. In this way will motion be propagated, from any source whatever, into the sphere, however large, in which may be thought to be the greatest circle, whose diameter is AE, and the solid shells AB, BC, CD and DE are equal to it.[6] Therefore at the instant that the point B starts to be propelled in the straight line AE, point C will be advanced, and also point D and point E. So at the same instant when any part of a shining object is moved towards the eye (which we suppose at E) the motion will impinge on the eye, i.e. the shining body will act on it. Similarly the action is also communicated to the inner regions of the skull, where the brain and the animal spirit, the organs of sight, are; so vision is effected at the same instant as the shining object begins to dilate.

FOL. 66v

FOL. 67

3. But it is impossible that the sun spreads out in all directions without being larger (if it could, it would be completely diffused) or, conversely, that it keeps diminishing. There must always be in the sun, therefore, a systole and diastole, or a contraction and dilatation, alternately, such as happens in a man's arm lifting some heavy weight; the arm weakens and gathers strength in turn.

FOL. 67v

---

[5] A hit at Descartes?
[6] 'It', the circle (or orb) whose diameter is AE; i.e. the four solids severally equal in volume the greatest sphere.

4. The action of a shining object, when propagated to the fundus of the eye and thence to the brain, is the cause of the reaction by which a motion is transmitted back from the brain, through the eye, to the objects outside. The latter motion, however, is experienced not as motion, but as the fantasy or image of the sun or of some other shining body. This fantasy we call illumination or light. The greater the motion excited by the object towards the eye, the brighter the light. Also, the lesser the distance over which this motion is propagated, the quicker the motion; the greater the distance, the slower the motion. Now, given that solid[7] spheres [*orbes*] of radii AB, AC, AD and AE have volumes[8] increasing in order of magnitude as numbers starting from unity, i.e. 1, 2, 3, 4, etc. Since [the spheres] are one to another in the triplicate ratio of the same AB, AC, AD and AE, then AB, AC, AD and AE will be each to each as the cube roots of the numbers 1, 2, 3 and 4, etc. Thus AB:BC will be as the side of the single cube [i.e. the side of cube of unit volume] to that of the double cube [i.e. the side of cube of two units] less the side of a single cube; and BC:CD as the side of a double cube less the side of a single cube, to the side of a triple cube less the side of a double cube; and CD:DE as the side of a triple cube less the side of a double cube, to the side of a quadruple cube less the side of a triple cube. Hence AB exceeds BC and BC exceeds CD. Now, in the same time as [a point at] B moves forward to C, [a point at] D moves to E.[9] Therefore in equal times, the nearer the motion is to the shining object B, the greater the space traversed; and the further [from B] the motion, the less the space traversed. In other words, motion is slower when derived at a greater distance. Hence[10] it is obvious that the widely held view of the philo-

---

[7] Solid but yet penetrable; see note 9.

[8] For 'have volumes', text: 'are'.

[9] This sentence is added by Mersenne in the manuscript: it refers to the third sentence of Article 2 (but what is given is only a hypothesis by Hobbes). The present passage is paraphrased by Mersenne at p. 75 of the Ballistics section of his *Cogitata* (1644). Properly, for 'in the same time' Mersenne's annotation reads 'at the same time'. 'To (*ad*) C' means 'so as to arrive at C' rather than merely 'in the direction of C'. Hobbes's argument is that for equal increases in spherical volume the increase in radius is less at greater radii; and the times being equal, the motion will be faster along the radius of the inner sphere and slower along that of the outer shell. Hence light is propagated longitudinally at uniformly diminishing velocity.

[10] 'Hence' probably refers, not to the sentence immediately preceding, but

sophers, that 'action is weakened in direct proportion to the increase of distance', is untrue. Now, what has [just] been said concerning the degrees of speed in the action by which light is produced should also be understood as applying to the degrees of brightness of the same shining object at different distances. For illumination owes its origin to the strength of the action, but the latter consists in the speed of the motion transmitted. Moreover, what has been said [here] about the sun must also be said, and for the same reasons, of any other shining object that emits its own light.

But if the distance between a [point-source of] light and the eye is divided into equal parts, [the light] will traverse the first part at a speed, in relation to that with which it moves in the second, as the side of a single cube to that of an octuple cube less the side of a single cube: this speed is to the velocity in the third part as the side of a single cube to that of a 27-fold cube less the side of an octuple cube, and so on. For as AB is given as equal to BC, and BC as equal to CD, AC is to AB as the side of an octuple cube to the side of a single cube; so AB is to BC as the side of a single cube to the side of an octuple cube less the side of a single cube. But AD is the side of a 27-fold cube. Therefore CD will be the side of a 27-fold cube less the side of an octuple cube.[11]

FOL. 69

5. But light is directed not only according to the speed or the strength of its action, but also according to the size of the shining body. For what has been said [by me] of the whole sun, or of another shining object, as regards dilatation and shrinking,

---

to the foregoing argument. It will be noticed that Hobbes makes no reference to an inverse-square relationship between strength of illumination and distance. The paragraph following, beginning 'But if the distance...', appears next in the *Cogitata*, Mersenne omitting the intervening sentences.

[11] Properly, another diagram is required, consisting of a straight line divided into equal parts. The reasoning is:

$$\frac{\text{Veloc. from A to B}}{\text{Veloc. from B to C}} = \frac{x}{\sqrt[3]{8x^3} - x} = 1$$

and $$\frac{\text{Veloc. from B to C}}{\text{Veloc. from C to D}} = \frac{\sqrt[3]{8x^3} - x}{\sqrt[3]{27x^3} - \sqrt[3]{8x^3}} = 1$$

So $$\frac{AC}{AB} = \frac{\sqrt[3]{8x^3}}{x} = \frac{2}{1} \quad \text{and} \quad \frac{AB}{BC} = \frac{x}{\sqrt[3]{8x^3} - x} = 1$$

Therefore

$$CD = \sqrt[3]{27x^3} - \sqrt[3]{8x^3} \quad \text{Figure 12}$$

FOL. 69v

is also to be understood of each of its parts—at all events, of each part that we can perceive; so any shining object may be considered either as one large, bright thing or as many lesser, lustrous things. Therefore just as two lights that illuminate any opaque object increase its sheen, so a shining object twice as large [as another one] creates the greater brilliancy. We must therefore be able to explain the shining of a body in terms of one thing's being less capable of illuminating than another's. We must also consider the site of the body that is to be illuminated; its position is sometimes perpendicular to the motion of the light, sometimes oblique. Also, when two shining bodies cast light on the same opaque body we must examine their positions relative to one another; for whenever they are closer together the combined effect can then depend on rays sparser [than when the sources of light are distant one from the other].

FOL. 70

6. Bodies are said to gleam when, because of [their] great roughness and the diversity of the position of their innumerable facets, shining objects reflect into our eyes a broken and confused image.[12] When the sun, or another star, shines on the smooth surface of water at rest so that the image of the star is revealed to us, and this image is in one place only, the water itself is not illuminated but is invisible. If the same water is whirled round so that a mixing of images occurs, the water seems like a wall with light on it. Illumination, then, is nothing but the reflection of light from innumerable surfaces on to one and the same point where the eye is.

FOL. 70v

7. Now, where reflection takes place the angles of incidence and those of reflection are always equal. Given any surface AB, a source of light at C, and the eye at D.[13] Let the shining object act on the surface AB at the point E. Therefore the motion along the straight line CE tends partly from the side AC towards the side BD; but partly from side CD towards the surface AB perpendicularly. As regards the motion from AC to BD, the surface AB cannot be affected by BD because AB is not approached by this motion.[14] So the source of light C acts on the surface AB only by the perpendicular motion from C to

---

[12] Or '...facets, they reflect...confused image of a shining object'.
[13] Figure 13.
[14] I.e. two components at right angles are independent of one another; CD and AB are parallel.

A, and hence in the other, perpendicular lines parallel to BD. There is resistance to this perpendicular motion from the body whose surface is AB. Because of this resistance the motion from CD to AB is lost, and a contrary motion, from AB to CD, generated. For bodies that are struck yield first at the spot where they are hit and then, as they recover themselves, they implant a contrary motion in the things that strike them.

Hence motion [i], compounded of the motion lost from AC to BD and that gained from AB to CD, is equal to motion [ii], that by which the movement from C was transmitted to E. Motion [i] makes the straight lines CE and ED equal, they having been described by equal motions in equal time. It also makes CA equal to BD. But the right angle DBE equals the right angle CAE. Therefore the angle of reflection DEB equals the angle of incidence CEA.

FOL. 71

8. From this compounding of motions is also understood why illumination is strongest when effected in a perpendicular line. For when rays are emitted from C to E the action is greater when a larger recipient is placed opposite to the action, but lesser with a smaller one.[15] But AB is opposite to the motion in the straight line CE only insofar as CE moves towards AB, not insofar as [CE] moves towards DB. If we now suppose that a movement is created from C to E by two movers AC and CD acting together, so that CD reaches AB at the same time as AC reaches EF, point C must come to E in the same time also. Hence the velocity through CA, which alone activates the source of light at C [directed on plane AB], is less than the whole velocity through CE, i.e. in the ratio as the length of the straight line CA to that of CE. If, therefore, through point E a straight line DG is drawn perpendicular to CE, source of light C will bring to bear its whole velocity upon DG. That is, C will illuminate surface DG more strongly than surface AB in proportion as the straight line CE exceeds CA in length. Hence we learn the reason why objects [directly] opposite to the sun or to a candle are more illuminated perpendicularly than are those obliquely opposite, and why those obliquely opposite are more illuminated than those even more obliquely opposite.[16]

FOL. 71v

---

[15] Or, '...a more recipient object [*patiens*]...a less recipient one'.

[16] Hobbes means that when rays are emitted from C to E the area of illumination is greater when the illuminated surface is at right angles to the incident rays (in this sense it is a larger recipient) than it is when the surface is not

FOL. 72

9. If two sources of light cast their rays on one point, and another two sources do so on another point; and if the sources are equal [in intensity], and the distances equal, but the angle at which their rays fall in one act of illuminating is less than in the other, then when the angle is less the illumination will be greater than when the angle is larger.

FOL. 72v

Given four sources of light, A, B, C and D, all equal.[17] Let A and B shine at the same time on the point E; and C and D on the point F. If the distances AE, BE, CF and DF are all equal, and the angle AEB is less than the angle CFD, then I maintain that point E will be more brightly lit by the sources A and B than will point F by C and D. Let AH, BG, CK and DI be drawn so that the angle EAH equals the angle EBG, and the angle FCK equals the angle FDI. Let two rays meeting at L be drawn through the straight lines AH and BG, and let another two rays meeting at M be drawn through the straight lines CK and DI. From the meeting of these rays will be created a compounded motion [the resultant] through the straight lines LE and MF. Now, [i] the angle ALB is less than the angle CMD, and [ii], the angle AEB is given as less than the angle CFD, and the angles EAH, EBG, FCK and FDI are given as equal. So the angle ALB will be less than the angle CMD.[18] Therefore the motions through AH and BG are the less opposed, each to each; hence the distance to which their own velocity will carry them will be less.[19]

Therefore also the motion through LE, a motion compounded of the motions through AH and BG, will be swifter than a motion through MF, this latter motion being compounded of motions through CK and DI. Hence the action of the sources of light A and B on the point E will be swifter, i.e. stronger, than that of the sources C and D on point F. So point E will shine more brightly than will point F.

---

at right angles (i.e. the recipient then appears smaller). E.g. consider a rectangle ABCD pivoted in the vertical plane round its side DA. As the rectangle rotates through 90° its apparent area, as seen from a source facing it, decreases.

[17] Figure 14.
[18] This repetition follows the text. 'As equal': 'as two pairs'?
[19] I.e. they are not distorted as much from their own original paths; this means that the resultant EL is greater then the resultant MF. Such seems the best interpretation, even though it involves a slight emendation of the text. For the last two clauses the original has: 'They will mutually take away, by their speed, less of themselves'.

10. Once we have grasped this,[20] let us come to our author's reasoning. First, in order to prove that the fixed stars are easily seen, not because of the sun's illumination, but because of their own light, he bases his argument on the distance between them and the sun.

Astronomers say that the distance of the sun from earth, as compared with the distance of the fixed stars from the sun, is as 1 to 2160. The illumination of [a point on] the earth by the sun [even] when directly overhead is not really great. Therefore the illumination of a fixed star by this same sun at so enormous a distance cannot be so great as to be seen by us. This reasoning is wrong, however, in that it has not brought into the reckoning the sizes of the fixed stars themselves. That the earth, were it placed among the fixed stars, would be noticed little, if at all, may be true; but if we attribute to the fixed stars sufficient magnitude they will be visible, however distant they are, because of their [reflected] light only, without emitting light themselves.[21] For although the velocity of an action is always less at a greater distance, [the action] never completely ceases. If the earth were indeed situated among the fixed stars it would be illuminated by *some* degree of light; but this degree would, if the size of the earth were increased, continually become greater, so that at length [the earth] would become visible even from [a point in] a heaven of fixed stars.

FOL. 73

FOL. 73v

11. Again, he argues thus: Given a luminous cone, with the sun as its vertex and a great circle on the earth's sphere as its base. If the axis of this cone were 2160 times larger [than it is], so that it extended [from the earth] to the fixed stars, the base of that cone would [then], according to the duplicate ratio of the bases of cones to their axes, exceed a great circle of the earth by 4,665,600 times. Hence the sun's illumination, dispersed over such a space, could not make the star shine sufficiently for one to be able to read by its light.

The first flaw in this argument is that it does not suppose that a great circle, even of the star to be illuminated [by the sun], is 4,665,600 times bigger than a great circle of the earth. If the point is insisted on, however, that star would appear of the same magnitude to us on earth as the earth would to those

FOL. 74

---

[20] Cf. Mersenne, *Cogitata*, Ballistics section, pp. 77–78.
[21] Cf. Drake, pp. 290, 373.

viewing [this planet] from a distance equal to that between earth and the sun. We see that the moon, at the great distance by which she is separated from us, appears little less lustrous than a wall near at hand, illuminated by the midday sun, would be. Why so, unless because [the moon] appears much smaller than she [really] is?[22] For if, perhaps because of the weakness of the sun's action across the moon's distance [from earth], everything seemed to be as large as it actually was, the moon would not be seen at all.

The second flaw is this: he concludes that the illumination of the fixed stars by the sun will be less than is necessary if we are to read by the light of one of them. This is true, but it is not what he should have deduced, i.e. that [the light of a fixed star] will be less than that required to enable the same star to be seen.

Third, he does not correctly argue, from the increase in the base of a shining cone, a lessening in the strength of the light. For light is not body, nor is it spread in the same way that metal is increased in width by [beating with] a hammer. It waxes and wanes partly because of the velocity of [its] motion, which diminishes and increases in the manner explained above, and partly because of the decrease and increase in the size of the shining object itself.

12. Galileo has shown the reason why perpendicular transmission of light illuminates more than oblique does, as follows: Suppose any number of rays to be represented by any number of parallel lines, the limits of which lie between A and B [laterally]; and suppose that a line CD has been drawn perpendicular to these, but that DC is then inclined towards DO. It will be clear that fewer rays will be directed on to the oblique line than on to the perpendicular line of the same length.[23] (Further, Galileo supposes that DC equals DO from the very fact that he says, not that DO is to be drawn 'obliquely, in any fashion you wish', but that DC must incline towards DO.) Now, he supposed that the rays at this spot are parallel because of the distance of the heavenly bodies, the sources of light, from us;[24] owing to this

---

[22] Drake, pp. 336, 371.
[23] Figure 15.
[24] 'He', Galileo. The reason stated, however ('because...from us'), does not appear in this context either in White or in Galileo, though it is referred to elsewhere (e.g. Drake, p. 287).

distance the angle of illumination[25] lessens, as far as our perception of it goes. He could have said the same, however, of conical radiation. Given a source of light at E, let its rays fall on the line FG, as far as possible perpendicular [to E].[26] Then incline the line FG to GH so that it is more oblique. Clearly more illumination is brought to bear on the line FG than on the line GH. He takes it as noted that where there is more dispersal of rays there is greater light. I do not see any fallacy here.

FOL. 75v

13. Our author introduces, in the form of [his] third interlocutor, a truly paralogistical censure of this demonstration of Galileo's.[27] If, he argues, the straight line CD be divided into two parts at E, the rays falling on CE fill the space CO as perfectly as they fill CE;[28] but 'to be equally illuminated' is nothing else but 'to be equally filled with rays [of light]'; so the spaces CO and CE are 'equally illuminated'; but space CE is as 'equally illuminated' as space CD; therefore CO and CD are 'equally illuminated'.

In this argument he assumes that 'to be equally illuminated' is the same as 'to be equally filled with rays'. Yet Galileo did not say or suppose this. He said only that those areas are equally illuminated which, being seen to be manifestly equal, are struck by the same number of rays. The following, said Galileo, is less well illuminated: that which, remaining constant [in area], is acted on by fewer rays, or that which, if increased in size, is acted on by rays that stay unchanged. Now, surely [says White], anyone can see that CO, the greater line, is illuminated by the same rays by which CE, the lesser, is; and that consequently CO is always less illuminated than CE?[29] The censure therefore contains an example of false reasoning in that Galileo [indeed] supposed the perpendicular and the oblique to be equal; but he [White] substitutes CE the perpendicular, it being less than the oblique CO.

FOL. 76

14. Our author sees that this criticism [at the start of Article 13] is false, and he rebuts it, but with a censure even more

FOL. 76v

---

[25] I.e. the angle subtended at the source of light.
[26] Figure 26.
[27] DM, p. 88f. The speaker is the character Andabata
[28] Figure 17.
[29] Figure 17.

ridiculous. He denies that the whole of the line CO is, or can be, illuminated by rays passing through CE, unless the light is dispersed to infinity. Hence only as great a part [of CO] is illuminated as equals the line CE, the rest being covered in spaces of darkness.[30] Why? Because the parts intercepted on the line CE between the parallels that cut it are, each to each, smaller than the parts cut off on line CO by the same parallels, i.e. because all CE is smaller than all CO. Similarly, if the triangle CEO were of wood, there would be no wood anywhere on the line CO, as is the case with the line CE, unless the straight line CE equalled the straight line CO, which is absurd.

FOL. 77

15. At last, on his page 93, he shows the reason why the oblique CO is not as [brightly] illuminated as the perpendicular CE is, as follows: 'Do you not observe', he says, 'that rain falling in a straight line moistens a pavement over its whole surface, but that walls in a line parallel to the [rain's] fall are moistened by the rain with perhaps a slight sprinkling, as if with an accidental touch? Consider, then, that the light, etc.' This is correct, for a wall directly opposite a source of light is more illuminated than one set obliquely, for the same reason that a pavement directly facing falling rain is moistened more than is a wall facing rain at an angle. But when we ask for an explanation that fits both occurrences our author does not furnish any; he neither produces one nor tries to find one. [These matters] we have sufficiently treated of above in Articles 8 and 9 of this chapter.

FOL. 77v

16. Our author has expressed on his page 91 the surprising opinion that not all of the oblique line CO is illuminated, but only as large a part of it as is equal in size to the line CE, and that the remainder [of CO] is enclosed in spaces of darkness. Lest, perhaps, someone were to attribute a viewpoint so remote from common sense to madness, but rather [that it be attributed] to some supernatural knowledge, he makes an addition [to his argument]. 'Or', he says, 'note from this that

---

[30] ...*Reliquum interstitiis tenebrosis contineri*. *Interstitia* suggest segments of light alternating with segments of darkness, the sum of the latter being equal in length to the difference between CO and CE. Hobbes is therefore not accurate when he says that 'the rest' of the line is 'covered in'—or adjacent to?—the *interstitia tenebrosa*: this happens only with those segments adding up to make the length CE.

metaphysical considerations are of a different nature from that of physics or mathematics. These latter studies indeed supply very valuable evidence for metaphysicians, but are quite unsuitable for metaphysics, which must be studied because of its authority [*dignitas*].[31] Moreover, that compounding of quantity is to do with factors we must seek beyond nature; and how much of the purpose of metaphysics is more abstract than that of physics!' In these words he misuses the term 'metaphysics', because of his ignorance and his bragging alike, so I have here brought forward as needing explanation what the knowledge is that we call metaphysics, and whence it has been so named.

FOL. 78

When Aristotle was about to deal with the nature of the heavens and of the elements;[32] with coming- and ceasing-to-be; and with the rest of the things of nature, he first published certain books, which he called *Lectures on physics*, and in which he argued about matter, form, place, time, motion, and the other questions pertaining to body in general, i.e. to all bodies whatsoever. Unless one knew this beforehand one could not approach the book *On the Heavens* or the rest of his books on physics. (Yet he did not [in the *Lectures*] deal with all the topics he needed to deal with.) Realising, then, that all the most common things must be investigated before anything less common is examined, and that the concept of òn, i.e. a being, is the most common of all, he considered that the knowledge of being must be the first to be gained; from this, he thought, one should then come to the special kinds of being, such as heaven, earth, animal, etc. So he wrote several books in which he took being as his subject; this knowledge he called *sophia*, wisdom, for this science embraces all sciences, just as his subject, being, includes all subjects. This same knowledge he called *Philosophia prima* because, if anyone wishes to philosophise correctly, he must commence with this.[33] Here, therefore, [Aristotle] defined the notions or those names which are the commonest of all [kinds of] being and of the essences such as substance, *accidens*, quantity, number, unity, time, place and motion; and he deals with many others that he had written about earlier.

FOL. 78v

FOL. 79

---

[31] One assumes that White means that the whole of metaphysics deserves study; but his wording could mean equally well: '...for the metaphysics' (i.e. parts of that discipline) 'which must be studied', etc. His general meaning here is unclear.
[32] I.e. those of matter (*elementa*): not 'the weather'.
[33] Cf. fol. 5.

They are extant, almost in his own words, in his eight books of *Lectures on physics*, where as far as possible he follows up everything by natural reason and claims to be wholly unacquainted with matters the knowledge of which 'transcends nature'. (Such matters are inscrutable to all save someone granted divine revelation restricted to himself.) Those books on *Philosophia prima*, i.e. on the elements of philosophy, came to be called the *Metaphysics*. Either Aristotle gave them this title because they were written by him at a later period than that of his treatises on physics, or, as most authorities believe, his successors did so: they found these books to be without a title, and in the canon a position after his books on physics was assigned them. Indeed, even to one of quite a subtle turn of mind they were hard to understand from the start, and as time went on they became formidable through foolish commentaries and disputations, leaving aside the fact that the books his expositors have written have proved even more difficult [than the original texts]. Because of this problem, and owing to the title *Metaphysics* (since 'meta' means not only 'after' but also 'beyond'), the ignorant believed that certain supernatural doctrine was contained in these books, just as if those who applied themselves to metaphysics were to do so in order that, by means of their doctrine, they might step beyond nature's confines.[34] Nearly everyone who spoke in the schools or whose writing could not be understood because of its gross absurdity wanted to be called a metaphysician. If anyone confronted such people with solid reasonings, either geometric concerning quantity or with some from physics on the subject of motion, they could not reply and claimed, through arrogance and in order to hide their ignorance, that the mathematical sciences and physics were unworthy of their attention; yet they understood the metaphysical doctrine least well of anything. So our author writes: 'Physics and mathematics are quite unsuitable for dealing with metaphysics as it deserves'–as if the road to metaphysics lay with physics and mathematics, and not that metaphysics were the road to physics and mathematics! Also, 'the compounding of quantity is to do with factors we must seek beyond nature'–this is as though metaphysics were the knowledge of supernatural things, i.e. of those that cannot be known. Again, 'This is why great men, but learned only in

FOL. 79v

FOL. 80

[34] Cf. *EW* III. 671 (*Lev.* III. 46).

mathematics and physics, when they come to deal with the nature of a continuum, of rarity, of density, of infinity, of place, of time, and with similar topics, become entangled in snares from which they cannot extricate themselves, and allow the dictates of reason to be confounded'–which supposes that the examination of rare and dense has nothing to do with physics! Neither rarity nor density, however, is a bare declaration of being or of essence, as are place, time, a continuum, and other terms, but is of being as such. His examination, then, is to do with physics, [he says].

FOL. 80v

It is easy to see how ignorant these protestations of his are. In a word: those who profess mathematics know almost all of what they teach; those who profess physics know something; but those who profess metaphysics know nothing; yet, wishing to appear to know more than the others do, they rail against the sciences of the rest.

FOL. 81

CHAPTER TEN

❦

On Problem 11 [of the First Dialogue]:
That the telescope is at its maximum
development
(that is, it has reached
perfection)[1]

1. A description of the telescope. 2. The description examined. 3. The position of the image seen through a telescope. 4–9. Four [six, in fact] arguments of our author, by which he denies that the theory of the telescope is being developed, confuted. 10. That the cause of gravity is not some agent that brings it into being.[2] 11. What gravity is, and in what kind of motion it consists.

1. Whether or not there can be made optic tubes or telescopes less prone to error than those now in existence can be ascertained only by reference to the ones whose nature and construction have been well studied. So we must first describe the optic tube, and trace the lines of radiation from an object itself, through both [its] lenses, [convex and concave,] to the

---

[1] Text: 'It is at its greatest perfection'. With this pleonasm cf. fol. 360v, Article 4.
[2] For 'some agent...being' the text has *generans* only. The manuscript gives articles 10 and 11 the numbers 9 and 10 (articles 4 to 8 promising to consider four of White's arguments), but there are eleven articles in the narrative below. This suggests that, the copyist merely reproducing what he saw in front of him, Hobbes modified his original plan by adding, changing, or omitting material as he proceeded, forgetting to correct the titles. See also fol. 96v, note.

fundus of the eye. We must then examine, one by one, the reasons why it appeared to our author that an optic tube better than those now in being cannot be made.

Given, therefore, a convex lens ACB and any object placed so far from it that the length of the straight line AB has no perceptible ratio to [the lens's] distance from the object.[3] Then let two straight lines, drawn from the two limits of the object to two separate points of the lens, A and B, make the angles dAe, dCe and dBe; and let us suppose that the same be done for all the other points that may be taken between A and B. Now let dA and eA, crossing one another at A, be [both] produced; likewise dC and eC, that cross one another at C; likewise dB and eB, that cross at B. Let them all [three] be produced, I say, as far as a concave lens placed at any distance e.g. at DE, so that both lenses lie parallel—or not markedly oblique—to one another. This being done, it is clear that all the [rectilinear rays represented by the] lines that have been produced have been refracted, and that the closer they are to the concave lens the more each pair converges on another pair. Now let the [rays represented by the] refracted lines, as they pass through A, be indicated by broken lines; those passing through B by pecked (---) lines; and those through C by unbroken lines. Let FG be the eye; and let us imagine that such is the shape of the concave lens, and such its position relative both to the convex lens and to the eye, that all the lines [of light] coming from e are refracted [by the concave lens] on to the surface of the eye in such a way that they are in turn refracted by the eye itself, as is required by the latter's nature. Let them all meet on the fundus of the eye at one and the same point H. Likewise let all the lines coming from d be so refracted that they all meet on the fundus at point I. If this is done d will be seen [to lie] on the straight line drawn from I and parallel to the part of the line dAI or dCI or dBI which lies between the concave lens and the surface of the eye; and e will be seen [to be] on the straight line drawn from H and parallel to the part of the line eAH or eCH or eBH that lies between the concave lens and the surface of the eye, such that the visual lines from the extreme limits of the object are the straight lines HL and IM. Hence the

FOL. 81v

FOL. 82

FOL. 82v

---

[3] Figure 18. In his *Opticks* (Harleian MS. 3360) Hobbes gives (pp. 203-207, i.e. its leaves 174-176) an account of the working of the telescope similar to what follows here.

image of [that portion lying between] the extremities of the object has been multiplied by means of the force on HI, as many times as there are points on the convex lens AB; and for the same reason the points that can be taken between d and e were multiplied an equal number of times, through the force on the points between H and I.[4] Furthermore, the distance between the limits of the object, one from the other, is greater than it would be were it looked at with the naked eye. For owing to the refraction of the lines of irradiation emerging from the concave lens, the more these are made to diverge one from another, the more will the lines of vision parallel to them cut one another at an angle larger than dAe or dCe or dBe. Indeed, at this greater angle they would cut one another even if there were no telescope used.

2. That the above is what happens may be clearly realised from the phenomenon itself; for if one covers over any part, or any number of parts, of the convex lens AB, but leaves uncovered any one part, and this either in the centre or towards the edges of the lens, the whole object will nevertheless always be seen distinctly, but less well illuminated.[5]

From this we surmise, first, that the object is seen by means of rays passing through any one point of the convex lens, and, second, that the whole object is seen by means of rays coming from any one point of the convex lens AB. So if the two dotted straight lines come from A to H and I (for the latter [lie] on the lines of vision by which the limits of the object are seen through A); and again, if the two straight lines coming from B fall on the fundus of the eye at two points other than H and I, say at N and O, then N and O lie on the lines of vision by which the extremities of the object are seen through B. So if N and O are not coincident, both ends of the object will be seen by means of different lines of vision; and so not one object but two objects would then be seen, contrary to what is actually observed i.e. a single and distinct object. Hence all the rays coming from d fall on I, and all those coming from e fall on H. The same must be said of the rest of the middle points of the

---

[4] The meaning of the word 'force', here used twice and also in a similar context on fol. 88v, is unclear. Perhaps, as on fol. 52 and elsewhere (in Chapter IX), Hobbes sees light as being propelled by the air and actually striking the lens.

[5] Figure 18.

object, namely that they register on the individual places of the fundus that correspond to them. It also follows that those parts of the lines of radiation coming from the same point of the object and intercepted between the concave lens and the surface of the eye are parallel to one another. For otherwise [i], the point that they represent[6] could not be seen as one point; and [ii], if the point of the object were seen in the same line of radiation intercepted between the lens and the eye, the point would appear to be several points,[7] because it would be seen by means of several rays. Now, the point would be seen along [in] any one line drawn from the fundus of the eye parallel to one of the intercepted lines, but it would also be seen along the lines drawn parallel to the other intercepted lines. In this way, again, one object [alone] would be seen. It remains to say, therefore, that lines of radiation emitted from the same point of the object have their parts (intercepted between the concave lens and the eye) parallel to one another, and the true line of sight is that drawn from the point of meeting on the fundus of the eye and parallel to all those.[8]

FOL. 84v

3. But the position of the image seen through a telescope, together with the magnitude of its apparent diameter as compared with the magnitude of its apparent diameter [as seen] without a telescope, may be determined thus: Given some object, [e.g.] the moon viewed without a telescope, whose apparent diameter and whose distance from the eye we estimate by our sense alone; and let us suppose that a straight line be drawn near the eye so that it cuts the optic axis at right angles; such a line is PQ in the last figure. Let PQ be equal to the true diameter of the moon [as opposed to that of its image]. Parallel to this [line] let another straight line be postulated that is equal to the apparent diameter of the moon (which is, say, half a foot.) Let the distance between these parallels be as great as the apparent distance between the eye and the image of the moon–indeed, this image cannot be measured, but is gauged by our sense as being two hundred paces long. If now two straight lines are understood to be drawn through the ends of the said two parallel lines and to meet within the image of the moon; and if the

FOL. 85

---

[6] Presumably the point on the retina registered by each ray separately; or each ray's point of origin at the source?
[7] Hobbes returns to this theme at fol. 285.
[8] Cf. fol. 283 below.

FOL. 85v

straight lines HL and IM in the last figure are produced until they meet those [parallels], they will determine the distance and diameter of the moon that appear through a telescope.

For instance, let there be any object AB, which is so far from the eye (the latter at C) that it appears of a magnitude [equal to] DE, which subtends the angle DCE; let FG be equal to AB, and let FDH and GEH be drawn through the ends, D and E, of the image.[9] Now if the same object AB[10] subtends the angle ICK, which is greater than the angle DEC, then (I say) the apparent diameter [of the object] will be IK and will be greater than DE, and the apparent distance will be the straight line from the point C perpendicular to the straight line IK. In proportion as the object recedes from the eye it subtends a smaller angle, until this angle reduces to zero, say at the point H; and as the distance HK or HI is to HE or HD, so the diameter of the image IK is to that of the image DE. If, then, AB were the true moon

FOL. 86

and DE the apparent moon, no telescope being used, and the angle ICK were the only one that the rays of vision subtend through a telescope, IK would be the apparent moon [as seen] through a telescope. Let these therefore, in brief and sketchily, as befits the plan of this work [of mine?] be the remarks to be made on the nature and construction of the optic tube.

4. Of the arguments used by our author to prove that there cannot be made a telescope more free from error than those that have already been constructed, the first is this: René Descartes has shown[11] years ago a new way of making telescopes by which—unless the hands of skilled workmen fail us—he promises that we shall see in the heavens objects so small, and these as distinctly as the ones we are used to on earth, and yet the results do not so far measure up to what was promised.

FOL. 86v

This argument [of White's] is indeed less than philosophical. If, not having been demonstrated, the said method [for making telescopes] is not the true one, does this mean that no method exists for [accomplishing] the task?[12] What if that construction

---

[9] Figure 19.
[10] Text: 'ABH'.
[11] Pluperfect in the original (*DM*, 96–97). The reference is to Descartes' *La dioptrique*, and Hobbes paraphrases White's words up to 'on earth'.
[12] 'Demonstrated': by White? Or proved in practice by others? 'The true one': that actually used? Perhaps a reference to the *Discours de la méthode*, Sixième partie.

of the tube–a method shown above to be genuine[13]–should call for spherical lenses, and not hyperbolical ones? Or cannot spheres larger or smaller be made than have so far been made by the same craftsmen? Our author in fact wishes [to imply] that René Descartes has erred, not over method, but in respect of the technicians who can make lenses of a hyperbolical shape. In White's opinion we should be despondent that Descartes himself made such sweeping promises for the success of his hyperbolical glasses, not because he hoped it would be possible for something greater to be achieved than was achieved, but because (in White's words),[14] 'It is sometimes useful that claims of this kind are put forth with the intention of stimulating the craftsmen's labours: I am unwilling to believe he was merely in doubt about the subject: I say he knew for certain that the craftsmen's skills would not be adequate'. Indeed, he presents him [Descartes] to us as a facetious fellow who would have said those words in order to stimulate the craftsmen's labours when he knew for certain that these would be lacking.[15]

FOL. 87

5. The second argument is as follows: The result of [making] so long a tube would be that the sides converge [at the eyepiece end]. But, [I reply,] even if this were true, it would not prevent telescopes from being made twice as long as they are [now]. Although their degree of accuracy cannot be increased indefinitely in this way, nonetheless it might be increased; yet it is not true that, the longer the tube, the larger [in diameter] the convex lens needed, and hence [it is untrue] that the sides will always diverge.[16] Nor will they appear to converge, contrary to what he suspects: for he bases his idea on what usually happens to anyone looking at long colonnades with parallel sides.

6. The third argument is this: If a man is to be seen [by us who are at a distance] from the moon, he must be seen by means of the rays he emits to us when he is established there. [I answer:] We cannot increase the lunar illumination in the way that we increase the illumination of objects that we see through

FOL. 87v

---

[13] Here *genuinus* can have several different meanings.
[14] It is not clear from the text whether the two clauses of reason go with 'be despondent' or whether with 'made'.
[15] Cf. Drake, p. 292.
[16] *La dioptrique*, Discours VII.

the 'interpreters' glass' (otherwise termed the microscope); so the rays reaching us from the man placed on the moon are much too weak to enable him to be seen by us at such a distance. Therefore if a man were on the moon, no device could make him visible [at places far removed] from her.

Indeed, White could have inferred from this that, without [recourse to] a telescope, a man cannot be seen [by people distant] from the moon. Yet the use of a telescope consists in its gathering rays and leading them, when collected by means of refraction, to the eye; so the rays given off by a man illuminated on the moon will be adequate to see him by if [only] all the rays dispersed over a whole hemisphere be collected into one ray. White, therefore, should not have assumed that there are a few rays transmitted by the man to us, before first determining how many can be brought together and transmitted to us by a telescope–which is the subject of the present disputation. He is foolish to go on to say what he does: '...Unless we see the man in such a way that there is nothing besides [this] that can put him into shadow, we are wrong in believing the man to be visible; for the large objects that we can see obliterate our view of the smaller ones'. From the very use of telescopes we know that the same causes which magnify the image of an object reduce the hemisphere visible. This, however, is to be conceded him: it is useless to expect anyone to see a man on the moon if some other, brighter object has to be looked at as well as [this man, and at the same time], just as [one can see] neither the moon, if she is so close to the sun that the latter must be viewed together with her, nor even the sky illuminated by the sun. So if a man is to be seen [by those] away from the moon, his image must darken the whole orifice of the telescope.

7. The fourth argument, and the most alarming, I have transcribed word for word. It goes thus: 'What frightens me and compels me [to believe] that this art will not progress much is that, to me, things certainly do not appear larger unless, and when, they appear nearer, and that they appear nearer only if they subtend greater angles and cones at the eye:[17] for the angles of vision diminish with the same force [vis] by which the nearness of the object is increased. Therefore it is necessary to reach quickly a certain modicum of excellence'. I honestly wonder

---

[17] Text: '...only if they fill the eye at greater angles and cones'.

why he could imagine that 'the angles of vision diminish with the same force as that by which an object draws nearer'. For the contrary is true. I would suspect a printer's error did not our author's argument tell me it had to be read as I have read it. I do not know what to feel, unless, perhaps, that he thinks the nature of an angle consists in its possessing a sharp point, and, as it were, a kind of spike. Result: when the point is greater the angle is larger. [Such being assumed,] he could imagine that an object close at hand subtends a smaller angle than an object further off does, because [the former has] a smaller point; but in whatever way he gauges the size of an angle, surely to put forward, almost as if in the same breath, two propositions so contradictory to one another shows amazing forgetfulness. For he says: 'Certainly one is convinced that things do not appear closer unless they subtend greater angles', and immediately he follows with the opposite: 'In proportion as things appear closer, so the angles they subtend are less'.

FOL. 89

8. He bases his fifth argument on the construction of the telescope. 'It is equally true', he says, 'that the telescope is the result of [using] lenses of opposite type, i.e. concave and convex, in whose characteristics we note this difference of properties: the larger the sphere of which convex lenses are parts, the more blurred the image they cast of the thing viewed.[18] With concave lenses, on the other hand, the smaller the spheres from which they are hollowed out, the sharper the image they throw. So you can gather from these phenomena the same thing: we should compare the parts of the tube in a proportion which cannot have a great size [*latitudo*].'[19]

FOL. 89v

In the same way that, in his books on logic, he distinguished a true syllogism from an apparent one, so in his books on rhetoric Aristotle distinguished contrariety, or the opposition of propositions, into true contrariety and apparent contrariety. For instance, to the statement: 'I shall sit while you stand' is opposed, not in fact but only apparently, 'You will stand while I sit'. But [the first] is really opposed to another statement, namely: 'I shall stand while you sit'. We have, in the words quoted [by me] above, a further instance of the same false

---

[18] Cf. *LW* II. 84.
[19] This literal translation, which is repeated at fol. 91 and interpretation of which is uncertain, retains the text of the *De mundo*, which Hobbes accepts.

FOL. 90

opposition. Now if it be that, the larger the sphere of which convex lenses are parts, the more blurred the image they cast of the thing seen; then the smaller the spheres of which the same convex lenses are parts, the more distinct the image they throw. The latter property is, as White asserts, that of concave lenses. So the properties which he cites as opposites are the same, and are not opposed, but only apparently so, and he commits that absurd fallacy which, as we have said, Aristotle put on record in his *Rhetoric*.[20] There could, perhaps, be a printer's error here [in White's text]. Let us therefore say: 'With concave lenses, the smaller the spheres of which they are parts, the more blurred the image they cast'. But this is false, as is the other of the 'opposites', namely: 'The larger the sphere of which convex lenses are parts, the more blurred the image they cast of the objects viewed'. For lenses, whether convex or concave, cast images (of viewed things)[21] sometimes more distinct, sometimes more blurred, according as the lenses' position between the object and that particular eye by which they are viewed is suitable or unsuitable. If anyone with normal

FOL. 90v

eyesight without spectacles puts glasses on, over the same [given] distance he will see less clearly with them than without; but if he looks [at an object] from closer up [and still wearing the spectacles?] he will again see clearly. The larger the sphere of which the convex spectacle-lens is a part, the more blurred will his sight be over the same distance [as just mentioned], but he will see very clearly an object that is moved close up [to him]. I believe that our author could see normally without eyeglasses, but had nevertheless used them for the purpose of the experiment. He had noticed a difference of distortion [of the image] in the difference of the sizes of the spheres of which [the spectacle-lenses] were parts;[22] but he did not notice that he could have seen the same objects distinctly, even through the most convex spectacles, if he had viewed them from closer to. So he thought that a blurring of this kind was to be attributed,

---

[20] 'As we have said'; at fol. 39. The reference does not occur in Hobbes's *Brief of [Aristotle's] the art of rhetorique* (1637?), neither is it clear to which passage Hobbes refers: perhaps *Post. An.* I.32 or *Top.* II.7.

[21] The brackets are inserted to remove an ambiguity otherwise resolvable only by remodelling of the clause, as has had to be done with the sentence preceding.

[22] Interpretation is dubious, and matters are not helped by the possibility of textual corruption: '...in differentia magnitudinum sphaerarum cuius erant portiones...' (read *quarum* for *cuius*?).

not to the position of the lenses, but to their convexity. Yet the blurring that is seen through the concave lenses which are parts of a small sphere is not really a blurring, but a faintness of the image. For images are said to be blurred when [our] perception of the individual parts is removed through their common action (i.e. through the action of the whole [image]); but when our perception of the parts is lost because of the smallness of the whole [image], this is due, not to its blurring, but to the weakness of the action [of the illumination]; it is not a blurring of the images but a dimming.[23]

Now if it were true that the lenses of a telescope differ in the way that he says, he still does not show how we may infer from this the impossibility of advancing the art of dioptrics. For those closing words [of his]: 'You can gather from these phenomena the same thing: we should compare the parts of the tube in a proportion which cannot have a great size', are unsuitable and hard to grasp. Perhaps he meant to say this: The larger a convex lens becomes, the greater becomes the blurring of the image; so in order to produce a sharp image we must choose a smaller sphere [from which to make] a concave lens. Hence we shall quickly arrive at [the question of making] a concave lens from a sphere that is hardly perceptible; so a longer telescope cannot in fact be made. That is to say, the greater the sphere of which a convex lens is part, the smaller should be the sphere of which a concave lens is a part. This is not true: we see [used] in the largest telescopes, the lenses of which are convex, parts of spheres greater than [those from which are made] the convex lenses of shorter telescopes; the concave lenses of the shorter telescopes, on the other hand, come not from a smaller but mostly from a larger sphere than do the concave lenses of telescopes of middle size.

9. The sixth and last argument is based on the following claim: 'Catoptrics are the basis of dioptrics. Galileo had read Archimedes' catoptrics, but when he had done so he could not have constructed a telescope better than his own'. I pass over what Galileo, the greatest scientist not only of our own century but of all time, could do or wanted to do. I also pass over the catoptrics of Archimedes, which I shall admit are,

---

[23] The rendering of this sentence follows the original as closely as possible: Hobbes could have expressed himself far more clearly.

through the very genius of the man, faultless. Only this I ask: for what reasons can it be shown that catoptrics are the foundations of dioptrics? 'I have found', says White, 'that there is the closest connection between a convex lens and a concave mirror, and between a concave lens and a convex mirror; and that when certain rules are observed, almost always the same things happen.' But in what this connection consists (perhaps because the matter seemed self-evident) he has not said. So let us compare telescope-lenses with mirrors.

A lens which in a telescope is convex magnifies the image of the object (so long as the former is not inverted) by means of refraction, but makes this image, when inverted, smaller; a concave mirror does the same by means of reflection. So far, they are alike. But they differ in other respects; for in order that the image be magnified, [as] by a convex lens, the mirror must be moved away from the eye, and sometimes the object must be moved away from the mirror as well; but for the image to be magnified by means of reflection in a concave mirror the object must be moved up towards the mirror until it is between its centre and surface (*sic*). Moreover, when the object is nearer to the surface of the mirror [by more] than a fourth part of the diameter [of curvature?], its image again becomes smaller as the object is moved towards the mirror.[24]

FOL. 92v

With a convex lens nothing like this happens. The greater the sphere of which it is a part, the less does it diminish the image of a thing seen by means of refraction.[25] But even if there are any resemblances whatever [between them], this is not enough to prove that catoptrics are the basis of dioptrics. It was necessary at least to have brought forward some conclusion [met with] in dioptrics and to have shown that in no way could this be reached except by adopting some proposition from catoptrics or from some property of reflection. He could not do so; for the nature of refraction can be demonstrated even by those who have learnt nothing whatever about the laws of reflection. And if the similarity between images produced by refraction and those produced by reflection indeed suggested that the one science is the basis of the other, he could equally well conclude from this, his own argument, that dioptrics are the basis of catoptrics.

FOL. 93

[24] These obscure sentences are given as literally as possible.
[25] Mersenne's correction of a passage where the copyist runs parts of two sentences together.

10. When, in the next problem, he is about to pass to a consideration of the nature of the earth, our author here prefaces a kind of general notion on the causes of gravity and levity, namely that heavy and light bodies are moved, not of themselves but by some producing-agent [*generans*]. This, according to Aristotle, the originator of the statement, is not (declares White) any single, defined agent, but that series of causes which universally unite in order to produce coming-to-be. Let us suppose, therefore, a stone falling by its own natural gravity straight towards the centre of the earth. The reason for this fall, [he concludes,] is therefore that which brought that stone into being, i.e. not any single agent, but all the several agents that originally combined to create that stone. Even if this were true, we learn nothing more than that the cause of [something's being] 'heavy' is the cause of gravity. But we are not asking this; we are asking, rather, the reason why a stone is heavy, i.e. why it is moved towards the centre of the earth more than towards any other point. Next, though that universal creating-agent is the cause of gravity in one thing that is brought into being, and of levity in another, we wish to learn in what these two comings-to-be differ; for the investigation of the nature of heavy and light things is just that. Again, what about the heavy thing that was the first of all [heavy] things to be created? To what centre did it press? If to its own, it was not being moved, and was therefore not heavy; if to another [centre], then it was not [i.e. it could not have been] the first heavy thing. Furthermore, a stone cast violently upwards does not in any way descend when so borne; hence by what propellent force does it fall back? That the stone is created anew is unthinkable.

FOL. 93v

FOL. 94

11. It is sufficiently agreed that an investigation into the nature of heavy and light things is very difficult, because nothing can yet be gleaned from the innumerable writings of the philosophers that has not involved a thousand most ridiculous consequences. The things I think must be re-stated about this quest are few, and are not relevant to the pith of the enquiry; they are merely certain preliminary questions, [a scrutiny of] which may lead us a little nearer to a knowledge of the subject.

Gravity has been fairly generally accepted as being nothing else but the tending [*conatus*] of certain bodies towards the centre of the earth. This striving is a motion either of the entire body (as when the whole falls), or of the parts of that

FOL. 94v

in which lies the pressure by which the parts move forward, even if the whole body is not yet advanced. So those who ask 'What is the cause of gravity?' are asking: 'By what mover does a stone, or another body, fall through the air or press down on the hand or on something else that supports it?' I think, first, that it is now sufficiently clear to most philosophers that the stone in question does not move itself; so it has an external mover. Now every mover moves [a] by pushing, or [b] by pulling, or [c] by impeding the previously existing motion of the thing to be moved, so turning the thing to one side. Examples of driving and of attracting can be cited by anyone. Let the following be an example of the third type.

FOL. 95

Given any spherical body, such as A, and let us suppose the whole of it to be at rest, but that its parts are being moved according to the shape of the whole, say circularly or otherwise.[26] Now let a straight line BC press tightly on body A at the short part D,[27] upon which straight line another body [x] is moved. It is clear that the motion of the body [x] carried over BC will vitally affect the motion of the parts of body A. The result will be that body A is either shattered at D or that, if A is too hard for this to be done to it, it will withdraw itself so that it preserves its parts' own motion. This withdrawal is now the motion we are talking about.

FOL. 95v

Everyone can very clearly see such motion in spinning-tops. These are rotated by a fast, whirling movement, yet at every lateral contact[28] they recoil with a straight [*directus*] motion. We must also bear in mind that, from the time that their parts come-to-be, all heavy bodies contain a certain motion in which their essence consists, as shown above in Chapter V, Article 3.[29] Such motion is one not *of* the whole [body], but *within* the whole–in the parts that do not quit the whole.[30] Therefore the motion of their separate parts is a kind of turning within itself [*in se*]; for otherwise a straight motion would make them quit the whole, i.e. the whole [body] would be broken up, as occurs with putrefaction. These points being despatched first, it is seen that the motion of heavy bodies is either a pushing or

---

[26] Figure 20.
[27] Presumably a straight line traversing a plane surface; and the body presses on this rather than this on the body. Cf. fols. 175, 222v.
[28] With one another? With an immobile body?
[29] This is not exactly what Hobbes said previously; see also fols. 20vff.
[30] Cf. fols. 116, 399.

a pulling or some third motion of the kind we have described. Let us therefore suppose, first of all, that heavy bodies are *driven* downwards. The degree of velocity which the driver communicates initially to a heavy body ought to be that at which that heavy body should always descend; or less because of the resistance of the air, unless a new impulse approach the now falling body. Since, therefore, the motion of heavy objects is being continually accelerated as they fall, then if that motion be a driving, the driver must continually rest on them, and must be close at hand with the same increases of velocity in every part of the air traversed by a heavy body. This is inconceivable. So there is no single driver that perpetually follows and propels heavy bodies.

FOL. 96

Second, let us suppose that heavy bodies are *drawn* downwards. What draws should cling fast to the heavy body that is drawn: this is a difficult claim to substantiate about the air that adheres to a stone. Moreover, what draws ought also itself to be drawn by another thing that was drawn previously, and so on continually down to the centre of the earth. There, however, if anything draws, it will do so by rising,[31] which is against the nature of heavy bodies. So it seems that the descent of heavy bodies is not effected through traction. There remains [the possibility] that it is done in the third way, i.e. that [round the world] the air has a kind of motion that is everywhere circular. If any heavy body climbs strongly into this air, the motion of the body's innermost parts (a body's essence consists in this motion) will be thrown into confusion because of the discordancy among, and the diversity of, the paths [they follow]. This is why, unless a force intervene, a heavy object always recoils as it rights itself [after an impact] until it reaches the centre of the earth.[32] So it seems to me, at any rate, to be impossible to describe these motions of the parts in different kinds of bodies, i.e. to expose fully the nature of gravity.

FOL. 96v

---

[31] Because it can descend no further?
[32] This sentence, to do with impact, seems out of context here. A possible, though very speculative, interpretation is: 'A heavy object will always reach a maximum height from the earth's surface; then it orientates in the vortex and is carried back to the earth.' This of course presupposes the reference to be to the Cartesian vortices rather than a development of the preceding Aristotelian theme, but Hobbes's changing the topic is due to his following White.

FOL. 97

CHAPTER ELEVEN

On Problems 12 and 13 [of the First Dialogue]:

That the Earth is not a magnet.

And that its effluvium is made up of

large, individual solids

1. Why Gilbert considered the earth a magnet. Our author's arguments [to show] that the earth is not a magnet: (2) that a vein of iron is found near the surface of the earth; (3) that stones are lighter than metals; (4) the depth of the sea; and (5) that the earth possesses no other body to the poles of which it is attracted in the way that a magnet is attracted to the earth's poles. 6. A conjecture of our author concerning fire in the centre [of the earth]. 7. The points he makes about the earth's effluvium are self-contradictory. 8. What fluid and humid are. Our author's arguments FOL. 97v to prove (9) that our air is everywhere impure; (10) that the several globes of the world have their own effluvia; (11) that all air *simpliciter* is impure; and (12) that there is absolutely no impure air or impure ether in those places where the effluvia of the spheres of the world cannot reach.

1. As regards the material forming the heavens, our author has declared earlier that it is the same as the sublunar. Coming now to sublunar things, he distinguishes their material into [a] terrestrial, and [b] air. Of these materials, he thinks that the latter is the effluvium of the former, as being [the material] which an internal fire or the sun turns into air by heating it. FOL. 98 So the opinion of Gilbert and of Galileo,[1] who declare that the

[1] Drake, p. 403.

earth is a magnet, he first of all disposes of, while he also presses home his own view (this is Problem 12). Then, by asserting that the nature of the heavenly bodies is the same as that of the sublunar, he infers in Problem 13 that every solid object, i.e. a fixed or a moving star, has its own effluvium, or its air.

From this I might seize the opportunity of saying something about the causes of magnetic properties, but I shall not do so, because the views I think should be put forward are only conjectural and are therefore unsuitable for disposing of conjectures.[2] Meanwhile, I am required to show here that Gilbert who thought that, because of the similarity of the phenomena,[3] the earth is a magnet–although he used arguments that are far from incontrovertible–has so far not been adequately refuted by our author. In the course of examining the latter's belief about an effluvium, I shall [later, in Article 8,] state my own opinions about fluidity and dampness. [But first,] Gilbert had noticed [a] that a magnet possesses its own heavy bodies, i.e. iron and all the magnetic things which seem to be drawn to it without external aid, as heavy bodies towards earth; [b] that it has its own poles, as the earth has; and [c] that a magnet's poles have the same direction or disposition with respect to the universe as the earth's poles have. Gilbert conjectured from this that a magnet is a homogeneous sample of very pure earth, and called it *microgê*, *terrella*, or 'little earth'; and the [planet] earth he called a large magnet.

FOL. 98v

These are not trivial arguments. For on seeing wholly similar qualities in two bodies, of which one is the larger and the other the smaller, who will not think that the two bodies themselves are also similar as to their nature and that they differ only in size? For how else can the likeness and unlikeness of the natures [of bodies] be distinguished, except from the resemblance and the dissimilarity of their qualities and the ways in which bodies themselves may be acted on? If someone ponders, on his own, the motion by which a magnet always inclines towards a meridian of the earth, and [the fact] that by that daily motion, either of the earth round the sun or of the whole world round the earth, the whole earth must have acquired, after a great number of revolutions, a tendency to produce that motion in itself, what other conclusion can he come to except that those

FOL. 99

---

[2] Cf. Drake, p. 356.
[3] I.e. that the earth behaves in a similar way to a magnet: *De magnete*, I.17.

parts of the earth which have acquired the same tendency with the whole are the purest of the whole [earth]?[4]

FOL. 99v

2. For the earth's not being a magnet our author makes out as his first plea that, to a greater extent than with any other metal, a vein of iron (of the kind that a lodestone itself seems to consist of) is found to be embedded in the earth's crust, and that people who tear open the earth to quite a depth find metallic substances and congealed liquids–as though the minerals gold, lead, and copper-ore, etc., are not also found in the earth's crust; or as though anyone had penetrated so far into the bowels of the earth as to realise that no vein of iron is found at an even greater depth [than that]; or as though the finding of a magnet in the earth's crust were proof that it cannot also be found in the bowels of the earth even as far as the centre![5]

FOL. 100

3. He draws another argument [against the earth's being a magnet] from the fact that stones are not the heaviest of all terrestrial things, but that metals weigh more than stones do. This suggests only that a magnet is *not* created by the compression of heavy bodies when they tend to the centre of the earth. I do not know whether anyone has asserted this; Gilbert certainly made no such claim. He thought it enough that a magnet is pure and unadulterated earth, in no way changed and corrupted by its surroundings air, water and amber; and that it has the same origin as the nature of earth. We all believe, however, not that earth was created by a compression due to gravity, but that it possessed its own nature from [the time of] its creation. Earth, then, is not earth because it is compressed; but, because it is earth, it compresses.[6]

FOL. 100v

4. The third argument is derived from the depth of the sea. The latter (no-one knows how deep it is) perhaps reaches the innermost parts of the earth. The consequence will be that the earth has hollows and therefore cannot be a magnet. But say, as our author wishes, that the bowels of the earth are hollow and that these concavities are filled with air, or water, or fire.

---

[4] It was thought best to retain this involved sentence as Hobbes left it.
[5] Drake, pp. 245, 402.
[6] By *terra* in this sentence the element earth seems to be meant.

CHAPTER XI: 6

Nothing prevents the hollownesses of those caverns from being able to be magnetic. For if there are caverns within it; if, therefore, the whole [of the earth] is not a magnet (except if [it is] an actual lodestone), [then], if porous, and its parts heterogeneous and not magnetic, [the earth,] though it produces magnetic effects, will still not be called, *in toto*, a magnet.[7]

5. Lastly, he says that the earth has no magnetic properties. Why? Because a magnet is attracted to the earth's poles, but there is available to the earth nothing with poles to which it can itself be attracted. Yet, [he continues,] if a magnet inclines towards the earth's poles, there must also be available to earth something with poles to which the earth itself can be attracted, for the earth has to turn to the poles of a magnet. Now, Gilbert lays it down that the earth is a magnet. So two magnets exist: the [planet] earth, and the 'little earth', or a lodestone. They differ only in size–unless, perhaps, in purity also–so the [planet] earth is a less pure magnet.

FOL. 101

Let us suppose, therefore, two magnets placed close to one another. If they are equal[8] each will be equally attracted to the poles of the other. If one is the stronger, the [force of the] weaker will be overcome and it will be drawn slightly towards the poles of the stronger; also the stronger will incline quite an amount towards the poles of the weaker. But each seeks the poles of the other, though unequally; yet they are both magnets. So the earth, which is attracted a little, though imperceptibly, to the poles of a magnet, can [itself] be a magnet.

6. Here are the arguments by which he thinks it more likely that fire is enclosed in the bowels of the earth than that stones are. First, if this were not so, the earth would not appear to be of any use.[9] 'As we notice that in animals and plants the active warmth is directed from the heart and root into the whole body, [we notice that] some exhalations are emitted from the centre of the earth by Demogorgon, or the chemical Archaeus,[10] into

FOL. 101v

---

[7] Hobbes loses himself (and the reader) in this long and involved sentence, which it was thought best not to attempt to break into more manageable units.
[8] In strength? In 'purity'?
[9] Text: '...does not appear...' White argues that its internal fire renders the earth useful.
[10] The term Archaeus was used, and perhaps invented, by Paracelsus (*De natura rerum*, Bk. 1).

all the limbs of the earth through certain smoky conduits.[11] As I cannot assert this, likewise I have no reason to deny it.' Surely, in order that vapours can be emitted from a central fire, there must be in the bowels of the earth not only fire but also water or some other damp matter? But say we concede the point. Why shall we consider that the earth is useless unless his condition is fulfilled? Because grasses and plants will not be created in order to feed animals? Yet if animals and men can be born even if there were not fire at the centre of the earth, why may not grasses and plants also be created? But he will say that animals cannot be born under such circumstances; so the earth, [he infers,] would be useless unless man were born.[12] To whom, then, [one asks,] was it useful for man to exist, so that it was for his pleasure that in the centre of the earth fire was located? Indeed, I do not know what White is saying here, unless he means: 'It was useful to men themselves that they existed'; for it is sinful to believe that Creation was useful to the Creator. But if our author is saying: 'Men have been made because it was to their own profit that they were made', why may he not also affirm that the earth could have been made because this was useful to the earth?

FOL. 102

Thus it was unnecessary to postulate a fire at its centre. I do not say, however, that, were the fire removed from the centre, sterility of the earth would follow, for [the earth's] fertility is to be attributed not to an internal fire, but to the sun's. [I say that] fertility can be produced not by light but by heat.[13] The earth, which has been turned towards the solar fire for so long, may be considered as having derived thence enough warmth to produce at least plants and animals–if the generation of metals is not also to be ascribed to the sun. But, he says, mountains have been torn from their places by the force of fires bursting forth from the earth as, to common knowledge, happened at Puteoli and with Mount Vesuvius. Is this, [he asks,] not an indication that there is fire in the bowels of the earth?– Not at all, unless the burning of some buildings were also evidence of the same thing; for what was that fire of Vesuvius, and fires like it, but sulphurous mineral ignited just beneath the earth's surface? [The mineral] did not have to be kindled by

FOL. 102v

[11] Perhaps 'smoky' should precede 'exhalations'.
[12] Cf. Drake, pp. 61, 367, 402.
[13] The 'I say' has been added; in Hobbes's Latin this sentence continues from 'I do not say' in the previous one.

fire deep within [the earth], because, as we see, fire can be ignited in so many [different] ways.

7. He claims on his page 104 that 'air is a vapour of earth', i.e. some fluid squeezed forth either by a force lying within the bowels of the earth itself,[14] or by the external action of the sun. But as he defines neither 'fluid' nor 'humid'; and as he speaks inconsistently about their formation, he has made this context so confusing that I have no means of telling what his opinion is. First he says, following Aristotle, that 'a fluid is created by rarefaction', so that earth is changed into water and water into air.[15] Then, two or three lines later, he says that a fluid is 'that which has been squeezed out, having been mixed with multitudes of dry things floating in it'. These two statements contradict one another in a most glaring fashion. If indeed earth, which is to be changed into fluid, were mere earth and not yet fluid, how was a not-yet-formed fluid squeezed out of a non-fluid? For compression brings about, not the creation of the thing squeezed forth, but its separation from something that adhered [to it]. If by 'a change into fluid' he understands only 'the separation of fluid from non-fluid', in this case rarefaction takes place; for what was dense remains dense, and what was rare is not made rarer by separation. But granted that an effluvium is formed in some way or other, and that fluid bodies are emitted from the earth; he therefore calls these fluids air, i.e. as he explains the term, 'this undulating element provided for our use'. This last is the proposition he ought to have proved; but he leaves out the proof and passes to another proposition: i.e. air is not pure, but is mixed with dry particles floating about in it. But where do these particles come from, [he asks,] if not from the earth? So a vapour is the

FOL. 103

FOL. 103v

---

[14] White's words (occurring on his page 105, not 104) are '*vi viscitus latente expressum ipsum fluidum*', '*viscitus*' having (in the printed text only) the diacritic to denote adverbiality. Finding no other instance of the word, one translates according to the gloss here supplied by Mersenne: '*id est in visceribus*'. Compare Hobbes's description of the lead mines near Hassop in the Peak District (*De mirabilibus Pecci* (pub. 1636?), lines 96ff.; *LW* V.328).

[15] Hobbes's paraphrases of White would appear to end where shown, rather than at 'into air', because Aristotle does not say that earth's changing into water is a direct *consequence* of rarefaction, whereas White does. The original lacks quotation-marks; but the proposal draws support from the sentence next following.

emission not only of fluids but also of solids. This goes against what he declared previously.

FOL. 104

8. Everybody is in the habit of calling fluid that of which the parts are easily separated from one another. Hence not only water, oil and the like, but also molten metals–even ashes, dust, sand and other dry substances–are said to flow. No-one ought to object here that any one particle, considered individually, of ash, dust or sand, is not fluid, for by such reasoning water, whose particles may be so small that they no longer flow, will not be fluid. The definition of fluid that he adduces here (in Aristotle and others it is the definition of *humid*), namely: 'Fluid is that which has no limits of its own, but is bounded by that which is dry', is nothing but a sound of words and is devoid of meaning.[16] So a fluid has no bounds of its own? Then fluid is infinite, for that is to have no confines. Yet it is absurd that he should interpret[17] these [quoted] words as meaning that the fluid is bounded by the dry, i.e. that dryness

FOL. 104v

starts where fluid ceases; for the result would then be that fluid commences where dryness ends, and hence that dryness would not have its own limits but would be bounded by fluid. From this it would follow, according to the above definition of fluid, that dryness is the same as fluidity; and, as this definition [of fluid] is the same as that of *wet*, dryness will also be dampness.

Dryness and fluidity are not opposed, but dryness and wetness are. Many dry things flow, e.g. molten metals, ashes, etc. Other wet things do not actually flow, e.g. ice; we include among wet things only those that moisten or are in near-potential to doing so. Now, in order for something actually to moisten, it must be fluid and continuous, and must adhere to the bodies it touches–if not to all, then at least to most of them.

FOL. 105

Indeed, because it is not determined of what kind these bodies are, the term 'humid' cannot be defined, for not every name that is assigned and adopted signifies on all occasions the same definite thing, and in such a way that [this] can both be defined

---

[16] Hobbes is less than fair to Aristotle here. Aristotle's definition (*De gen. et corrupt.* II.2; 329.b.31), after saying that moisture or liquid (*hygron*) cannot be confined within limits of its own, then says: 'Dryness is easily confined within *its own* bounds' (translator's italics), and not that 'Dryness sets bounds on humidity'. But Hobbes's parenthetical gibe against White strikes home.

[17] Presumably Hobbes means: 'It is absurd to interpret...'

CHAPTER XI: 11

and is useful in philosophy; in favourable circumstances, however, a name suffices for unfolding the mind, inasfar as [this] is useful for everyday purposes.[18]

9. To prove that, of necessity, air is impure, our author pleads as follows: A fluid is, by definition, that which does not possess limits of its own, but is bounded by the dry. (We have just shown that this definition is worthless.) What is bounded by dryness clings to dryness (untrue; for dust in a dry vessel is bounded by, but does not adhere to, it). Air is not damp (now he changes the *fluid* boundary to a wet one), so air adheres to dry objects floating in it. Next, for 'the smallest bodies to adhere to [other] bodies' is the same as 'to be mixed'. Therefore, [he says,] air is mixed, i.e. impure.   FOL. 105v

Before this conclusion could [fairly] be drawn it was necessary to prove that air is damp; for only a thing that moistens is seen to be wet. The fact that air dries rather than moistens is an argument why, normally, moistened things are exposed to wind and air in order to be dried. So for us to conclude that air is wet, it ought to have been proved that air contained other dry vapours to which it could cling and with which it could be mixed.

10. Under Problem 13 he tries to prove that large solids, e.g. the stars, taken individually, possess liquid of their own. All he has to offer is contained in the following words: 'What has become useful on earth we must also expect [to find] in the heavenly spheres if, as we have said, they are equal to earth'.[19] But [I say,] they are not equal, for it is not agreed [that they *are* equal]. Even he does not claim that this can be true except in the planets. For because the sun and the stars shine of themselves and are consumed in flame, they cannot resemble terrestrial nature.   FOL. 106

11. The sequel, in the same strain, aims at proving that nowhere is there pure air. 'Therefore', he says, 'nowhere will there be found air serene and unpolluted, save where vapours

---

[18] 'Can be defined' and 'is useful' both lack a grammatical subject. The reference is not to 'thing'; perhaps, '[a name]'. 'However': one expects cumulation ('moreover') rather than concession. The whole sentence is unclear.

[19] 'Equal' seems here to be used in the sense of 'similar'.

from some place do not reach it. Even in such a place it will not escape the sun's rays, will it? But if those rays are flame, then they are also dry and hot, and thus they are attracted to air. So [the air] will be mixed with them, will be changed by them, and will be deprived of its purity.' Those vapours which he imagines to flow from the earth he has declared to be, all of them, fluids, i.e., as he believes, damp. Likewise, [he says,] if the rest of the heavenly spheres resemble earth, they are all damp also. (Thus the place from where he fetches those dry bodies by which [the spheres] are polluted is not one that these bodies can come from.) So, [he infers,] air is not rendered impure in any way, except by the sun's rays. These, he declares, 'are dry'—indeed, they are those that 'are flame and are attracted to air', so the air will adhere to them, and out of air and the sun's rays will be formed a mixed body.

FOL. 106v

What light is, and how radiation is effected, has been stated above in Chapter [IX].[20] From what is said there it can be realised that radiation [i.e.] what most people call rays [of light], is nothing but a motion continually propagated from the sun's body through the air, i.e. that the sun moves the air that is close to it, and that this latter air moves other adjacent[21] air, and so on, until the [ensuing] motion, directed to the eye, creates [a] a mental image that we call illumination, and [b] radiation, or the sun's rays. I am surprised that he who had read and approved of René Descartes' *Dioptrics* could have clung to the idea, widely held by the common people, that the sun's rays are flame.

12. He will have it that beyond this impure air exhaling from earth and from the other globes of the world there is neither air nor ether of any kind. 'For', he asks, 'what use is air in spaces so far apart? What living creatures does it benefit?'—as if God had to create nothing from which some benefit did not reach the animals, or the wisdom of God were to be measured not only against human wisdom but also against human self-esteem!

FOL. 107

---

[20] Hiatus in the original. The rather abrupt transition—and repetition—is explained by the last section of the article.
[21] Text: *sibi continuum*. Perhaps *contiguum*, as in 'close to it'?

CHAPTER TWELVE

❦

On Problems 14 and 15 [of the First Dialogue]: That the same individual body cannot naturally return [to its original state]; and that another universe would result if any single thing were changed [in this][1]

1. The state of the question concerning the principle of individuity.[2] 2. When the principle of individuity applies to matter and when to form. 3. When, by means of names that signify matter only, one asks if anything is [a] the same thing *simpliciter*, or the same being, *ens*, or [b] the same body, and the like, then the principle of individuity applies to the material *simpliciter*; but when the question is asked by means of names that signify matter determined through form, the principle of individuity applies to the amount of material sufficient to [make] the form. 4. When the question is asked by means of names by which the form alone is determined, the principle of individuity is from the form. 5. The possibility of the universe's naturally returning [to a former state]. 6. That the principle of individuity is not a result of the unity of non-essential qualities (*accidents*). 7. Our author's argument, seeking to prove that something individual cannot naturally recur [to its former state], refuted. 8. The universe shown not to

---

[1] Parts of this chapter are closely followed in Ch. 11 of *De corpore*.
[2] Hobbes's rendering of *individuitas* as 'individuity' (as in *EW* I. 135–136) is here (as far as fol. 113v) preferred to 'individuality' (but see fol. 314v). The question is: What makes something what it is and not something else? i.e. that of identity: cf. '*Ichheit*' (fol. 353v) and see also *O.E.D.*, 'individuation'.

FOL. 108

become something different as a result of change. 9. That the consequence of a change or motion of any particle of the universe is not that the whole universe is changed, but only that its separate parts change place[s]. 10. Whether there can be created something the same in number as something now in existence.

1. The contents of these two problems can be referred to that part of philosophy where the subject for disputation is the principle of individuity. In such a disputation we are concerned only with what we must properly call 'the same' [as something else, or the same body at different times], and what we must call 'something else', i.e. how 'the same thing' and 'another thing' are defined. Now, by 'the same' we do not understand—as the common people do—'the same in appearance', which is merely 'similar', but we understand 'the same in number', i.e. that two things when equated[3] are not two [separate] things but are one thing only. For if we say that what we have seen today is the like of, or the same as, though only in appearance, what we saw yesterday, then the things we have seen are understood to be two. If, however, it is the same in number or is the same individual thing, then there are not understood to be two things. Further, since only a miracle can put the same, individual thing in more than one place at the same time, the question of individuity cannot possibly be raised in respect of things close at hand; for if anyone sees, as in Plautus's comedy [*Amphytryon*], that Sosia is onstage and that another Sosia is also onstage at the same time, such a person cannot doubt that the Sosias are two in number, not one only.

FOL. 108v

In the question argued among philosophers and concerning individuity, therefore, two times are equated. It is asked whether what now exists as a single thing in number is the same as any one of the things in existence previously: for instance, 'Was a ship refitted every year at Athens the same in number as the ship that once carried Theseus to the Island of Crete?' Moreover, this question can be asked only about bodies or about material things, because bodies that have been changed do not, on that account, perish at all, and for the same reason they may at a future time be the same thing[4] as they were in the past.

---

[3] Text: 'both things'. For 'equated', perhaps 'compared'.
[4] For 'may', perhaps 'can'. 'The same thing': so the text.

But since non-essential qualities (*accidents*) are created and destroyed by the action of body upon body through continual change,[5] it is impossible for them to be, over a period of time however short, the same as they were. Therefore there is nothing to be said on the individuity of non-essential qualities.

2. Some ascribe individuity to the unity of the material [of a thing], some to the unity of [its] form,[6] and some say that individuity is the sum effect of all the non-essential qualities (*accidents*) that act together [on a thing]. Our author is of the third group. A case for [claiming that individuity stems from] the material is that a piece of wax is spherical or cubical, but is always the same cube[7] in number; a case [that something's individuity is due to its] form is that from infancy to old age a man is the same person in number, although the material [of his body] is not the same [in his old age as in his youth]. As for the acting-together of non-essential qualities I do not know whether any example can be given except insofar as such an aggregate [of accidents] is necessarily understood to include form. But in questions like these I see that we must examine by what name we speak of that body about which we ask: 'Is it the same [as another body]?' That is, there is a great difference between the question 'Is Socrates, in number, the same man who [once] existed?' and the question 'Was Socrates the same man?' i.e. the same body or the same being, *ens*, that existed.[8] Now, the name of any body is either a very general one—such as body, *ens*, matter, substance—or one restricted and limited by a certain quality by which [the body] is distinguished from some other body, e.g. [the names] wax, ship, river, man. Those general names indicate a definite and determinate material; but of the restricted ones some signify a material identified by its quality or qualities (e.g. wax); others are for indicating the form when no explanation of the material has been considered—except to the extent to which an explanation of the form is required (a ship or a bench is such a name). Now if we expressly enquire about some thing whose name signifies determinate

FOL. 109

FOL. 109v

FOL. 110

---

[5] 'Through continual change' seems to relate to 'are created and destroyed' rather than to 'the action of body upon body'.
[6] 'Unity': either 'consistency' or 'sameness'. 'Form': in the Scholastic usage; not the subject's shape.
[7] 'Cube': Mersenne's emendation of Hobbes's word 'form'.
[8] The second question may end at 'that existed'.

material, no explanation of its form having been considered, the identity of the thing is taken from the identity of [its] material. If the name signifies a determinate form, and no explanation of the material has been taken into account except[9] insofar as some explanation is necessarily sought as to the form, then the identity of the thing is gauged by the identity of the form. This will be made clearer in what follows.

3. Say, therefore, with respect to any body such as a ship, we ask: 'Is it the same being, *ens*, or the same body as it was before?'—for nothing but the material is determined by the word 'being' or 'body'. Then if the material is the same as formerly (no part of it having been removed and no new material added), that material will be, in number, the same being and the same body, as to number, that existed before; but if some part of the first material has been removed or another part has been added, that ship will be another being, *ens*, or another body. For there cannot be a body 'the same in number' whose parts are not all the same, because all a body's parts, taken collectively, are the same as the whole. Now if we ask, concerning a ship: 'Is it, as regards number, the same ship as existed before?' [then,] because its form is determined by the word 'ship', namely that it may be suitable for sailing, that which remains [after the removal of some of the ship's parts?] will always be a ship, and that ship the same as before, because the ship previously existing was never destroyed. But if some part of the ship is removed so as to make the ship no longer fit to sail, what is left will not be a ship at all, and therefore not the same vessel [as existed before]. Hence if another plank is substituted for one that has been removed the result will be that a different form of ship (the first having ceased to exist) is imposed on different material. From this material, which is now not the same [as that of the earlier vessel] a ship will be created that is not the same in number [as the earlier ship], but is the same in appearance, i.e. similar. Besides, the fact that a ship of which one plank alone is changed—this plank being essential for the form of the ship—is not the same in number as it was before, may be shown thus: If it is the same ship after a single plank has been changed, then by the same criterion it

---

[9] Between 'account' and 'except' the copyist has repeated 'The identity of the thing...of its material' in the Latin. In 'the identity of [its] material' the last word is Mersenne's correction of Hobbes's word 'form'.

CHAPTER XII: 5

will be the same ship when a further plank has been changed, and when yet another; and so on until all the planks are changed. So if anyone believes that when planks have been removed they join up again in the same pattern as before and will make up a ship just as previously, then there will be two ships the same in number; so two and one are the same number, which is impossible.

4. Now if we ask concerning a river, 'Is it the same being, *ens*, or the same body, as previously?' the reply will be 'No'. For the term 'being' or 'body' signifies material alone, and waters move along and pour themselves into the sea, [being replaced by waters, different as to number, that are discharged from a spring, and these latter move forward in the same way. Further, a river is, in number, not a body];[10] it only resembles body and is of the same appearance. So if we ask, 'Is it the same river?', because the term 'river' determines material no more than as being fluid,[11] the reply will be: 'It *is* the same river,' for a single river is classified by the continuity of its flow, which is one unbroken motion. Therefore, since the motion and the flow are one and the same, the river will also be one and the same.

Likewise if one asks: 'Is a man, when old and young, the same being, *ens*, or matter, in number?' it is clear that, because of the continual casting of [existing] body-tissue and the acquisition of new, it is not the same material [that endures],[12] and hence not the same body; yet because of the unbroken nature of the flux by which matter decays and is replaced, he is always the same man. The same must be said of the commonwealth. When any citizen dies, the material of the state is not the same, i.e. the state is not the same *ens*. Yet the uninterrupted degree [*ordo*][13] and motion of government that signalise a state ensure, while they remain as one, that the state is the same in number.

5. Having grasped this, we must next see in what way it is true that the same individual body cannot be re-created.

FOL. 111v

---

[10] A hiatus involving the loss of a complete line or lines is here filled by Mersenne, probably conjecturally.
[11] The question may continue as far as 'than as being fluid'.
[12] Perhaps, for 'it is not...endures]', 'he is not the same material', i.e. of the same material, or that his material does not change.
[13] Cf. fols. 238, 341v, 348.

FOL. 112

First, if one asks: 'Can being, *ens*, the same as to number, exist again?' [We answer:] 'Clearly it cannot'. Hence in order for an existing thing to return, i.e. to exist again, we must imagine that there first took place a ceasing-to-exist. But matter cannot naturally cease to exist. For although a ship, or a plank, ceases to be a ship and a plank, it nonetheless never naturally ceases to be an entity; because *ens*, unless it be reduced to nothingness, does not cease to be *ens*. But to reduce to nothing is a supernatural task, and one for God. So the question is: 'Can, for instance, any existing ship, if it ceases to be a ship, be made a ship again, and the same in number as existed before?' I do not see why this cannot be done. If anyone has broken into pieces an existing ship, this ship will cease to be a ship; but if one has reassembled, with the same nails, the separated planks into the same structure as before, the ship will again possess material the same in number as it had previously, and it will also be a ship. A ship that has material the same in number [now as at other times] is surely the same ship; for although

FOL. 112v

the form has not been re-created the same in number, this does not cause the ship (when the same material is existing) to be another ship: it causes the ship to exist again.

6. But my author will deny me that the same dispersed planks can be reassembled in the same order.[14] The artisans will blunder; or some circumstances, necessary if the ship is to be the same in number, will be absent or have been neglected. For he says on his page 111: 'Individuity depends on a mathematical unity of circumstances; but this unity wholly supposes an infinity of variety'. According to our author, then, a man who sometimes blushes or sometimes goes pale; who sometimes sees, hears, touches, and does these things and at other times does something else, *will become*, in succession, as many individual things as the different acts he performs.[15] How, therefore, was Orestes in Sophocles recognised by Electra, and by

---

[14] Presumably, 'in the same order as was followed in the original building-process, i.e. in the same positions as before'. The inconsistency 'reassemble'/'re-create', permitted by *O.E.D.*, seems justified here: spelt as a single word, the latter term has, even in the present context, distracting associations.

[15] The examples given, e.g. blushing, are not in White, but are Hobbes's (cf. fol. 45v). 'These things': awkward as the expression is, it seems best to leave it. *Will become*: translator's italics; cf. the reasoning at fol. 309v.

means of his father's ring, as being that very Orestes to whom, in actual fact, he bore no resemblance? Because, I think, the limits of continual change from infancy to manhood were represented by a ring, the same in number [as a ring] which Electra had noticed both in Orestes's childhood and when he was a man. But perhaps, says White, [Orestes] was thought to be, not the same in number, but the same in appearance; i.e. [it was thought that] not the same man, but one like him, had arrived in order to avenge a father's death. Likewise White could say that, when someone has committed murder or theft, it is not the same man in number, but someone resembling him, who is punished—which is to violate all human laws and observances. But such views are absurd, as are the countless other ones propounded by those who believe that the individuity or the numerical unity of a thing derives from the unity of chance circumstances (*accidents*). [It is] the ambiguity of the term 'other' [that] has deceived these people, I think. For something which is changed or altered is very often called 'another thing' (although the subject of the alteration is not a different thing but is always the same in number), because [the thing changed] is in a different state and appears to us differently. Hence when the non-essential qualities (*accidents*) are not the same in number as they were, people say that the subject is not the same.

FOL. 113

FOL. 113v

7. But if we concede that individuity depends on the numerical identity of all the causes working together at a given time, let us see whether he has rightly inferred from this that it is impossible for the same individual body to be naturally recreated. 'If a game is played with three dice', he says on his page 108, 'I see that three sixes will certainly set aside [*obviare*] some total arbitrarily composed.[16] The same thing will happen if six dice are used and six sixes are required [*posci*]. But a player will throw a set of sixes far less frequently than his opponent will throw other numbers; so the infrequency of this occurrence [i.e. of throwing six sixes] is increased according to a

---

[16] I.e. three sixes will always defeat a set of three other numbers, however comprised; and so with any number of sixes relative to the same number of other scores (see the next sentence in the text). White prefaces the quoted words with: 'The greater the number of things mixed, or the greater the number of turns [*vices*] equally combined, the less frequently can any one [combination of numbers] occur' (the word 'one' is repeated, erroneously).

FOL. 114

fixed law.[17] I see perfect infinity only in the nature of concurrent causes.' A little further on, he concludes thus: 'If, therefore, it is necessary that, the greater the variety of non-essential qualities (*accidents*), the more rarely the same result occurs, then it is equally necessary, where multiplicity surpasses every measure [*omnem modum excedit*], that it is impossible [for the effect] to return a second time'.[18] But who will grant him that 'there is perfect infinity in the nature of concurrent causes'? On his own admission, he believes he has shown the world to be finite. Assume, then, [I say,] that each of the atoms which can fill the entire world, and that every part of those atoms, has been divided into as many secondary atoms as there may be primary atoms in the whole world; and that those secondary atoms are again subdivided as often as you please. The atoms will still be finite in number, and the combinations [*complicationes*] of things finite in number will also be finite,[19] as anyone with an average knowledge of arithmetic will readily admit. Assume next that effects were produced as in a throw of sixes.

FOL. 114v

The number of sixes thrown may equal the number of dice used, but the same sixes may again appear at another throw. Then however many the particles of things be, and however many their combinations, an effect produced on one occasion may also be produced on another. So his argument based on the multiplicity of concurrent causes (this multiplicity he wrongly calls 'infinity') is worthless.

Yet the same effect has many proximate causes, of which any one in turn has many further causes of its own, and causes are thus continually multiplied. He infers from this that eventually there will be an endless [number of] causes of the same effect; but in fact it is impossible for a number of causes in the finite world to be augmented infinitely. Perhaps the multiplication of causes in such a manner may bring it about that an effect the

---

[17] In the first part of this sentence the original says: 'But there will be a far smaller number of sixes of repeated picking-up'. White appears to argue (wrongly) as follows: A player is less and less likely to achieve six sixes at one throw, as against throwing any other set of six numbers (however comprised), the more and more he tries. Though he may succeed at his first attempt, his ability to repeat his success recedes as he goes on.

[18] White may mean: 'A cause cannot return as something ancillary', i.e. that a given cause cannot recur with reduced potency. Cf. Galileo (Drake), p. 407.

[19] If indeed White saw the present manuscript, or this part of it, he must have remained unmoved by Hobbes's pleadings, for in *Exercitatio geometrica* (1658) White's position is little changed.

same in number can be produced only if at the same time there were also produced all the factors resembling those that accompanied the earlier state [of the body concerned].[20] This is not to deny the existence of Plato's Great Year,[21] but only to postpone it a little, until through all [nature] some effect is produced [exactly] similar to a foregoing one.

FOL. 115

8. It is easy to prove that the statement 'The universe will become a different one if any single part of it is altered' is untrue. Granted that there is a universe, however regulated yesterday, and that upon it is wrought any number of changes; in this case, chance happenings in the universe are not the same today as they were yesterday. Now, these happenings are inherent in the universe of today; but the subject of the happenings (*accidents*) which were present in the universe yesterday either remains or has perished. Yet the universe or a part of it cannot perish, unless it has been supernaturally annihilated by God;[22] so the universe remains what it was yesterday. Either, then, there are two universes today, and this at the same time and in the same place, or else both of them are one universe. But two universes cannot be in the same place at the same time; so there is one and the same universe, just as much yesterday as today, whatever the ways in which different changes befall it.

FOL. 115v

9. If we were to concede that individuity depends on the unity of concurrent causes, I would not have denied the consequent: that any slight movement or change alters the world. For, [he says,] if its parts are indeed all dissimilar, then the whole (which is none other than all its parts in conjunction) will not be the same [as it would be were all the parts similar to one another]. What he is trying to prove here is as follows: If any one part of the universe is changed, all the remaining parts will be severally changed. This–if he calls 'change' the local motion of every part[23]–is true, for when one part is changed a different part will take its place and another part will replace the last-

---

[20] Perhaps, (1) '...can be produced only if there are also produced all the factors jointly resembling...' or (2), reading *effectus* for the text's *affectus*, '...all the effects that accompanied the first effect'.
[21] Cf. *DM*, p. 115 and Bacon's essay 'Of vicissitude'. Hobbes had assisted Bacon in the translating of some of the *Essays* into Latin.
[22] Drake, p. 244. Cf. fols 112, 334, 453v, 458.
[23] Or, '...if he calls a change of any one part "local motion"...'

FOL. 116

named part, and in this way will be effected in succession, according to the cohesion of the parts, a disturbance–though a slight one–throughout the separate parts. But, [I reply,] the local motion of parts is a change, not of the parts, but of the whole, and not always in the whole but only when this motion of the parts makes the whole behave differently from before. When, for example, something cold is made out of something hot, something black out of something white, or something round from something square, [the thing] is then said to be changed; but when all the parts of the earth are carried round, or when all the parts of a stone that is hurled forward by the hands fall, or fly forward, no-one, I think, will say that the earth or the stone is thereby changed.

FOL. 116v

10. If something is understood to be reduced to nothingness, and if a restoration [to its original state] is effected by means of creation, one may ask: May we equate numerically what is restored with what was previously destroyed? I think that this question is unanswerable; for created material is not supposed to have existed previously merely because we say 'It has been created'. The material [of something created] is therefore not the same in number as the material of something that has been destroyed; so there are two [kinds of] being, *ens*, not one.

CHAPTER THIRTEEN

※

On Problem 16 [of the First Dialogue]:
That there is no need [to propound the
existence] of a world that contains
all other [created] things

1. The components of the present problem. 2. What *tending* [*conatus*] is. 3, 4. Of our author's two arguments that the surface of the world is uniform, the first, based on the *tending* within the world, proves the contrary; the second, inferred from the winds and from circular movement, is refuted. A rebuttal of his argument against (5) the roughness and (6) the flatness of the earth's surface. 7, 8. His arguments for the roundness of the world are found wanting. 9. An argument against an increase in the size of the world's surface is shown to be invalid.

1. Our author thinks in Problem 2 [fols 9v*ff*] that the world has now been shown to be finite. It follows from this that the outermost form [*figura*] of the world's surface is a fixed one; we shall next see what this shape is.

In this question he thinks he has first to show that the shape is always the same and is constant, i.e. that parts of the outermost surface[1] of the universe are neither driven forward nor come to rest again. (Such movement may easily happen with the surface of bodies that are violently agitated from within.) Lest, then, the earth's surface be thought to be rendered irregular by reason of its inward movements in the way that the surface of boiling water is continually disturbed and, because of the boil-

---

[1] This is not pleonastic: see later.

FOL. 118

ing, is ever on the move, one of two points must be laid down: either the outermost part of the heavens is very hard, or every movement and striving [*conatus*] is directed towards the inner parts. Following, in this question, Copernicus, Galileo[2] and other very well-known scientists, White commendably rejects as false the [belief in the] hardness of heavenly bodies. Therefore the reason why the outermost surface of the world always remains the same must, he thinks, be that no motion from the internal parts either restricts or presses on it. That no motion from the inner parts presses it has, he believes, been proved by the fact that winds do not extend upwards that far, and that the motions originating in the effluvia of the earth and stars are borne circularly.[3] He [then] says that circular motion is 'harmless' to the things surrounding it, i.e. that things carried circularly do not press on the bodies adjacent to them. In the end he denies that the world can be naturally enlarged.

FOL. 118v

2. It is here that we must ask what tending [*conatus*] is. Now, everyone knows that motion is nothing but the loss of an initial position and the continual[4] acquisition of a second one, and that *conatus* is therefore the same as the principle of motion. Moreover, it is clear that every part of a movement is motion and that the principle of anything at all is its primary part.[5] It follows from this that all *conatus* is motion. What if someone says: 'A principle is not the same as a primary part. It consists in an indivisible point in such a way that that which is now at rest may have within it, while at rest, the principle of motion'? It will follow that what is at rest can be moved of itself and will not require the approach of another moving body to set it in motion.[6] This is against White's own opinion and that of nearly all philosophers and, moreover, against observed fact. Or if someone said, 'The principle of motion is the potential to motion, without the act [of motion]', it will follow that

[2] Drake, p. 69.
[3] The text requires emendation. Perhaps we should render: '...and that the wind originating in the effluvia...is borne circularly'.
[4] Here and later, perhaps 'continuous' (*continuus*).
[5] This sentence suggests that the clauses preceding ('and that...of motion') are not dependent upon 'everyone knows'; but the text is translated as it stands.
[6] Grammatical considerations in the Latin make it clear that this sentence is not part of the question asked but is Hobbes's own inference. On 'principle' cf. fols 38v, 289v.

conatus[7] is not an action and does not achieve anything, whether the *conatus* be inward or outward. *Conatus* is therefore motion in actuality, even though the motion be very small and indistinguishable by the eye. This may also be shown as follows: Let each of two bodies be moved towards the other until they touch, say an iron ball and another iron ball, so that they come to rest in pure contact (as understood in mathematics); neither will move or restrict its fellow or the part of it which it touches. Why? Because, while it is at rest, neither has *conatus*. If, during a movement, or while it was being moved, the same sphere collided with the other[8] and propelled it, why was this? Because, in this instance, had it not collided with the sphere resisting it, it would have moved further on; but in the first example it would have stood still. So *conatus* lies in the fact that a body that tends [*conatur*] is being moved.[9] Likewise, heavy bodies lying on the ground may be said still to tend downwards, because when the impediment is removed they actually descend; but if they did not descend they would not be said to tend previously. Likewise a strung bow is thought of as trying to restore its parts [to their former state of non-tension or less tension], because, when the impediment is removed[10] it actually restores itself, and when it has straightened itself it no longer possesses *conatus*. The removal of an impediment does not constitute an action, however, but an action is needed if things at rest are to be moved. It remains, therefore, that the principle of fall of heavy bodies, and of the return [to straightness] of a bow, is that some mover [*movens*] has actually been moved in a heavy body that tries to fall and in the same bow that tries to straighten itself. So *conatus* is nothing but an actual motion, either of the whole body that tends, or of its inner and invisible

FOL. 119

FOL. 119v

---

[7] Translated below by words such as 'force', 'impulse', 'effort'. The (incorrect) tense-sequence here is Hobbes's.

[8] '...was being moved': by a third mover? '...with the other': the latter being initially at rest?

[9] 'Tends': Hobbes would say: 'endeavours'. '...is being moved', perhaps: 'is moved', i.e. habitually; but the sense is little altered: 'When a body in motion endeavours (towards something), this is *conatus*,' i.e. *conatus* is *conatus*. And (fol. 199) Hobbes chides White for explaining the precession of the equinoxes as the precession of the equinoxes!

[10] By *tensus*, here given as 'strung', Hobbes seems to envisage the situation before the bowstring is further stretched when the arrow is inserted just prior to firing; hence the tension will cease if the string is cut. In his other instance, shooting of the arrow would reduce but would not remove altogether the tension of the string.

FOL. 120

parts. But, [I say,] the presence of motion in the inner parts of all hard bodies and of those whose visible parts cohere and resist an agent is argued from the fact that all resistance is motion: for resistance is a reaction; a reaction is an action; and all action is motion.

3. Now where our author thinks that, because the *conatus* of the outer parts of the world tends towards the inner parts, the result is that the outer surface is nowhere dilated or compressed, but that it always remains smooth and uniform, as part of the same uninterrupted process, the contrary is true. For as the force [*conatus*] of any part of the outer surface tends toward the interior, that part must actually be moved toward the interior. So an irregularity [in the surface] must result from the downward pressure. But he imagines this impulse toward the interior to be an actual motion, for he supposes that the impulse [*conatus*] is directed inwards because of the abhorrence of a vacuum. It is understood, therefore, that because of this force [*conatus*] a descent takes place to fill places which would otherwise be empty; but this can be done only through actual motion.

FOL. 120v

4. The argument he deduces from the fact that winds do not reach this far [i.e. to the limits of the world? folio 118] conflicts with his own doctrines stated earlier; for there may also be winds wherever there are effluvia and vapours which are emitted not only from the earth but also from every other star [*sic*].

But he thinks that the measure[11] of the universal world must be determined from all consistent bodies taken together, such as the earth and the stars, together with their effluvia.[12] It follows, then, that there may be winds in any part of the world, because there are effluvia of this kind in every part of it. Movements [in being] towards the outer surface of the world are, he declares, circular; but in circular motion there is no impulse [*conatus*] towards the outer parts since, he asserts, 'It is only when the proximate air is thickened that circular motion has no effect on its surroundings'. Now this seems, at a first glance, likely; but anyone who thinks about the matter and examines it closely by means of experiments sees it to be false. As regards

FOL. 121

---

[11] Possibly the text's *modus* should be *motus*.
[12] For 'consistent', perhaps 'that constitute [the world]'.

an experiment, if somebody whirls round himself [in the horizontal plane] a stone or other heavy body fastened to the end of a string he will find himself drawn against his will out of his position, i.e. towards things 'external to' him.[13] Furthermore, if any particle of a rapidly rotated wheel falls off during rotation the particle will be borne, not in a circular motion, but (except inasfar as gravity hinders)[14] along a tangential path. This also is in accordance with reason and with the very nature of motion.

Given a circle of radius AC, let a straight line BC touch it.[15] Suppose that some body moves from C to B in a straight motion. The body in motion will therefore always endeavour to be moved away from the centre A. Imagine, next, that whilst this moving body tries to progress from C to B another moving body approaches that urges the first body towards A, continually and with an effort exactly equal to that by which the first body previously tried to move away from A. This occurs if the mobile in C has been connected to A by some string AC and is then struck a blow aimed in the direction of B. So there is created a mixture or compounding of two motions, one being along the tangent and the other from the tangent towards the circumference. Thus the circular motion from C to I is compounded of two straight motions across CD and DI; the circular motion along CH [is compounded] of the motions along CE and EH; and the circular motion along CG [is compounded] of the straight motions along CF and FG. Force [conatus] is still found, however, in a mixed motion, although the force of either mover is reduced; so a force along the tangent, i.e. the force of moving away from the centre, is still present in circular motion. Even though, then, the motions round the outer surface of the world were all circular, some force would still tend outwards. White has not, therefore, proved that the surface of the world is always uniform, as he claims. On the contrary, the very nature of motion and of the effluvia (which he supposes to be on the earth and in all the stars) shows that of necessity the surface of the world is diversified through a perpetual inconstancy. Finally, this is also shown in what he has laid down in the

FOL. 121v

FOL. 122

---

[13] I.e. towards the path described by the stone.
[14] By pulling the particle 'downward'? Will this apply to any plane of rotation?
[15] Figure 21.

previous problem, namely that at every motion, even the slightest, of any part [of a body], all the other parts are moved.

5. In the next place he tries to prove that the earth's surface is not rough, and he uses the argument by which he earlier disposed of the possibility of a vacuum.[16] 'For if', he says, 'some *swellings* stand out [on the world's surface], there may be drawn from the top [of a hump] to another part of the [world's] surface a shortish straight line which passes through the foot of the hump. But if there are any *hollows* these must be covered by straight lines; so you have an expanse that is not filled, i.e. a vacuum.'[17]

So that we may understand this more fully, let the spherical surface of the world be represented by an [unbroken] curved line ABCD that is not circular, but is gibbous at A and concave at F.[18] Now if the straight line AB be drawn, it will be shorter than [the concave curve] AFB; so the space contained, AFB, will be empty. Likewise the straight line CD is shorter than the curve CED, so the space contained, CED, will also be empty. But he says, from the arguments adduced above, '[The concept of] a vacuum has been demolished, so there cannot be humps or hollows of this kind'. This reasoning is[19] invalid because, in Chapter 3 above, we refuted the arguments produced to destroy [the concept of] a vacuum.

In order, however, to prove that there is no vacuum within the world, must we not also say that there will be none outside it? That is, if the world is indeed finite and a vacuum does not exist, will there be a plenum outside the world? 'There will be', White will declare, 'nothing: neither plenum nor vacuum nor space.' So both AFB and CED, which are supposed to be outside the world, are for the same reason neither plena nor vacua nor spaces, even. So the straight lines AB or CD, which he supposes can be drawn, cannot, nor can any lines at all. In order to destroy [the concept of] a vacuum, therefore, it is unnecessary to consider the surface of the world as level.

What lies outside the world is surely able to contain body. We must call this ability either space or absence [*privatio*] of space, although the space inherent in a body has been created

---

[16] Fol. 23.
[17] Translator's italics.
[18] Figure 22.
[19] Mersenne alters to 'is rendered'.

together with the body's subject. Before the Creation, however, it was possible to contrive a world and to place it in a spot different from that occupied by our own world; so there exists outside the world an ability to receive body–an ability that no arguments have removed. Hence, as what is enclosed by AFB and CED is the ability to contain as much body as nature allows, the surface of the earth need not, because of this, shape itself into a spherical figure.

FOL. 123v

6. Then he denies that the world's surface could have been composed of planes, e.g. [that it is] cubical. He adds as his reason the fact that such a shape will be full of corners; but that corners presuppose a protuberance that serves no purpose. He declares, furthermore, that no power in nature can do such a thing, because her influence [*momentum*] is exerted inwardly. As regards utility, my reply is only this: those philosophers who assess the shape of the world, or other works [of creation], by its usefulness to God fix their own prudence as the yardstick of God's prudence and are therefore seriously in error; for utility and non-utility concern purpose. [Our author] should say first with what intention the shape of the world was determined by God, and afterwards he should show how a cubic shape is more contrary to such an intention than a spherical shape is. If the world has been created with the intention that the whole be moved round some axis, a spherical shape seems more useful than a cubic one. If, on the other hand, God wished it to be at rest, and the internal parts only to be moved, a cubical shape would, as far as men can judge, be more suitable than a spherical one. What White adds on the inward-drawing force [*conatus introrsum*] of nature has already been sufficiently refuted.

FOL. 124

7. Some authorities have considered that the world's surface is spherical, and for this reason: the world, the noblest of bodies, merited a form also of the worthiest; and they thought the noblest shape to be a sphere.[20] Our author, in spurning such reasoning, is of my own opinion; but his reason for holding this view is not the same [as mine], for I do not understand what 'nobility of shapes' is. But he, taking 'nobility' as meaning 'beauty', therefore denies the consequence from 'nobility' to 'a

---

[20] Cf. Galileo (Drake), pp. 84, 209. Also in *Il saggiatore* (1623) in a pass of wit at the expense of Lothario Sarsi (see Drake and O'Malley, *The controversy on the comets of 1618* (1960), p. 279).

spherical shape' on the grounds that outside the world there are no observers. He has no reason for doing so: that God surveys His own work is not implausible. What White then concedes–that the world's inner surface may be spherical because it *can* be observed [from the outside]– is ridiculous, for no surface of a sphere is concave. Again, if something is brought to our view and appears spherical, a surface need not on that account be termed spherical. For whatever the shape of the world, over so huge a distance vision will still be limited [*determinari*] equally, i.e. spherically.[21]

FOL. 124v

I do not follow what he says next: 'The portions set apart', (i.e. the convex surface of the world), 'have been assigned to usefulness inside. Therefore you [one of White's interlocutors] favoured the stirring up of wild dreams, sounding through imaginary space, to arouse secondary purposes. So we shall ensure that the outer surface of the world will flourish because of innumerable kinds of beings, *entia*, in order to fulfil the wish of these'. Even if behind these words lay some quality which advanced the author's design and which it was vital for me not to ignore, I still must censure the obscurity of his wording: it is so metaphysical that I have thought it necessary to pass over this last extract, just as I have done with several others like it that occur from time to time.

8. He thinks that the roundness of the world, as regards the outer surface, is to be inferred from its usefulness rather [than from its symmetry of form]. He argues as follows: 'If the remainder of the universe [*mundus*] (the suns [*sic*] and the paths of the planets being in various circles) is similar to our own world, the one we inhabit and behold, then it follows that the whole world has been framed of spherical mists. Now, I do not accept, [he continues,] that spheres can be enclosed with less wastage of space than [occurs when they are enclosed] within a sphere. So if the world is made up of spherical things alone, it does not seem unreasonable [to assume] that its surface is also drawn out into a spherical shape.'[22] Even though it be allowed him that 'the whole world is framed of spheres of vapours', he

FOL. 125

---

[21] Maximum range of vision is the surface of a sphere with the observer as centre, irrespective of the shape or size of his surroundings.
[22] This sentence is also part of what is conceded to White, and is a mere repetition of the sentence preceding it; the sentence that follows it states the point yet again.

might not readily accept that those spheres are therefore enclosed 'with less wastage of space than [occurs when they are enclosed] within a sphere'. Let me demonstrate this so that others may believe it.

Given any [finite] number of spheres, e.g. a, b, c, d, touching one another.[23] Let a sphere ABCD be drawn that touches and encloses them all externally. Let the several planes EF, LM, KI, and HG be subtended by the separate parts—AB, BC, CD and DA—of the sphere that encloses the remaining spheres. It is clear that this figure, made up of surfaces partly spherical, partly planar, is smaller than the spherical surface ABCD, and yet that all the smaller spheres are contained by it. It is obvious, moreover, that less empty and useless space is contained [in it than in the sphere ABCD]; furthermore, from that description of the spheres it appears that the world cannot possibly be made up of bodies, the surface of every one of which is spherical, unless he revives the concept of vacua, which he claims to have done away with. So the world is not composed of spheres, and it is not (as our author thinks it is) very convenient for spheres to touch a sphere.

FOL. 125v

9. Finally, he declares that 'the world cannot be enlarged through natural causes'. How we are to take this is uncertain; if, in fact, it is understood of the quantity of the whole world, i.e. that he denies that the quantity may be made, by natural causes, larger than it is now, I readily agree with him. He who can create the whole can create part of the whole, namely new matter not yet existing; but creation is a supernatural action, not a natural one. Even if White understands the proposition as 'It is impossible for the outer surface of the world to be enlarged', he has proved it by no reasoning. As he admits,[24] his argument does not suffer from the fact that there is no space outside the world into which the latter may expand, for, he says, 'space is indeed required for local motion, but not for quantitative motion' (i.e. the motion of increase). This is false, however: for in quantitative motion a change of place occurs, since the larger is not contained within the same place it was in when it was smaller; hence the place is different from before; so [this] motion is local.

FOL. 126

---

[23] Figure 23.
[24] Hobbes's verb is *confiteor*; but White is stating, not conceding.

But, he declares, 'my argument suffers from the fact that, if the world is enlarged, either its surface must be increased all at once and from some cause or other, in order that a larger sphere may be [formed]; or else its shape must be changed. Nature abhors either alternative'. But why nature abhors the latter, i.e. why the world cannot be enlarged by natural causes —which he had proposed to show—he does not say. Perhaps he considers that this may be sufficiently understood from the fact that it cannot happen without one's conceding the possibility of a vacuum. Since, then, he has not shown that there cannot be a vacuum he has not shown that the world cannot be enlarged.

The topics I have dealt with so far are those that, in my opinion, call for treatment in connection with the First Dialogue—which concerns the material of the world. The next Dialogue is devoted to the form of the world, or the motions of its principal components. We turn now to consider these.

CHAPTER FOURTEEN

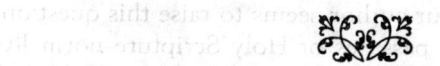

୫୨୧୬

On Problem 1 of the Second Dialogue:
What is meant by the word 'motion'

1. Questions concerning the meaning of terms, especially the meaning of the word 'motion', are not unworthy of examination. 2. That by the term 'motion' is properly signified, not the changing of position relatively, but a change of place absolutely. 3. That bodies are said to be changed, not because of a new 'relative essence' [*esse*] but because of a new 'absolute essence' [*esse*]. 4. Definitions: Relation, Related Things, and the basis of Relationship. Also that an absolute and relative non-essential quality [*accidens*] is the same non-essential quality in number. 5. A new name is given to things, on some occasions, without their being changed. 6. The author's argument disproved. 7. The points he has here adduced on the interpretation of Scripture are confused and doubtful.

1. As it consists in the perception of differences, true philosophy is clearly the same as a faithful, correct and accurate nomenclature of things. Now the only person who knows the difference between things seems to be someone who has learned to assign to separate things their own correct names. Moreover, right reasoning, which philosophers need, is nothing other than the correct combining of true propositions into a syllogism.[1] A valid proposition is formed out of the correct coupling of terms, i.e. of the subject and predicate, in accordance with their proper and adequate[ly-defined] meanings. From this it emerges that there cannot be a true philosophy that does

---

[1] Cf. fol. 346v below.

not lay its foundation on an efficient nomenclature of things. So the question of the correct meaning of this word 'motion' is worth looking into, because, if a knowledge of motion is lacking, nothing certain can be laid down about motion, and hence (because whatever is done by nature is done through motion) about nature. Our author seems to raise this question [solely to show] that the passages of Holy Scripture normally cited in negation of the earth's motion will seem capable of explanation only if we attribute to them an unsuitable and figurative mode of expression.[2] He wishes, therefore, [to demonstrate]–and this properly speaking–that as often as the position of two things, each to each, is changed, both things move, and consequently that when the earth and the sun occupy given successive positions with respect to one another it may be said that (both statements being correct and accurate) [a] 'first the sun moves', and [b] 'then the earth moves'. We shall show this to be false.

FOL. 128

2. As we stated above in Chapter V, Article 1, everyone explains this term 'motion' or 'to be moved' as 'the continual quitting and acquisition of place'. Something is said to be moved, then, that continually quits one position and gains another, and something is said to have been moved that has left one place and has reached another. All declare, however, that what is in the same place it was in previously [at a given time] is at rest. This is the correct and accurate meaning of 'motion'. So when our author decides that the sun stays in the same place but is moved because it has changed its position relative to that of parts of the earth, he does not take the term 'motion' in its correct significance,[3] as he claims to do. The sun when motionless may indeed seem to be moved because of this continual change of position; but to attribute motion to those things that stay in the same place, and that only seem to be moved, is to speak inaccurately.

FOL. 128v

3. Yet he thinks that whatever undergoes some change is moved;[4] but [I say that] a change of position is a change of some kind, namely a change of relationship. We must therefore

---

[2] 'This question' is: What does the word 'motion' mean? A double negative in the text makes interpretation dubious.
[3] That is, White does not take the term 'motion' as Hobbes wants him to do!
[4] Or, 'whatever is moved undergoes some change'.

examine what this change of relationship is–also if, and when, it is the reason why a related thing may itself be said to be changed.

Now, when the position of two bodies is not the same as before, we rightly say that one of them has been moved out of its place. Likewise when some object appears to our senses differently from before, we at once say either that the object itself has been changed or that our senses have; but it is not yet agreed to which of the two [statements] we must ascribe change. For things that were at first white and then appear yellow, or that were at first sweet and then seem bitter, appear so because it is sometimes they that are changed, sometimes we. Either (properly and correctly speaking) the only thing that is said to be changed is [a] one in which some new absolute essence, *esse*, is produced, or [b, one in which] is present that which was not in it previously.[5] For instance, if some body has contained a white essence, but the whiteness later disappears and blackness is produced in the same body by some agent, so that a black essence is now present in it, then that body is correctly said to have been changed. Again, if a different [*novum*] body has earned some absolute title, i.e. a new, absolute name, the body must have been changed; but with relative names a different thing happens. If John's son dies and his father does not know, then, although John ceases to be the father, it does not follow that John is changed. So that this may be better understood we must say something on the nature of related things.

FOL. 129

FOL. 129v

4. Related things are those which are compared with one another.[6] Two bodies are compared in respect of the non-essential qualities, *accidents*, of each, e.g. 'father' relates to 'son' in regard to the non-essential qualities according to which they are compared, namely that 'to have begotten' signalises a father and 'to have been begotten', a son.[7] Thus

---

[5] Reading *inest* or *interest* for *intereat* in the text's...*producitur, vel intereat* (why the subjunctive?) *in eo non erat prius*. To retain the latter supposes dropping of the *non*: '...or that [absolute essence] which was in it previously may perish'.

[6] *EW* I. 133. See also fols 28 and 216v.

[7] Changes in the English language since the seventeenth century necessitate a translation such as this. Clearly Hobbes does not mean 'A father and a son are kinsmen' (or the modern 'relatives' or 'related'), but rather 'A father is to a son as X is to Y'.

FOL. 130

a white object and [another] white object, when compared in respect of colour, are related things and are said to be alike; but white and black, when compared in respect of colour, are said to be unlike. So relationship is the same as 'a thing's having been compared with something else, or with itself at a different time', e.g. to be a father, to be similar, to be equal, to be the same, to be different, to be larger, etc. Now, things are compared in respect of [their] non-essential qualities (*accidents*), and these latter are called the bases of a relationship. For instance, 'to have begotten' is the basis of a father's connection [with his son]; the whiteness of a white object that is compared with another white object is the basis of the similarity between the two objects, and the whiteness of a white object compared with a black is the basis of the dissimilarity. So the root of the connection, namely 'whiteness' or 'being white', is the essence or the absolute, non-essential attribute; but 'to be like or unlike' is *compared accident* or *compared essence*, i.e. relation. These two expressions: [*a*] 'to be white' and [*b*] 'to be similar to other white objects' relate, not to [two] different things, but to the same thing, [*b*] compared, or [*a*] not compared

FOL. 130v

but taken solely for what it is. So 'to be white' and 'to resemble something white' are the same concept, but are different expressions, since what we call 'white' when we are not making a comparison we call 'similar' when we are.

5. From this it follows that a previously non-existent relative name can be acquired even though one of the objects related [one to another] undergoes no change at all. Given some white object [*a*], let some other object [*b*, of a different colour] be made white. The one that was white first of all, [*a*], is unchanged, but yet it gains an additional title, namely 'similar to the one made white afterwards'. Indeed, we may perhaps say, not incorrectly, that the object that was white all the time, [*a*], has now become similar to the white one that exists at present, [*b*]. So however we frame the statement–correctly or incorrectly–it is clear that 'to resemble a white object' and 'to be white' are the same accident. So it is unnecessary that, in every new rela-

FOL. 131

tionship, i.e. with every new comparison we draw between two things, both be considered changed: only one need be.

6. Our author uses the following argument (his page 131) to prove that we may correctly and properly say that the sun,

although it stays in the same place, is moved: 'Motion is continual change. But even though either the sun or the earth be changed in position, the following is agreed: each is, with respect to the other, continually varied. If, [he goes on] this is true, both statements are made with absolute correctness and accuracy: that the earth has been continuously moved with regard to the sun and that the sun [has been moved] with respect to the earth'. To see that, in his reasoning, he can be relied on to ignore the logical handling of philosophy, note the offences against logic he has committed. The first is this. In the second figure[8] he has deduced an affirmative conclusion, for he reasons thus: 'Motion is continual change; the sun undergoes a continual change of position; therefore the sun is moved'. Perhaps he will say that the first of his premises should be taken in reverse, so that a syllogism is made as follows: 'All continuous change is motion; the sun undergoes continuous change; therefore the sun is moved'. But either (in his doing so) the major [premise] is false, i.e. if he thinks that a change which is not a change of place is nevertheless local motion; or the minor [premise] will go against his own hypothesis, i.e. if by 'change' he understands 'change of place', for he believes that the sun stays in the same place.

FOL. 131v

The second offence is his saying that change, or 'variation of position', is [that] kind of change [which occurs] in the sun; for it is obvious, as was shown in the last article, that there may occur a change of position, i.e. of relationship, even though only one of the related things is changed. So nothing may be inferred from such change except that one of two things is moved: it may be the earth, it may be the sun.

FOL. 132

Third, where he should have concluded simply 'The sun is moved', he adds, 'with respect to the earth', i.e. that the movement is seen by those dwelling on earth or that [the sun] has, in relation to earth, a different position from the one it had before—which is one of his premises.

Though at the beginning of his treatise Copernicus affirms that the sun is at rest, he later speaks of the sun as if, in common parlance, it were moved.[9] So from what White says about

---

[8] 'The form of a syllogism as determined by the position of the middle term in the premises'—O.E.D.
[9] Hobbes's words 'at the beginning of his treatise' are not very accurate. In the Preface of the *De revolutionibus* and the Dedication to Pope Paul III, Copernicus says merely that the earth moves; he does not mention the

FOL. 132v

Copernicus, 'His plea that he is addressing the common people seems superfluous and harmful to itself', I fear that White has misunderstood him. At the opening of the *De Revolutionibus* (this part of the book is wholly to do with physics) Copernicus openly declares that the earth is moved and the sun motionless. But later, when he had to approach the calculating of movements (this section is to do with astronomy), it seemed right to him to speak in a different way, not because he believed he was addressing the common people–for the book is not for the general reader–but because he was unwilling, by [using] a new way of speaking, to divert the minds of his readers from astronomical calculation.

7. What White here disputes concerning the opinion of theologians is quite confused. As a philosopher he wishes to uphold the motion of the earth, and as a Christian he wishes to leave the interpretation of Scripture to the Church. So he is torn between two obligations, and I do not know what view he is putting forward. He says on his page 132: 'The Scriptures must be understood [i] according to the Church's tradition and [ii] in the light of the disciplines of the arts or sciences'. His linking of 'tradition' and 'disciplines' is like Odysseus's yoking two draught-beasts of unequal strengths.[10] What if they should pull in different directions? I should like to know which he intends following: [i] or [ii]? Again, 'To wish to elevate to [the status of]

FOL. 133

a rule of faith and a law of Christian doctrine over the heads of all the faithful', he avers, 'that which, from your ignorance or from gross ill-manners, you can only foolishly talk about, has the mark of a great offence'. What is the purpose of this? Suppose that at any time the Church should decide that those texts which seem to assert that the sun moves, and the earth stands still, be understood in such a way that anyone holding the contrary opinion would be considered a heretic. Will not such teaching about the immobility of the earth and the move-

---

sun. And Chapter 5 of the First Book, asking: 'Does the earth have a circular movement?' does not say: 'The sun is motionless', but rather: 'The apparent motion of the sun is only relative'.

[10] Ulysses, in order to avoid leaving Penelope when going to the wars, feigned madness. He yoked a horse and a bull, ploughed the seashore and sowed salt. The device was exposed by Palamedes, who put Ulysses's son Telemachus in the path of the animals, and Ulysses would not run Telemachus down. The story is not in Homer.

ment of the sun be established by authority over the heads of all the faithful, as if it were the law of Christ? Yet if our author thinks he *has* proved the earth moves, he must have imagined that those who have taken the contrary view have been blinded by ignorance; so if by chance she has decided otherwise, he apparently intends to find the Church guilty of a great crime.

FOL. 134[1]

CHAPTER FIFTEEN

※

On Problem 2 of the Second Dialogue:

That the wind in the torrid zone

is caused by the sun

1. The source, according to Galileo, of the wind that blows unceasingly from east to west in the torrid zone [of the earth]. 2. Our author's views on the origin of this same wind. 3. Though he thinks he is refuting Galileo's view, he is refuting another one he substitutes for it. 4. Why the sign—from the motion of the sea—of an impending storm precedes the gale. That the explanation is not the one our author gives, which is that the seas move more quickly than the wind.

FOL. 134v

1. Those who have sailed the Atlantic Ocean between Africa and the southern part of America have noticed that navigation within the torrid zone from east to west is easy because a certain wind, blowing from the east without interruption and at a constant rate, urges forward their sails;[2] return from west to east within the torrid zone, on the other hand, is impossible because of the same wind.[3] Moreover, those who undertake a voyage from America to Spain leave behind the area of the tropics, intending to take advantage of any chance winds. I have heard that some English sailors call this [east-west] wind 'The Breeze' and that, unless interrupted by unpredictable gales, it

---

[1] Fol. 133v is blank.
[2] Galileo: *Two chief world-systems* (Drake), p. 439; *Il saggiatore* (1623), as in Drake's *Discoveries and opinions of Galileo* (1957), p. 260. Subsequent references designated 'Drake' are to the first-named work. Cf. fols 170vff.
[3] Drake, p. 440.

blows unceasingly not only across the Atlantic Ocean but also across the Pacific.[4] On page 434 of his *Two World-systems* Galileo adds that it has been established from experience that even in the Mediterranean a voyage from east to west takes a quarter less time than one from west to east.[5] Furthermore, when all other winds are stilled, a wind from the east blows quite hard everywhere over a wide expanse of the oceans of the earth and over the tops of mountains. Not without cause does Galileo seize upon this wind as evidence of the earth's rotation, the wind being a necessary consequence of the supposed daily motion of the earth; for, this motion being from west to east, creatures carried simultaneously by the earth in its rapid twenty-four hour movement must face the air that opposes them and so feel a wind. Indeed, whether the air strikes the sails, or the sails strike the air, the result is the same, whichever way the wind is caused.[6]

FOL. 135

2. Our author wishes to take issue with Galileo. This wind [from east to west] is, he says, caused by the sun in the following way: 'The sun continually raises mists from places under its influence; and air, driven from its own position by the mists that have been raised, takes their place. These same vapours, left behind by the sun as it advances, and condensed, again drive away the air that replaces them, and for this reason the vapours sink at an angle and with great force. The air thus driven forward is the wind we are talking about that follows the sun continually, i.e. from east to west'.

FOL. 135v

In this account of the origin of the wind several things are impossible. First, if the wind were formed in this way, the vapours that were raised would form a single integral body, so that the air taking their place would have to move towards them in one great mass. Without a vacuum this cannot happen; for if, as the single atoms of a mist rise, the separate atoms of the air touching those of the mist will simultaneously fall, no more will a wind be created by the air falling in one region of the world than it will by the vapours rising in another. Second, even though all the mists left behind by the sun and

---

[4] Drake, pp. 440–441.
[5] *Ibid*. See fol. 169, note.
[6] Cf. fol. 137v, and with what follows compare *EW* I.88–89, there cited from 'a certain treatise' (unnamed, but clearly the *De mundo*) as an instance of 'a false cause'. See also *EW* I.469–470 and VII.100.

FOL. 136   condensed by the cold were simultaneously to sink as one unbroken body, they would not sink faster than they rose. In particular, they would be gradually condensed, as the sun moved away, at the same rate as that by which they would have been thinned as they rose; they would gradually fall nevertheless. From this we again see that the wind formed by a falling vapour raised in the east would not be greater than that formed by vapours in the west. Third, the wind we are discussing drives ships westwards at every hour, no matter when they are sailing: in the morning, at noon, in the evening, or in the middle of the night. So in the morning the sun succeeds the wind, not the wind the sun. Therefore this wind was not generated by the sun. Fourth, this wind also blows in the middle of the night, at a time when we must consider these vapours to have subsided, namely when they have not seen the sun for a period of time; the wind is therefore not due to the descent of vapours generated by the sun. Lastly, because the sun raises vapours from any given spot, the vapours will also fall at any spot (except to the west) in the path of the sun. From this it will follow that the wind thence generated is set in motion not only from east to west, but also from north to south and from south to north.

FOL. 136v

3. The opinion of Galileo, who considers that this wind is produced by the motion of the earth, White turns to refute on his page 137, but he does not come to grips with it; for the view he does refute is not Galileo's or anyone else's. The path of the daily motion [of the earth], to which Galileo ascribes the origin of this wind, is, according to him and Copernicus and those who maintain that the earth moves daily, from west to east. But our author's arguments are against an east-to-west motion–a motion no-one has dreamed of applying to the earth. He sets out his first argument in these words: 'Common motions are not felt; but if this wind has its origin in the motion of the earth, the motion of the air will be that which is called "common" and imperceptible to the senses. How, then, does it urge forward the waves?' According to Galileo and all the Copernicans, [I reply,] the earth is indeed moved from west to east. The air, however, moves neither from west to east nor from east to west but, as far as these [two] movements are concerned, it can do either, or else it is at rest; yet it can *appear* to be carried from east to west, especially because we are borne in the oppo-

FOL. 137

site direction. When, at a given time, the air is motionless but the earth is moved, this is not a 'common' movement but a movement of the earth alone. As a result of such a movement the same breeze is felt, whether the earth is at rest or whether the air is moved from west to east. Similarly, if someone bangs his head against a door or a door bangs against his head, the effect is the same.[7]

FOL. 137v

White brings forward another argument as follows: 'Again, either [a] the air is driven more strongly than the earth, or [b] the air is driven equally strongly, or [c] the air is not driven. If more strongly, how is this effected by something slower [than itself]? If equally, how does [the earth] drive it? If [the air] is not driven at all, those going in one direction will experience it equally with those going in the reverse direction'. But, [even] supposing a [rotary] motion of the earth from west to east, we do not understand the air as being moved towards, any more than being turned back from, the east. Galileo does *not* believe that the air is driven forward by the earth; he believes that the air drives the waves back and that we feel the air because of the earth's impact on the resisting air. Air when resisting forms a wind, [I say,] just as much as air when advancing does; clearly, therefore, our author's arguments dissuade us from finding a possible origin for the [east-west] wind (the cause of which we are investigating) in a [west-] eastward motion such as Copernicus and Galileo ascribe to the earth. We concede that the wind cannot originate in an [east-] west motion of the earth–but no-one assigns to the earth such a motion [anyway].

FOL. 138

4. It has been noticed very frequently, and I myself know from experience of a great calm at sea when no noise or breath of wind was discernible, that the waves have gradually swollen up into an amazing configuration; that soon, agitated by a light breeze, the sails have wavered to and fro, seemingly because of a wind to come; and that, warned by these signs of an imminent storm, the seamen have furled the sails and prepared to face a gale. That is, the men know by experience that the waves which swell up when there is no wind are the sign of a future tempest. They do not, however, trouble to learn what causes this effect, which seems to be due to the following: As often as the sea is

---

[7] There are many statements on relative motions in the *Two World-systems*, e.g. Drake, pp. 116, 163, 171.

FOL. 138v

moved but no wind is in evidence, a storm, though not in sight, hangs over the sea. The wind caused by the descent of a laden rain-cloud does not run along the surface of the water but, as a result of its perpendicular, or almost perpendicular, descent, presses down on it. This causes the waves to be far bigger than they would be if the wind did not exert a pressure in the direction of the sea-bed, but merely (as it moved) skimmed the sea. The first waves rapidly generate the second, the second the third, and so on across an extensive area, especially when a cloud follows them. Their rapid extension is due, not to one wave's being carried from its position, but to the motion generated by a continual pressure of wave upon wave. The reason why sailors notice the propagation of motion through the waves in this way more quickly than they notice a motion of the air is this: air or wind which, as we have said, sinks on to the ocean and presses towards the sea-bed is repulsed upwards by the surface of the sea and so is spread over a very wide expanse. As a result, its motion slackens; and then the air glides over the ship. It is not perceived, however, until a storm-bearing cloud

FOL. 139

hangs over very close.

Our author gives a different explanation of this phenomenon. He says it occurs because 'the sea is moved faster than the wind' (he states this in the margin of his page 141). But later, on page 142, he changes his wording, saying that the thing happens because 'motion is propagated more swiftly in water than in air'. Either opinion he advances without giving a reason. [a] For the first, it is clear that even a laden ship is carried faster than the water on the actual surface of the sea because, as we sail, objects floating on the surface are left behind; the wind driving a ship, however, will move far more swiftly than the ship itself. I wonder, therefore, on what grounds he will say that the surface-water moves more swiftly than the wind, or even than the ship. When he declares he remembers waves at sea beating against the stern of a ship, either his senses have deceived him or else the ship must have been motionless whilst the sails were being re-set. [b] His second

FOL. 139v

point is that motion is more swiftly propagated by water into adjacent water than by air into adjacent air. Here he is proved wrong by the fact (among others) that, if it were so, even sonic propagation, which takes place by means of motion transmitted through a medium, would be swifter in water than in air. Let us not, however, pass over the reason he gives to prove his point: 'The viscosity', he says, 'of a liquid morass also does this,[8] with the result that a part quite close to a [source of] motion is reluctant to leave behind a part further away, and in consequence drags it with it. The tenuousness of the air is more easily penetrable and eagerly invites the delivery of a blow by a fairly direct path'. This context is obscure, and similar to one noted somewhere above,[9] where he speaks of the arousing of 'secondary fancies'.[10]

---

[8] I.e. the fluid propagates motion.
[9] At fol. 124v.
[10] Possibly White's meaning is as follows: Imagine that we are looking down on a fluid and elastic medium. Point M is a source of motion (say a vertical rod) moved towards two points A and B: say that MAB is a straight line. Either, as M advances, A is driven towards B by the compression of the medium between M and A; or, as M advances, the fluid that moves to fill the void created by M comes in part from a direction forward of A, leaving a partial void between A and B. The fluid in advance of B urges B towards A. On motion in fluids see *EW* I.334.

FOL. 140v[1]

CHAPTER SIXTEEN

On Problem 3 of the Second Dialogue:

That the Sea is moved from

one coast to another

FOL. 141

1. What experiments are essential for an investigation into the reason for the ebb and flow of the sea; and what the ones we do possess are. 2. A description of the vertical circles travelled by the moon in the course which always corresponds in time to the ebb and flow of the sea; and the names of the same [circles] given them by sailors today, translated, as far as they can be, into Latin. 3. A point-by-point description and refutation of our author's views on the reason for tides. 4. A discussion and refutation of his opinion about the cause of the ebb and flow of the sea every six hours. 5. That it is probable that the seas are driven back from the coast of America. 6. A censure of his opinion about the reason for the increased tides at the equinoxes and, (7), at new- and full-moon. 8. A denial of his opinion about the reason why times of day precede times of tides.

FOL. 141v

1. In this problem and the one that follows it our author looks into the cause of the ebb and flow of the sea. Many have written on this, and have suggested different reasons for it; yet they have made it no clearer at all–indeed, I do not think it can be determined by the experiments so far performed and now in our possession. So the explanations sought for the tides of the sea are not to be broached before clear agreement is reached on the effects, i.e. at what hour tides have been re-

---

[1] Fol. 140 is blank.

170

corded at any given coast or inlet, and what their height has been. Our aim is not only to study the tides with respect to the position of the stars, but also to compare the tides one with another.

This has not yet been done, and there is in existence no account of the very great tides that occur on the opposite shores of the Atlantic, in the Indian Ocean, and in the Pacific, or of those occurring 'in the north ocean of Asia';[2] so I do not see by what means we can light upon a reason that explains them all. The characteristics of the tides on the coasts of Spain, France, Britain and Lower Germany, however, are but one: that the hours of the tides correspond absolutely with the times of the moon, with two exceptions: [a] At new- and full-moon the [rise of the] tide precedes [the phase of] the moon; [b] When the moon is in a northerly latitude the times [of moon and tides] again begin to vary[3] rather more quickly than when she is in a southerly.

When the tide is very high at London the moon is midway between south and west, or between north and east; and this is always and invariably true except, as I said above, at new- and full-moon, when the tide occurs more quickly than the moon moves to that position.[4] My words 'between south and west, or between north and east' are to be understood as meaning: 'If a tide occurs during the day, and the moon is between south and west, then, after midnight, the tide will occur at the time that the moon is between north and east'. What I have said concerning London has also been seen to apply to other seaports, namely that tides always occur when the moon is in the same verticals.

2. A tide, I say, corresponds with the movement of the moon in vertical circles, not with her movement in the meridians. A tide rises and falls not as the moon moves from meridian to meridian but as she moves from one vertical to another. These circles are called azimuths, and cut one another at the zenith and the nadir, i.e. at the poles of the horizon. Where [the circles] cut the horizon at right angles they indicate

---

[2] The north Pacific rather than the Arctic? (Translator's quotation-marks.)
[3] *Impleri*. Of dubious meaning.
[4] I.e. 'The occurrence of the tide precedes the arrival of the moon in a north-east or a south-west position.' Or the words 'that position' [*eum locum*] may refer, not to the position, but to London.

on it the areas where there are winds. Sailors normally reckon these areas as thirty-two in number and determine them thus: Suppose the horizon ABCD to be divided into thirty-two equal parts, and that the meridian, which is always one of the verticals, be AC. Let the diameter DB be always at right angles to AC. Now let A be the area of the north wind. The other areas are marked out in order, as on the accompanying diagram.[5]

In the sketch the circle ABCD is understood to represent the horizon, and the point E to be elevated perpendicularly from its plane in order to signify, not the centre, but the pole of the horizon. Let the thirty-two radii here drawn be understood not as radii but as quadrants drawn on the spherical surface from the pole E to the circumference of the horizon. Thus you will understand you have drawn thirty-two azimuths, or sixteen verticals, each vertical halved.

Now suppose the moon is in any one of these verticals, say in that drawn from the north-east to the south-west; and suppose a tide occurs at a given and definite place, say London. Then there will be a tide at London every time the moon is in the vertical passing through the north-east and the south-west, except, as I have said, at new- and full-moon. The tide at London will occur when the moon is on the vertical passing through the north-east to the north, and through the south-west to the south, which is, they say, to precede the moon by one point of the nautical compass.

The foregoing is, I assert, accepted as established by popular opinion and as proved by experiment. Hence I will believe that someone has explained the reason for the ebb and flow of the sea only if he has also explained the phenomenon under discussion. And since not even Galileo has considered this correspondency between the tide and the moon, there is no reason why one should agree with what Galileo has written on the subject [of tides]. He thought, of course, that the moon has some effect in general and brings about the unequal heights of the tides; but this has no bearing on the relationship of the tides with the azimuths of the moon. I have little hope, however, that the cause of such an ebb and flow can be discovered, and even less when I see that, although these are admittedly due to the moon, it is a very tricky matter to explain her mode of working, since hers is a uniform daily motion not from azimuth to azimuth, but from meridian to meridian.

[5] Figure 24.

CHAPTER XVI: 3

3. So I do not propose to give an explanation of the tides of the sea; I plan to set out briefly our author's reasons and to examine them. The essence of his theory is this: When the sun is over the Pacific Ocean, the wind, he says (which he imagines to be continually stirred up by the sun), drives the waters of the sea westwards, and pushes back the waters of the Indian Ocean that resist them. The resistance of the Indian Ocean brings about a swelling and a to-and-fro motion of waters (chiefly beneath the daily arc of the sun), because the wind generated by the sun should be strongest where the sun is directly overhead. This movement, which commences in mid-ocean, produces a rising of the waters on either coast [of the Pacific] until, when the Indian Ocean moves back and the tide falls, the motion reverts to mid-ocean from either shore. This is continued over the whole sea, until the Pacific Ocean recovers the water it poured forth. Such is the reason he gives for the tides on the shores of the Pacific and Indian Oceans.

FOL. 145v

FOL. 146

I have many things to say against this. First, imagine that, at a given moment, there were no ebb and flow of tides, but that later, when the sun was over the Pacific (incidentally, the assumption is absurd), an ebb-and-flow commenced, and that this generated a motion of the waters as far as the Indian Ocean. It would not follow that the latter must be raised so high as to produce a noticeable swelling. Water that remains at rest, and with no perceptible swelling, yields the instant it feels the lightest of impulses from the first water to approach it. This is because if the descent of a sudden and great wind creates a recognisable swelling, this quickly subsides owing to gravity.[6] Consequently, as gravity and the wind act against one another, the surface of the sea is merely rippled into small waves, and no tumescence takes place.

FOL. 146v

Let us also consider–if our existing tides are caused by a moderate wind said to be aroused by the sun–how great should be the tides raised by those powerful winds that rise unexpectedly from time to time and brood over the sea for two or three days without cease. Again, if a wind stronger than that brought about by the sun were to stream across the Pacific Ocean, the consequence would be either that no tides would result or that tides would extend in the contrary direction [to

---

[6] Or, 'owing to its own weight'. Cf. fol. 150v.

that from which the wind came]; for a wind may be presumed to set waters in motion not according to its distance from the sun but according to how hard it blows. But what if the wind arose when the sun was over the Atlantic? The water it propelled would beat against the shores of America, and not against other water.[7] Should one have to wait until [the water] reached the Pacific before one conceded that sea-tides had 'begun'? We must also ask: What wind does the sun raise (when borne across the tract of dry land which stretches uninterruptedly from farthest Asia to the Atlantic) at that time? How and where are seas raised?

But since he holds that the initial upsurge of the waters takes place midway between the two shores of the Pacific, he should declare accordingly how wide he imagines the middle of this ocean to be, i.e. the extent of the wind generated there by the sun. For if the width of [the area of] wind is equal to the width of the Pacific, in no sense can we say, 'A sea is moved from coast to coast'. On the contrary, even if the waters start to dilate in the middle of an ocean and move to the shore on either side, not even in this sense can we say that 'the movement is propagated from coast to coast'; it is propagated from the coasts to the main.

We must not ignore, either, a fact proved by the most reliable trials: that a wind blows across the whole circumference of the earth from east to west, everywhere simultaneously. Therefore, [if we take this with what White says,] at every degree of longitude alike there should be a tide that is the same and uniform, i.e. no tide at all! If a tide is created by the wind, the following question may also be asked, since a tide is formed anywhere the sun is overhead: 'By what wind is a tide generated in the same place twelve hours afterwards, when the sun is now in the opposite hemisphere?'

Lastly, why he should say this tide of his returns to the Pacific Ocean while the [earth's] movement [as he believes] from east to west still continues, passes my comprehension—unless I thought that the seas make their way through hollows in the land-mass of America (for we know that America extends from north to south through a whole semicircle), or that as much water passes through the Strait of Magellan as is set into motion [in], and rolls forward out of, that vast ocean, the

---

[7] Presumably the waters of oceans other than the Atlantic.

CHAPTER XVI: 5

Pacific.[8] We see, therefore, of how little account are his general thoughts on the swelling of the sea.

4. In the next place he lays down in these words a reason why the tides occur every six hours. 'Ereunius: "Only be sure you want the sun, in a given sign of its course, to have some effect. For how long previously has it been acting in that direction?" Andabata: "The sun completes its whole orbit within 24 hours. Therefore it must strive for six hours to reach a given spot (according to the law we make it subject to) and to move away from the spot for six hours. So at a given place and during a six-hour period the sun makes the wind stronger, and for [another] six hours, weaker."' Ereunius is asking whether it is at a given point on the sun's path that the sea must dilate the most, and for how long previously the sun should act on the sea and cause the waters to commence swelling. [To the latter question] Andabata replies: 'Six hours. For the sun reaches that place within six hours as it carries forward the tides, and takes six hours to quit the same place as it carries them back'.[9] But, [I ask,] when the sun moves away from any spot, is it not more and more remote, until it is a half-circle distant? That is, it moves away within twelve hours; but does it not then, for a further twelve hours without pause, draw closer to the same spot? What reasoning is this of his? 'If the sun is moved round the earth in twenty-four hours, then it must always approach a given spot within six hours exactly, and for the other six it must always move away.' I believe he thought of the hemisphere of the earth as being a mountain of water, and that the place in which a dilatation of the tide was to occur appeared as the top of this mountain; but when he saw that the sun had to ascend this mountain for six hours and descend it for another six, then, I think, he accepted this as meaning that the sun approached for six hours and moved away for six; this he seized on as the reason for the tides' changing every six hours.

FOL. 148v

FOL. 149

FOL. 149v

5. Because it suits his thesis to do so, he denies that, after

---

[8] Alternatively: '...unless I believed that...as much water passes through the Straits of Magellan as is set in motion and rolls forward from that vast ocean, the Pacific'. Two absurd propositions, the absurdity of the second being accentuated by the contrast between the narrow straits and the vast ocean.
[9] In Andabata's answer Hobbes is paraphrasing White's words.

some shores have been covered by a westward motion of the seas, the waters move away again towards the east; yet he offers no argument to the contrary. He does, however, suggest *some* reason to explain how this can happen in a northern sea but not in the Indian Ocean and the Pacific, namely that a northern sea is nothing else but a bay of the Indian Ocean; but why the same thing cannot happen in the Pacific or the Indian Ocean he does not make clear. I should like to know where, as the wind caused by the sun drives the Pacific Ocean to the shores of China and the Indies, the waters raised in those areas go; for if they do not move backwards towards the east, their path must be divided and they must be directed partly to the north and partly to the south. Hence they must move partly through the Indian Ocean and partly through the Arctic Ocean until they meet in the Atlantic, continuing their flow westwards and reaching the shores of America. If they do this, where (I again ask) do the waters raised near the shores of America go? Do they all make for the Pacific by way of the Strait of Magellan? How great then, will be the force of the flood passing westward through this strait![10] Experience suggests the contrary, however, and so they must return towards the east. Again, why tides cover sea-coasts twice a day has never been more feasibly explained than as follows: the waters are poured back from the American coast towards the east by their own weight.[11]

6. Why are the tides that occur at about the times of the equinoxes greater than those at other times? Our author thinks he has supplied an explanation that fits his own views closely enough when he says that [at the equinoxes] the sun is in its greater circle,[12] i.e. that it is borne faster than at other times. He forgets, however, what he has assumed earlier: namely that the *wind* creates tides and that a wind is started by vapours set in motion and is not due to the speed of the sun. Yet everyone knows that vapours are raised by heat, but that they warm fast-moving things less than they warm slow-moving things. From this it should follow that, the greater the circle in which the solar fire is borne, the lesser the wind, and in consequence the smaller the tides.

---

[10] Perhaps correctly, the manuscript punctuates as a question.
[11] Cf. fol. 146v, note.
[12] See fol. 152, note.

7. He declares, however, that the reason why tides are greater at new- and full-moon than at intervening times is that at full-moon the moon's light is greater. But what *is* 'burning-light'? Further, does it raise vapours to the limit of its power?[13] If it does, isn't the amount very small? And what are we to make of the same effect at new-moon? 'The light of the moon', he says, 'is brighter at the conjunction, when the whole [moon] is nearer the sun [than at other times and] by the diameter of her sphere.'[14] So, [he continues,] the light which the moon receives from the sun and which is directed away from the earth when the moon's opaque body intervenes [between sun and earth] may have some effect on the sea.[15] Or if, [I ask,] this light had such an effect, would not the latter necessarily be less at full-moon than at any other time whatsoever? For the moon is always equally illuminated, and is nearer the sun at every period than she is at full-moon.

FOL. 151v

8. He says nothing on the connection between the coming of tides and the motion of the moon in vertical circles. That the tides on the morrow of a given day should, in due order, follow closely on those of the day before, as does the position of the moon, he explains at follows:[16] 'Lastly', he says, 'consider this: the presence of low tides on every single day can be due to one cause only: that the motion of the sun is swifter than the motion of the sea'. I do not know what he here intends 'the motion of the sun' to mean. If he wishes it to mean 'the sun's body's own [rotary] motion' we shall certainly grant him that this motion is swifter than that of the seas in the ratio by which the radius of the [earth's] greater orbit is larger than the radius

FOL. 152

---

[13] The Keplerian 'burning-light' is not in *DM*, which has: 'The smallest part (of the moon) in no way performs to its full ability' at the intervening times. But Hobbes may mean merely that the moon raises vapours in her own way or according to her ability. Perhaps: '...greater. But [he says, this] light is of something on fire, and therefore it raises vapours to the limit' etc.

[14] 'When' could properly be 'since', though the sense would seem to require the rendering given here. Alternatively, 'whole' could precede 'diameter'.

[15] This, which should perhaps be phrased as a question, is not quite what White says, which is: 'During the moon's hour-angle or (perhaps) horary circle, the moon's effect is weakened. We see this effect daily'.

[16] The construction is cramped. Apparently Hobbes means: 'The positions of tides on one day are to those of the day before as, at a given hour, is the position of the moon on one day to the same on the previous day'.

of the earth.¹⁷ Astronomers believe this to be over 1000:1. Again, I do not think he takes 'the movement of the sun' as meaning 'variations in the hours of the tides', for he considers that not the sun, but the earth, is borne in a daily motion. Perhaps he understands the phrase to mean 'the daily motion of the earth' from west to east (indeed, he must intend it so); and perhaps he is saying that this motion is swifter than the movement of the seas from east to west. If so, this difference in velocities cannot be considered a reason,¹⁸ for it constitutes in fact the *effect* which we are investigating. When people ask: 'Why does the hour of the day precede the hour of the tide?' they are anxious to know why, though the earth rotates within twenty-four hours, the seas' circular passage takes twenty-four hours plus about two-thirds of an hour. So either his opinions are meaningless, or else he has indicated a given effect instead of its cause.

FOL. 152v

---

[17] Perhaps we are to understand the word 'orbit' at the end, reading 'than the radius of the earth's orbit'. Can the expression 'the radius of the greater orbit' refer to the difference in the radial vectors when the sun is said to appear to describe an orbit eccentric to earth, or at any rate not circular?

[18] I.e. a reason for the fact stated in the second sentence of this article.

CHAPTER SEVENTEEN

On Problem 4 of the Second Dialogue:
That a sea-tide is not due to the
motion of the earth

1. Galileo's opinion about the cause of the to-and-fro motion of the tides at sea. Demonstrations: (2) How any part of the sea is moved irregularly;[1] (3) Why this irregularity is greater at new-moon and at full-moon than at other times; and (4) Why, on the other hand, it is greater at the equinoxes than at any other time. 5. [Hobbes's claim] that Galileo's theory has not been adequately proved. 6–16. Refutations of our author's arguments against Galileo.

1. In the Fourth Dialogue of his *World-systems* Galileo ascribed the origin of the sea-tides to a combination of the annual and the daily movements of the earth.[2] In that our author has set forth his view in the previous Problem, he has considered that in the present one he must refute its contrary–needlessly indeed, if he has proved his own opinion valid, as he claims to do; for the same reasoning that shows the soundness of any proposition also shows at the same time the falsity of its contradiction. Let us see, therefore, whether White's refutation of Galileo's case is sound.

So that even the slower-witted can the better comprehend any kind of tidal rise and fall, and its cause, Galileo presents as it were before his readers' eyes an open barge filled with water that flows freely, such as the barges in which [the people] at Venice convey fresh water from the surrounding district into the town.[3] He shows that in such craft, at every stroke of the

---

[1] See fol. 154, note.   [2] Drake, p. 446.   [3] Drake, pp. 425, 457.

oar or jerk of the boat, i.e. at every change in speed, the water rises at the bow and at the stern alternately, but that amidships it merely moves to and fro without any perceptible rise or fall. Next, it is supposed that any given part of the earth rotates daily about the centre of the earth itself. To prove, therefore, that there must be risings and falls in the tides on seacoasts similar [to those of the water in the barge] we need but a single argument: It must be shown that any part of the earth has an unequal[4] motion, i.e. one that is sometimes rather fast, sometimes slower, and that this irregularity is greater at those times when the ebbings and flowings of the sea are usually greater.

FOL. 154v

2. In the first place, therefore, Galileo demonstrates that the true motion of any part of the earth, i.e. the motion compounded of its annual and its daily motion, is irregular;[5] second, that its irregularity is greater at new- and full-moon than at other times; and third, that, on the other hand, its irregularity is greater when the sun is over the equator than at other times. The first of these points will be clear from the following: Let the circle ABCD be the daily motion of the earth; on this circle let any part of the earth, A, be marked. Now, A is carried round, in its daily circle, 365 times a year; if, therefore, EF is the 365th part of the annual circle, it is about three times the whole circumference ABCD (for the radius of the earth to the radius of a great circle, i.e. the radius of the daily to that of the annual circle, is about 1 to 1095; but 365 to 1095 is also 1:3; so the daily circle is to the 365th part of the annual circle as 1:3).[6] Next, after the quadrant AB has been

FOL. 155

---

[4] Henceforward translated as 'irregular'. See fols. 283v–284 on the production of one motion compounded of two different ones.
[5] Drake, p. 446.
[6] An incomplete sketch (on fol. 155v) is shown in Figure 25, together with an explanatory figure not found in the original but comparable with that in Drake, p. 426. Here Hobbes is saying that

$$\frac{\text{radius earth}}{\text{radius earth's orbit}} \sim \frac{1}{1000}$$

$$\frac{\text{circle of earth}}{\text{circle of earth's annual orbit}} \sim \frac{1}{1000}$$

$$\frac{\text{circle of earth}}{\text{length of earth's daily arc}} \sim \frac{1}{1000/365}$$

$$\text{hence } \frac{\text{the circle ABCD}}{\text{the arc AF}} \sim \frac{1}{3}$$

divided into any number of parts, Aa, ab, bB, and the remaining quadrants are likewise divided, let EF (the arc of the annual motion described in one day) be divided into as many parts as the whole circle ABCD has been, i.e. Ec, cd, de, etc. Hence when point A moves forward to point a in its daily motion, and at the same time the centre E moves forward to c in its annual motion, the radius EA will be in cg; so point A, which has advanced by the length Aa, will be at α. Likewise when the radius cg has moved forward to dh, then point A, which has advanced by the length Ab, will be at β. In the same way point A will be at γ on the third occasion, at δ on the fourth, at ε on the fifth, at ζ on the sixth, at η on the seventh, at θ on the eighth and, at last, at ξ on the twelfth (the daily circle is divided into exactly the same number of parts as the annual arc is). If now the line $\alpha\beta\gamma\delta\varepsilon\zeta\eta$ is drawn, its longest part will be Aα, the next longest αb, and then bγ; but $\zeta\eta$ and $\eta\theta$[7] will be the shortest parts and equal to one another, but they will again increase as far as ξ at the same rate [*gradus*] at which they decreased. So part of the earth A will be swifter between A and α than between ζ and η. It is clear, therefore, that the motion of any part of the earth–a motion compounded of the annual and the daily motion–is irregular, i.e. it is sometimes quite fast, sometimes quite slow, yet on any given day it is very swift on two occasions, and very slow on two.[8]

FOL. 155v

FOL. 156

3. The second proposition, i.e. that the irregularity of this motion is greater at new- and full-moon than at other times,[9] we demonstrate thus: If the sun is at A and the earth is borne in its annual path through the circle BC, but the moon is carried round the earth in the little circle DFEG, a new-moon will occur when the moon is at E, and a full-moon when she is at D.[10] In the middle period [between new- and full-moon] the moon will be between D and E, say at F and G. Now both the earth and the moon are held to be carried round by the force of the sun (at A), so the force of the sun will not be hindered from carrying the earth circularly and at a uniform rate when the moon is at the side [of the earth]–it does not matter whether at G or at F–because the sun acts on each [i.e. earth and moon]

FOL. 156v

---

[7] So the text.
[8] Or, '...it is at its fastest twice a day'.
[9] Drake, p. 453.
[10] Figure 26.

in the same manner and from the same distance. Say, however, [i] as happens at new-moon, the moon is at E, directly between the sun and the earth, or [ii] [that she is] at D, so that the earth is directly between the sun and the moon, as at full-moon. The sun's force is [now] so weakened that it cannot move the earth as quickly as before, for the moon when at E takes away part of the sun's moving-power [actio], and when at D she increases the task of [the sun's] moving the earth circularly. So the irregularity in the earth's annual motion is in evidence both at full-moon and at new-moon except for the irregularity of the motion due to the compounding of annual and daily motion. In other words, at a time when the sea's tides are increased, the irregularity in the motion of any part of the earth is also greater.

FOL. 157

4. The third proposition, that the irregularity of [this] same motion is greater when the sun is at the equator than when outside, he [Galileo] proves thus: Let the sun be at A, and let BC represent the circle of the great sphere in which the earth is borne circularly.[11] Let B be the centre of the earth at the [celestial] equator and C the centre of the earth at the solstice. Now when the earth is at the equator let DG be the terrestrial equator that will cut the ecliptic BC in a common section BK, which is the same straight line as AB. The elevation of the equator above the ecliptic is $23\frac{1}{2}°$, which may be represented by the arc LD, whose sine is ED. It is obvious, then, that point D is directly over the plane of the ecliptic, i.e. over the plane of the paper, perpendicularly at E, and that point F, representing the same inclination, is perpendicularly over the same plane

FOL. 157v

at M. So when any part of the earth is carried in its daily motion through DKF, that motion matches the annual motion through EM. Hence [the daily motion] increases the speed of the annual motion; the daily motion through FND, on the other hand, is opposite to the annual, and so reduces the latter's speed. The amount of the increase or reduction is EM. Next, when the earth is at the solstice, the common section of the terrestrial equator and of the ecliptic is the tangent HI, so that

FOL. 158

the points H and I are in the same plane of the ecliptic. When therefore, any part of the earth is carried through IOH, its daily motion will match its annual and will increase the latter's

[11] Figure 27. Drake, pp. 458ff.

speed; [if it is carried] through HPI, on the other hand, it will be opposite to the annual motion and will diminish the latter's speed: the amount of the increase or decrease is the line HI. But HI is greater than EM. A greater amount of the daily motion must therefore be added to, or subtracted from, the annual motion when the earth is in the solstices than [the amount applicable] when it is on the equator. So far, then, Galileo's demonstration correctly goes on to prove that the irregularity in the motion of parts of the earth is not the same at the solstices as it is at the equinoxes. It does not follow, however, that [the irregularity] is greater at the equinoxes than at the solstices; but the contrary is true:

5. When the earth is at the solstitial point C, the space HI is crossed twice by the earth's motion at one part of the day: once by the annual and once by the daily motion; at the other part of the day the space HI is crossed once, minus the same HI, i.e. no space at all is crossed. When the earth is at the equator, however, at one part of the day [the line] LG is crossed once and EM once; but at the other part, (LG−EM) is crossed. So the irregularity between 2HI and (HI−HI) is greater than the irregularity between (LG+EM) and (LG−EM). The irregularity of the earth's motion is therefore greater at the solstices than at the equinoxes.

FOL. 158v

Further, in that comparison [Galileo's] between a barge and the earth as water-carriers I find a certain deficiency, and because of this the effects of the to-and-fro motions of the waters ought not to resemble one another.[12] [I say that] the irregularity of the motion of the waters in the barge occurs because the barge is indeed halted or driven forwards, but the water [in it] is neither brought to rest nor driven forward together with [the barge]. With the irregularity of the motion of any part whatsoever of the earth[13] this does not happen, because the same force that moves the earth forward, or halts it, also drives and halts the sea together [with the earth], by acting upon the water no less than upon the earth; but an oar, or any obstacle [struck], acts on the barge only, not on the water [within it]. Such is one point[14] in Galileo's theory that I find less accurately worked out.

FOL. 159

---

[12] 'Waters'/'water' (159): so the text.
[13] Or, 'of any motion of a part of the earth'.
[14] 'Is one point': plural in the text.

6. We shall now look at the questions (in the same theory) that our author censures. On his page 162 he considers that a sea-tide is more correctly compared to [the motion of] water whose level rises in any [unmoved] flask when we pour in more water than to an irregular motion [we may impart to] the flask [even] if we pour no [more] water in. This is most patently false. Where would the waters to be distributed over the whole sea come from? A little earlier he said that a wind [generated] by the sun rests upon the ocean in order to raise tides; surely such a wind has not now been transformed into water? Perhaps he will say: 'It is not the tides of the Pacific or the Indian Ocean that are caused by a pouring-on of more water, but those of the Mediterranean, of France and of Britain'. But the question discussed here is of a tide, not in [separate] parts of the seas, but in the sea as a whole. Galileo does not deny that the waters which happen to enter an inlet can sometimes cause a tide; but [he declares] that in those tracts of sea that stretch between east and west a tide is always due to the irregularity of the earth's motion. So Galileo will not deny that the Mediterranean Sea close in to the African coast is enlarged because it meets the waters from the Atlantic and from the Black Sea. The rise and fall of water at Venice, however, he will think are to be ascribed not, as our author ascribes them, to the water's being turned back and made to flow in a closed circle, but to the situation [of Venice], because the Adriatic, which stretches for 1000 miles from east to west,[15] is so affected by the earth's motion that it is acted on in the same way as the water in the barge is.

7. Next, he takes from Galileo the example of the barge and attacks it. He claims that the motion of the water contained in the barge resembles not the motion of the sea-water contained in a hollow of its own, but rather the motion that water would assume in a basin whirled round its own centre [in a horizontal plane]. This is also false. The earth's revolution causes the shores lying on the western edge of any sea to propel the waters lying in front of them, and the coasts' west-east motion is transmitted to these seas. Hence if the earth collided [with something] but

---

[15] An extraordinary statement, but Hobbes is quite unequivocal: '*Mare Adriaticum...extenditur per 1000 milliaria ab oriente in occidentem*'. Does he mean the Black Sea?

the water did not, the latter would be swirled towards that shore which faces [the motion]. But, in a basin turned in the way he supposes, nothing could communicate the movement of the basin to the water except the bottom of the bowl. Yet the only water the bottom can draw along with it is the portion clinging to it. So then, if the basin collided [with something], the water being carried, [itself] at rest, would not be put into motion through that impact.[16]

8. Third, having granted the earth to be carried in regular motion, i.e. at a constant velocity, and then suddenly to be brought to rest, our author denies that the result will be a forward rush of the seas. He argues as follows: 'Imagine', he says, 'that the earth is *not* carried at an irregular rate [*inaequaliter*]; what impetus will the water have? Will part press closely against part, or will a part change its position by means of an imperceptible fall? Not the first, etc.[17] We must not therefore suppose that the water acquires any impetus from the motion of the earth. But if [the water] does not acquire impetus, it will not experience any change due to the resistance brought to bear on the earth itself [which would retard or stop the earth]'.

FOL. 161

If the sphere (one body, including the earth and the water) were indeed moved uniformly on its centre, it is correct [to say] that one part does not press against another, because things moved in uniform motion do this no more than do things merely in contact. There is no doubt, however, that of things borne uniformly each individually changes its place, for motion exists only as change of place. But look at the argument he uses to prove that the water, even were it moved together with the earth, does not change its position! 'Suppose', he says, 'the drift of the argument is that all water will change its place in respect of its parts relative to those of the earth. Then it is unbelievable [that the water will not change position]. Again, if water changes its place relative to the whole universe

FOL. 161v

---

[16] If the basin is being whirled round, are we to understand that the impact hinders, or that it accelerates, this circular movement? Or that the collision has some other effect? Hobbes's words 'yet the only water...' may refer, not to the water's clinging to the bottom of the basin and being drawn along with it, but to the total water-content of the basin.

[17] Or, 'Not before, etc.'. The 'etc.' is Hobbes's. Here follows in White the passage quoted on fol. 161v below.

as well as to the earth, this is *denominatio* rather than motion;[18] for motion, unless it occurs wholly [*specialiter*] in that body that is said to be moved, has no effects of its own.'[19] These words, though very obscure, convey his meaning to me adequately enough, but with the help of his philosophical tenets that we know already. He believes the following: Change of place is the same as change either of position or of the relationship between [a body's place] and the place of a body [situated] elsewhere.[20]

FOL. 162

Now when, [he says,] the earth is carried round at a regular speed, any one part of the water rests always on the same part of the earth and always, [as it were,] looks down over the same part from the viewpoint. No part whatever of the water, therefore, changes its place with respect to the earth. One could object here that the parts both of the water and of the earth change their position or situation. That is, they change their place with reference to the sun and stars and to all bodies in general save those attached to the earth; therefore parts of the earth are moved as well as parts of the water. He answers [by saying] that such a change of place is not motion, but is only *denominatio*; he adds as his reason that such motion is not the motion which is restricted to [*specialiter in*] a moved body. So, then, [he goes on,] the waters are not moved by the earth's rotation sufficiently to change their place, or, indeed, to take on the impetus [imparted] by the motion of the earth. And if

FOL. 162v

they do not take on impetus, then there will be no forward movement [of the waters] if the earth pauses.[21]

Who does not see here, first, that any droplet of water does change its position or site relative to the sun and the stars, and

---

[18] White seems to mean: 'This is an argument about terms. Motion is meaningful only in terms of the relationship between the mobile and some single point of reference in successive instants of time. If an activity relates to two or more points of reference simultaneously, we cannot call it motion'.

[19] The words cited occur in the passage previously quoted at fol. 161. Hobbes omits White's opening phrase: 'Not the first, because all the parts are carried uniformly, but the second'.

[20] I.e. given two bodies, A and B. 'Change of place' seems to be, according to White, either a movement by one of these bodies, independently, to another site; or the altering of the distance between A and B. Alternatively, the text may be in error in retaining a diacritic, and we are to understand: '...or of the relationship to the place of another body'.

[21] Perhaps, 'Neither will the earth move forward, checking itself the while'. If, however, the translation in the text is correct, one understands at the end 'i.e. if the motion is now fast, now slow'.

CHAPTER XVII: 9

that, according to White, true motion is thus created wherever you will? [Again,] who does not see that our author supposes there to be present in the earth itself, or in the sphere comprising the sea and the dry land, the motion by which that [the earth's] position is changed? Hence who fails to grasp that that motion actually exists within a body said to move? Next, who can deny that that motion to be found in the parts both of the earth and of the water is the same thing as impetus? For what is 'impetus' if not another term meaning the same as 'motion'? Lastly, who does not see the following? When parts of the earth and of the water move forward all together and meet something which resists the former but not the latter, [then] the parts of the earth that are torn away from the parts of the water will be left behind [as the water moves on].[22]

FOL. 163

An absolutely absurd belief which, at the start of Dialogue II, our author takes to be an accurate one is that the nature of motion is relative and consists in a change of aspect–or, as he puts it–of [a body's] respect to other bodies. A 'relative' is not that which we can visualise in one thing alone (for in every relationship we have to consider *two* correlated things); yet it is possible to imagine that the world itself moves,[23] and, according to our author, the world is *one* thing. So the nature of motion is not relative.

9. In the fourth place, even if we grant that both the water and the earth are moved simultaneously and have impetus, he still denies that that irregularity arising from the compounding of the daily with the annual motion can produce that effect which impetus towards a [solid] obstacle would produce. Moreover, this applies, [he says,] because it proves that, [as seen] on bowling-greens, the little wooden balls which roll forward on the surface of the earth spin at an even rate [*aequabiliter*], even though the earth is moved, as he supposes, at a varying speed.[24] The comparison fails, however, in that the balls

FOL. 163v

---

[22] To divide a single sentence of the original between paragraphs may seem dubious practice; but clearly the words 'Who does not see?' (used elsewhere by Hobbes) make clear that what follows them is Hobbes's own opinion and not an 'absurd belief' that motion is relative.

[23] That we have come to see that White is a heliocentricist does not absolve Hobbes from writing ambiguously here; his Latin can equally mean: 'Yet possibly he (White) imagines that the world moves'. A similar instance occurs at fol. 352v, note 45.

[24] Presumably the balls are without bias. Perhaps one should translate: 'skittle-alleys' or 'marble-playing'.

are thrown forward by the hand in such a way that, through this motion, one by one they leave behind them the parts of the earth that lie beneath them. Water moved by the earth as a whole, on the other hand, always accompanies the same part of the earth, except inasfar as the above mentioned irregularity [of motion] prevents it. The comparison is also at fault in that the ball thrown forward by hand is hard and compact, so that it cannot be raised by impact were the earth to collide [with something],[25] but can only acquire or lose a degree of velocity.

FOL. 164

Whether it does so or not our senses cannot tell us. Lastly, tides—even those of the sea itself—cannot be perceived over so small a distance as that across which the ball is cast.

10. Fifth, he compares the irregularity of sub-equatorial motion, and of the motion to the circles [of latitude] near the pole, with the irregularity of motion near different parts of the same equator. 'That the polar areas are moved more slowly than those situated close to the equator', he says, 'is not considered an impediment to the motion of the [terrestrial] sphere. Similarly the variation in [the rate of] spin constitutes no obstacle [to the earth's motion].' In this comparison there is nothing relevant to the proposition, for a part of the earth that lies under the Arctic circle is not the same as one lying close to

FOL. 164v

the equator. So in this case there is no irregularity in the motion of the same part of the earth, or of the same soil, but from the compounding of the earth's annual and daily motion there arises an inequality in the motion of the same soil, of the same stone, or of some part—any one you choose—of the earth, as was shown above in Article 2, relevant to the part A [on the diagram to folio 155]. I omit the fact that Galileo does not say, as is objected in this context, 'This irregularity has the nature of an obstacle'.[26] He supposes that an irregularity of this kind, deriving whether from an obstacle or from the compounding of motions, still has the same effect in raising the waters.

11. Sixth, if the rise and fall of the sea were brought about in the same way as that of the water in a barge moved at an irregu-

---

[25] 'Raised by impact' (*elevari per impactum*) seems to mean either 'thrown upward from the surface of the ground' or 'bulge outwards', as do the seas to different degrees at different times, as White argues.

[26] I.e. White alleges that Galileo sees the varying rate of motion as equivalent to [?impact with] a material object, *obex*.

lar pace, then, he says, it would follow that '(a) the sea would roll forward both at its shores and on the high seas, and (b) that first the shores and then the deep would take on [*concipere*] a slowness. So a tide will not be produced by [its] approaching the shores from the deep, and then abandoning them, and vice-versa'. I do not understand these words. I know well enough what the water in the barge does. Its flow occurs both near the sides [of the barge] and amidships, depending on the length of the barge, without a noticeable rise or fall; but at the bow and stern there occurs a rising and sinking of the water, with no perceptible forward motion. Galileo says that the same thing happens across a whole ocean because, owing to its great extent, the latter reaches from east to west. If it follows, as he [Galileo] infers, that tides are *not* brought about by a motion from the open sea towards the shores, and from the shores towards the deep, then we are very likely to arrive at the truth; for though in the whole preceding Problem our author has striven to prove a tide is created in this way, he has not succeeded in doing so.

FOL. 165

FOL. 165v

12. Seventh, having drawn a diagram DEFG to represent the earth moved along EG, the ecliptic of the great sphere,[27] he argues against Galileo thus:[28] 'By the amount, therefore, by which the sea is forsaken by the land at [*ex*] one part [of the earth], i.e. at F, the sea from the remaining part, at D, will sink back at G.[29] So a tide will occur at [*ab*] both points near G, and these tides will be simultaneous.[30] Tides will occur face to face with one another, therefore. And those who are sailing at E and G will see that kind of frothing-together of the seas'. To understand this better, we must imagine that a part of the earth at D is moved towards E and thus proceeds through E, F and G till it returns to [*in*] D.[31] Since its motion is slowest

FOL. 166

---

[27] Figure 28.

[28] In the notes that follow, a tentative explanation of a difficult passage is offered, the text being given as literally as possible. Moreover, there is doubt about the prepositions used by White and by Hobbes: except where shown, *versus* in White seems to require 'at'; in Hobbes, 'towards' (for 'at' Hobbes's word is *in*).

[29] For '...at one part...from the remaining part...' White may intend: '...on the one hand...on the other...', but this does not greatly affect the sense.

[30] Presumably, at D and F the motion of the earth exceeds that of the superincumbent water, which then swirls towards G from both directions.

[31] By reason of the earth's course along a great circle of which EG is an arc–ignoring the earth's rotation?

when it is at F,³² the water [then] resting on it [that part of the earth] will swirl forward, and, towards G, will leave it behind. [But] since a part of the earth which at that time [while what precedes is occurring] is at D is carried more swiftly than before,³³ the water [resting on this part] will fight its way back towards the same G–not such that it rushes back from D towards G, but [rather] that it moves less from D towards E.³⁴ In this way a swelling-up of the waters will occur at G. There will not, however, be two opposite tides and a frothing-together of the seas, as he imagines. No-one will infer from his diagram that there will be, except someone who forgets the path by which the earth moves forward in its daily circle. The same person will also think that, because of the slow movement at F, the water situated there divides, part being distributed towards G and part towards E; also that the water at D pushes forward towards G and towards E at the same time.

FOL. 166v

13. Eighth, [our author] rejects the comparison between tides and the water in the barge because of the amount of water [in the barge relative to that on the earth]. The motion of a whole sea, he thinks–a motion created by an agitation, such as that of the small quantity of water³⁵ in the barge–will be so great that the tides caused by that reason would rise not only more than ten foot high but also above towers and mountains. Correct–if we suppose that in each instance there were not only a similarity between, but also a ratio and proportion in, the increase or the lessening of the speed.³⁶ But this is not so. The reason for that considerable to-and-fro motion (owing to the vessel) that we noticed in the [water in the] barge is the great and sudden difference in motion; yet the motion of the earth increases or diminishes little by little and by degrees, and in addition it unloads much more water into the Atlantic and into other great bays of the sea than

FOL. 167

---

³² The word 'its' seems to refer to the part of the earth under discussion, rather than to the water lying upon it. Why should this motion be slowest at F? One takes D and F to represent respectively north and south.
³³ I.e. than it was when it was at F?
³⁴ In this closing qualification Hobbes seems to have in mind relative motions: 'The water does move from D to G, but only in that this motion is slower than that from D to E'. Throughout, the motion takes place in the direction DEFG, not the converse, But the clauses here translated as clauses of result could be clauses of purpose.
³⁵ *Aquula*. Perhaps ripples are meant.
³⁶ Or does Hobbes mean that one instance (*ratio*) i.e. that of the barge, resembles the other, that of the seas at large?

it raises [as tides] on the shores. In a barge, from which water cannot escape because of the sides, this cannot happen.

14. Ninth, he asks: 'How is common motion imperceptible, if it has an effect such as this?' If he returns to [the question of] the barge, he will see that the barge's motion is in common with [that of] the water it carries, provided that [the water] is moved at a uniform speed. But the motion is not common when the barge is halted on colliding with land and the water [within it] flows forward; or when [at sea] the barge increases speed at a stroke of the scull, but the water [inside] does not move faster. Likewise the motion of any part of the earth and of the water that rests upon it would be common motion if the parts of the earth were carried at a uniform velocity; but on this occasion it is not common, because, owing to the irregularity of the movement, one part leaves another part behind.

FOL. 167v

15. Tenth, to [Galileo's explanation of the] monthly increments in the tides our author has, in the first place, only one objection: Given that the moon is at her greatest distance from the sun, as at full-moon, or very close, as at new-moon, in each instance there occur tides greater than the normal. He thinks this should not happen, but why it should not he does not say. Galileo, supposing–on the firmest grounds–that the motion of the earth is a result of a property [*virtus*] of the sun, shows adequately enough that that force is weakened as often as the sun, the moon and the earth are in the same straight line:[37] whether the earth be between the sun and moon, as at full-moon, or whether the moon be between the sun and the earth, as at new-moon, [is immaterial]. So in each case the force of the sun is weakened and the daily motion of the earth is further slowed down. Hence the irregularity of the [earth's] motion will be greater and, in accordance with these assumptions, the tides will be greater, each time.

FOL. 168

16. Lastly, [White] attacks Galileo's opinion about the reason for the increment in the tides at the time of the equinoxes; here, although he attempts to explain the [Galilean] demonstration of the compounding of the daily with the annual motions, he does not understand the conclusion, but infers one

---

[37] E.g. Drake, p. 326f.

FOL. 168v

totally different from Galileo's. In Article 4 above I have shown that the said demonstration is valid: the conclusion there [drawn] is that at the solstices the earth's daily motion is closest to its annual motion, i.e. by half a day, but is also furthest from it by, again, a half-day. At the equinoxes, however, it is least close and least distant. At last our author explains the matter thus: 'Ereunius: "So you see that the earth's daily course at the equinoxes is close to the ecliptic. At the solstices, however, [the earth] straightway veers in the opposite direction, drawing near [the ecliptic] by degrees that are lesser or greater according to whether it is steered closer or more distant to these points." Andabata: "Obviously". Ereunius: "So if these two movements are the result of the two causes, then the earth's

FOL. 169

motion at the equinoxes will necessarily be faster, and that at the solstices slower. Such was Galileo's opinion."' Whether indeed Galileo reasoned thus, let the reader decide from Article 4 of the present chapter or from Galileo himself, at his page 452 [of the 1635 Latin edition].[38] Here Galileo clearly declares and shows that the daily motion at the solstices agrees most closely with the annual in the one semicircle, but greatly differs from it in the other. At the equinoxes it also agrees, but to a lesser extent, in the one semicircle, but in the other it disagrees, again to a lesser extent. That is, at the solstices the additions and subtractions of the daily motion with respect to the annual are to the additions and subtractions at the equinoxes as is HI to EM in the diagram at Article 4 of the present chapter.

---

[38] Rather, pp. 451–452 of Galileo; Drake, pp. 458–459.

CHAPTER EIGHTEEN

❧❀❧

On Problem 5 of the Second Dialogue:

That in the earth there is

no resistance to motion

1. Our author's opinion on the cause of the earth's motion, both the annual and the daily. 2. The following is not made sufficiently clear: that the earth has no resistance to motion, not because its magnitude hinders this, but because the earth does not gravitate. 3. That the resistance of bodies at rest is not due to their being at rest. 4. For the earth to take on a motion that is definite and of determinable velocity it is not sufficient that the earth exert no resistance. That the motion of a wheel described by our author is neither (5) perpetual nor (6) applicable to the earth. That he ought to show the reason not only why the earth is moved but why its motion is [from west to] east, and why so swift. 7. That the earth's motion cannot be due to a violent wind in the way he supposes, i.e. to one caused by this same motion of the earth. 8. That an easterly wind cannot make the earth rotate in an easterly direction. 9. That [the rate of] such a wind, many degrees slower than [the earth's] daily motion, cannot be the cause of this latter. 10. That his explanations of the seas' pouring out of the Pacific into the Atlantic are inconsistent. 11. That, according to our author, the origin of [the earth's] daily movement antecedes the wind which he says causes it. 12. That the reason for the annual motion cannot be the flow of the seas from the Pacific into the Atlantic. 13. That the passing of the seas through any strait cannot be the reason why the earth is always carried beneath the

---

[1] Fol. 169v is blank. See fol. 202, note.

ecliptic. 14. A highly plausible reason for the annual motion, and (15) a likely explanation of the daily.

FOL. 171

1. In order to give a reason for the earth's motion, the annual as well as the daily, our author has supposed in Problem 1 of this Dialogue that in the torrid zone the sun raises a wind which follows the earth continuously from east to west and drives forward the waters of the sea, particularly the ones on which [the wind] immediately bears, i.e. those nearest the surface and furthest from the sea-bed. [(a), daily.] Now, those waters near the surface, he declares, are moved more quickly than those that are deeper; the former then move ahead, sink by their own weight, and drive the latter in the contrary direction, namely from west to east. He believes that as a result of the movement of waters resting upon it, the sea-bed, which is actually the earth, is moved together with them from west to east. Such, he says, is the explanation of the daily motion. [b] As regards the annual, he bids us recall a certain strait, the Strait of Majorca, which is so situated as to be near enough parallel with the circle of the zodiac (through which, according to the followers of Ptolemy, the sun travels, but the Copernicans say the earth does). Hence, because the water resting on the bed of this strait must flow in accordance with the topography of the place, the earth will travel on the same principle. This, he asserts, is the cause of the annual motion.[2] The impossibility of these explanations seems to me, at first glance, so obvious that I need not make it even more so by [offering] criticisms—had I not made up my mind to propose a theory of my own on the matters in question rather than to refute so erroneous a view.

FOL. 171v

2. First of all we must consider whether, or to what degree, it is true that, if at rest, the earth has no resistance to motion. Now, one understands that there are two ways in which an object is at rest. Either [a] the whole is indeed at rest, but its several parts are moved in the same way as those in a human being when in complete repose—yet [in this state] some movements of the parts may be understood, as of the blood, the spirits, and the internal organs on which life depends. Alternatively [b], the object resembles the things said to revolve about a centre of spirits (among which the earth is held to be included).

[2] An argument taken up again at fols 182, 183v.

Although the whole structure is not moved, i.e. it does not quit its place, the single parts are moved, and they describe circles. Let us understand the earth as being at rest in sense [a], so that the whole keeps its own place; the parts (whether they are moved or not) are completely disregarded; and that immobility is not due to a surrounding body which can occupy the earth's place by pushing the earth out of the way. [Such being granted,] what our author assumes seems to be true: that in the earth there is no resistance to a mover about to approach it.

The reason for this fact, however, does not appear to be the one he gives, which is that neither weight nor magnitude resists circular motion; for although no weight is present in the earth as a whole, who can doubt that the force that moves the earth will not move a smaller earth more easily than it will a larger? (The mover having to impart motion to the entire earth, it seems that the force that is adequate for activating a small earth is insufficient to activate a large.) Nor does it at once follow that, because the tendency [affectio] of a body is to be mobile, all bodies great or small are equally movable or are movable by the same force. Nor does it at once appear that, because White declares that nothing can be shifted but body, and hence that body, as body, is not opposed to motion, the same mover can impose its motion on a large body in the same time as on a small. Incidentally, in the same context I censure those words 'nothing can be be moved but body' as being contrary to what he states in his Third Dialogue at page 400 where, speaking of the will, he says that it moves itself through the intervention of the passions. 'The will', he declares, 'by moving any passion round the heart, transmits to the brain spirits. These excite the clear notions by which the reasoning-faculty will be set in action; this faculty returns the motion to the will.' Either this is not true, or else the will, ideas, passions, and fancies are reasoning bodies–or it is untrue that 'nothing is moved except body'.

3. The first reason why bodies [where] at rest [are those that] offer least resistance is that every resistance is counter to any defined and certain motion–counter, that is, either directly or obliquely. Now, rest seems to be not the converse of any defined motion, but the absence of all motions equally; but because someone might imagine motion and rest to be con-

traries, and that two bodies, one acting, the other being merely acted on, conflict, I add a second reason [why bodies at rest resist least]. It is this: If resistance itself were to consist in rest, it would follow that things in equal states of rest offer equal resistance. But all bodies at rest are equally at rest, for rest does not admit of greater and lesser [states of rest]. So all [instances of] resistance would be equal. But it is clear that one resistance is greater than another. Therefore resistance cannot consist in rest; so it consists in contrary motion. Hence we prove that there is no resistance present in the earth when at rest, [as follows:] If the earth possesses resistance to any motion [x], then it contains within itself a motion that is contrary to [x]. Therefore the earth possesses motion, i.e. it is not, as is supposed, at rest.

FOL. 173v

4. Taking as granted his first proposition, i.e. that the earth does not resist circular motion, he turns from this to propose a way of explaining the daily motion of the earth. For such motion to be generated, he says, only the following is required: the earth is to possess some force of compression originating in all its parts and borne along the same path [as they are].[3] It seems to me to be also necessary, however, that that body which exerts pressure is moved [versetur] not only along the same path but also by the same force by which, and at the same velocity at which, the earth itself is moved. This is a speed of ninety English miles an hour,[4] a velocity equal to, or greater than, the speed of a ball discharged by a piece of ordnance. Whether the source-of-energy [comprimens] by whose force the world is turned has this velocity we shall consider below [at folio 177].

FOL. 174

5. To explain more clearly the motion from which the earth receives the last-named motion, he brings forward by way of illustration a certain machine, which he says he has heard was in fact made. There was suspended between two very high walls, that supported its axle, a huge wheel fitted all round with containers in place of a tyre. In these were placed small bags

---

[3] The translation given of this sentence is that of White's own words (DM 177). Hobbes's version is confused: 'Anything which exerts pressure from all directions possesses the property of being forcibly moved in the same path, because the earth, which has no resistance, will necessarily follow along the same path.'

[4] Cf. fols 177, 185v, 220.

CHAPTER XVIII: 5

of sand which descended impetuously as the wheel moved downwards; but as the wheel rose they were carried gently upwards. 'Imagine', he says, 'that this wheel is moving in the air. Don't you see that this ceaseless descent of the little bags of sand is continued into (*sic*) the circumference, to become the source of the motion? You see, therefore, what must be done to the wheel and what to the earth to make it move in a circle.'

It seems to me that by this little tale (for he says merely that he has heard it, and admits that he quotes it not as a proof but as an illustration) White is unfair when he asks me: 'Don't you see that that descent of the sacks will constitute a source of perpetual motion?' He assumes that what he has heard tell about must be taken as having actually been witnessed by the reader, and not least by those who seem to take the opposite view to his. But he knew that craftsmen have made no less efforts–and with hopes no less fervent than those of the geometers trying to find the squaring of the circle and the division of an angle into a given number of parts, or of the chemists seeking that panacea commonly known as the philosopher's stone–to devise a machine that, once set in motion, could, from the nature of its construction, move for ever even when the mover ceased. He ought not, therefore, so carelessly to have passed over, supplying neither argument nor diagram, the explanation of this mechanism, as if, on seeing the wheel and the sacks, one must immediately foresee that a motion imparted once [only] can be perpetual. It seems to me that we can show that the opposite holds.

Consider, [he says,] a wheel suspended from its centre at C;[5] all round the circumference let there have been constructed cavities or small, partially-spherical holes with their openings plugged. In these let there be placed sand- or mercury-filled bags of the type he requires, at A, D, B, E, etc. It is clear that, if these bags have been set in their chambers so that they can take up [*suscipere*] a motion other than that of the wheel, they will prolong the wheel's motion no more than they would were the wheel of the same weight without them [as with]. In other words, the impressed motion will, after the motive power is removed, gradually cease when the weight of the device, and the resistance of the axle it presses on, gradually take over. This is because the weight's force [*vis ponderis*], which is in the

FOL. 174v

FOL. 175

FOL. 175v

---

[5] Figure 29.

sacks, is compensated, i.e. opposed and nullified, by a contrary and equal force diametrically opposed [to it]. We must conclude, therefore, [he believes,] as follows: The bags are loose [in their sockets],[6] so when any one of them passes over the top of the wheel, say at point A, its looseness will cause it to fall sharply as it moves towards D, and it will strike the bottom [of its socket] with a force greater than that of its weight alone. For the same reason the sacks opposite also rise more quickly than they would by their weight alone.

This, I say, cannot happen. Just as a bag strikes the more strongly against the base of its slot owing to its fall, so, during the actual fall, it weighs less *because of* the looseness, i.e. because it exerts less pressure on the wheel than does the bag opposite, at E, which presses upon the base of its socket with [its] whole weight. As a result, when the bag falls from the top to the base of its socket D, the wheel is moved backwards by its opposite side, which is [now] the heavier. Therefore nothing results from the looseness of the bags [in the sockets], save that the [wheel's] motion is rendered uneven, which the more easily annuls the motion imparted in the first place. Moreover, if the circular motion of the wheel due to its *conatus* (this remaining downwards) is to be reduced, and finally nullified, the same motion cannot conceivably be accelerated or maintained through an increase in *conatus* caused by the sharp fall of the bags.

FOL. 176

6. Applying these data to the motion of the earth, he asserts: 'Consider merely that the wheel is the earth itself; that the sockets are the coasts and the sea-bed; and that those bags of sand or mercury represent the currents of water we discussed at such length earlier. (The latter, we said, are driven from the Pacific into the Indian Ocean and thus, having completed the remainder of their circuit, are returned to the Pacific.) And, given that the earth has no resistance to motion; see: what will happen? I also add, [he continues,] two supplementary facts. (*a*) With the [rotating] wheel, [at any instant] one group of the moving weights is in motion, but the other group is resisting motion. To the earth, which draws from every direction a principle of motion, but from nowhere a principle of resistance, this does not apply. (*b*) After a little while the person who first set the wheel in motion may go away and let it rotate on its

FOL. 176v

---

[6] Or, 'are loosely filled'.

own; but the world's mover, i.e. that wind which we have considered as being widespread from east to west, is ever-present and blows unceasingly.'

As regards the effect of the sockets,⁷ I have already shown this to be non-existent; so it is useless to think of the wheel as the earth, the concavities as coasts and the sea-bed, and the bags as floods of water. I shall not deny that these two circumstances have the force sufficient to rotate the earth, even without the help of concavities. As the wind exerts constant pressure, or even blows only once, then the earth (which has been demonstrated or is supposed unable to resist even the smallest thing) must be moved with [the wind] and must retain perpetually its movement once acquired–unless some other object moving equally strongly drives it in the other direction–and this ever at a constant velocity and in the same direction to which it was driven initially by a wind that was blowing.⁸

FOL. 177

But the question raised in this context is not about *any* motion of the earth, but about its motion from west to east; not about any velocity the earth may have, but about that by which any part of the earth close to [*subiectus*] the equator traverses 90 miles every hour, or fifteen miles every first minute,⁹ i.e. with a velocity no less than that at which an iron ball travels when discharged from a cannon. Therefore he must show why, when the wind blows from east to west, the earth (conversely) is borne from west to east; and why, since this wind lacks the power to carry a feather or other very light object at three or four miles an hour, it can at the same time propel any part of the earth near the equator at 90 m.p.h. Indeed, he had to show first of all that such a wind existed before the earth's or the sun's motion began. Let us first ascertain, then, whether the only conclusion to which his suppositions lead is not this: that such a wind cannot be the cause of the said motion.

FOL. 177v

7. Consider, therefore, whether both earth and sun are at rest and whether vapours have been drawn up from the sea by

---

⁷ Presumably answering the question asked by White immediately before he introduces circumstance (*a*).
⁸ A near-literal version of this difficult passage seems best. Perhaps, for 'smallest thing' Hobbes means 'smallest wind'. He seems to envisage something on the turbine principle.
⁹ Text: '900 miles'; fol. 185v makes it clear that 90 is intended. Cf. fols 174, 220.

FOL. 178

the sun's heat. If (as is most consonant with reason) either this same heat will raise the vapours, or they will sink down by the same path [as that] by which they rose, then when both bodies [earth and sun] are at rest no wind will arise. Our author is well aware of this, for in the first Problem of this Dialogue he supposes, or asks us to admit, that 'in its daily path round the earth the sun is moved from east to west'. Then, because that postulate may seem to conflict with what he is later to deduce, he contends that we must consider the following [two proposals] as amounting to the same thing: either the sun is borne round the earth, or the earth is carried circularly round the sun. That is, he declares that motion means nothing in fact [*res*], but is meaningful only as regards the relationship of one position to another. From this he deduces that the sun leaves behind it the vapours drawn up by its heat and that, owing to either its own movement or that of the earth, these vapours are chilled and sink down. Hence they follow the sun; and this is the origin of that wind by which, in his opinion, the earth is moved. So, [I say], that wind would never have existed unless the sun or the earth were moved. That these events are impossible will become instantly clear from a single query. I ask, therefore: If the sun was moved then,[10] why is it now at rest? If the earth moves, why does its motion not antecede the wind, i.e. the result the cause?

FOL. 178v

8. His declaration that a wind blowing from east to west brings about a motion of the earth in the reverse direction, i.e. from west to east, is indeed a great paradox. How the thing can occur he shows as follows: 'One realises', he writes, 'that the earth is directly moved by the water resting on its sea-bed. The upper water is carried along more swiftly than the lower. Therefore in its flow it will precede the latter. It will not, however, float on the top of it but will (he says) force its way through the water close beneath it. If it does so, i.e. if the water that was near the surface sinks, [this water] will actually thrust forward that part [of water] in front of it, but will drive backwards the one behind it. Hence the water [whose flow] has been reversed in this way will be borne from west to east'.

Here I enquire first about the water that sinks right from

---

[10] Hobbes is criticising White for assuming, in order to prove some theories, that the *sun* moves (e.g. Article 6 of Chapter XIV); and for assuming also that the *earth* moves, as here.

the surface to the bottom: whether it sinks [a] perpendicularly; or [b] at an obtuse angle [to the eastern horizon], i.e. inclining more to the west than it would do were it to sink perpendicularly; or [c] at an acute angle, where its descent towards the sea-bed would be in an easterly direction. [a] Say it sinks perpendicularly. The reasons why any part of the sea should move downwards are the same for all its parts, so no motion, to east or to west, will take place either in the surface-waters or in those lying much deeper. [b] If the descent is in a westerly direction both the surface waters and those near the sea-bed will all be moved towards the west because the descent is similar at every place. [c] For the same reason, if the descent takes place with an inclination towards the east, both the surface-water and that much deeper will move towards the east. So in no way can the upper part possibly move in one direction and the lower in another.

FOL. 179

Suppose we concede what he lays down, i.e. that the deepest water is in part thrust forward, in part driven backward, by the topmost water as it sinks. It will follow that the earth is turned no less towards the west on the one hand than towards the east on the other, and hence that the earth is not moved at all. This objection, though he raises it himself, he does not resolve at all. His reply runs as follows: 'Suppose the water is carried further westward by the whole current, and more quickly than a lower part is carried. The lower part will necessarily be turned back towards the east by the whole current'.[11] This is no more than a re-statement [*repetitio*] of the conclusion to be proved, except that the words 'further...by the whole current' contain either a marked contradiction (for 'to be carried further' and 'to be carried by the current as a whole' are not well combined) or, at least, a value-judgment based on the knowledge of an opinion insufficiently explained. So, to supply a better explanation, he turns to an example furnished by the water under London Bridge.

FOL. 179v

As in a very powerful river over which a very strong bridge has to be supported by close-set and very stout piers, the water is impeded in its rush, so it swells up and passes at high speed

---

[11] Consider the analogy of a fluid that moves longitudinally with another fluid inside it. That the outer fluid may move the more swiftly relative to the inner (cf. White's remarks on the motion of earth and of the sun, fol. 178) does not imply that the inner fluid moves in the reverse direction relative to a static background, or that the outer drives it backward.

and downrush through [the arches]. But because it cannot push aside the water moving less quickly in front of it, those waters in the centre of the current [above the bridge] are very much compressed and therefore retire towards the banks and flow [only] gently towards the bridge.[12] That is to say, both the water near the surface and the lower water have the same motion: forward in mid-stream but ebbing near the banks. This example is not relevant to the question at issue, then, for that water on a river-bed does not flow backwards; nor is it valid [to infer], from his analogy, that the waters of the sea do so. If we are to argue from the example, whoever asserts that the waters on the sea-bed must move backwards every time their direction is changed by the flow of the waters near the surface must give a reason why the water on a river-bed is not, by the same reasoning, hurled back to its source by the impetus [it receives] from lofty places that slope down sharply into more level ones, since in elevated places the surface-water moves ahead of that lying deeper.

Lastly, suppose that a gentle wind (as is that beneath the sun) is the cause of the [earth's] daily motion by acting first on the surface-waters,' then on the depths, and then on the sea-bed. If, in addition, a great wind from the east brooded over dry land, it seems to me that this must [by White's reasoning] make the earth revolve toward the east and, at least, that every wind will either accelerate or retard, and palpably, the [earth's] daily motion. This could not occur without being noticed; but as no such thing has been recorded there is no reason why this motion should be ascribed to the wind.

9. The most damning of the objections one can raise against his theory is the great speed of the [earth's] daily movement. Because both the water and the earth complete their circuits in the same time (we disregard the deduction to be made for the tide), the speed of the water will be the same as that of the earth. This velocity is therefore communicated [a] directly by the land to the water, or [b] by the water to the land, or [c] both have it from a common source. If, then, [b] the earth receives its velocity from the water, where does the water draw its own velocity from? 'From the surface-water propelled from east to west as it sinks to the sea-bed.' And the surface-water?

[12] Cf. Pepys, 8 August 1662.

CHAPTER XVIII: 9

'From the wind that blows from the east.' But this cannot happen when the speed of the earth exceeds [that of] the wind immeasurably. I find that our author has proposed nothing to resolve this difficulty, save when, perhaps, he reports Galileo as saying: 'The slowest cause can, by repeating itself, furnish the source to the swiftest motion, just as it is an accepted thing in mechanics to show that any weight can be moved by any force'.[13] Whoever (either from Galileo or from a close study of mechanics) considers the reason governing this fact to be familiar will know only too well that it cannot be applied to the daily motion [of the earth]. The reason why a motion that at one moment is slow generates, with the aid of machines, a very swift motion in another [body] consists in the fact that the force of bodies that exert weight [*ponderantia*] is compounded of the magnitude divided by the speed. Therefore just as there is 'quantity' of surface, so weight [*pondus*] consists in a twofold measure: the body's velocity and its magnitude. Of these, one divided into (*sic*) the other gives the force [*vis*] by which [the body] weighs. That is why two heavy bodies whose masses and velocities are proportional to one another exert the same force and are thus of equal weight when placed on a balance. And in order that the weight suspended at the lesser distance from the fulcrum, hence describing the smaller arc as it moves, may have only the correct magnitude [to counterpoise the magnitude opposite it] it compels the body of lesser magnitude to describe a larger arc in the same time, i.e. [the greater body, *a*] imposes on the smaller, [*b*], a velocity greater than that of body [*a*]. But this cannot be applied to the wind that moves the earth. When, according to Galileo, a very slow motion can, by redoubling itself, produce a very swift one—one at least much swifter than itself—this is due not only to the force [*vis*] of the slow motion but also to the increment of motion [arising] from natural gravity.

FOL. 181

FOL. 181v

Galileo suggests an instance of this in heavy bodies hanging from a cord. These, if displaced from the vertical by a puff of breath, not only return to the perpendicular but also cross

---

[13] Galileo's *Dialogues of motion* is the only reference given (in the margin, DM 184) by White. For the idea, compare Galileo's *Two world-systems* (Drake), p. 215; *Two new sciences* (ed. Crew and De Salvio), pp. 98, 291; and *Le meccaniche* (c. 1600, as in I. E. Drabkin and Stillman Drake, *Galileo Galilei on motion and on mechanics* (Madison, 1960), pp. 67n, 157–177). Cf. Aristotle: *Mechanics*, ch. 6, and fol. 226v below.

it and come back again;[14] and when they begin to swing back, then (because their motion is very slow) they can again be made to move faster by blowing. There is nothing, however, resembling this in the earth's continual rotation, caused, as [White] supposes, by the wind directly beneath the sun. So no reason has been given by our author for the existing daily motion, a very swift one.

FOL. 182

10. As regards the cause of the earth's diurnal motion–if indeed the earth *is* moved–personally, I shall postpone revealing my own opinion for the present, till I have also considered what our author was about to say concerning the cause of the annual movement. Let us therefore proceed with the remaining questions.

'My curiosity', he says, 'is satisfied. Why, after the flood of its deeper waters turns eastward, the Pacific hurls itself with such violence into the Atlantic, I do not ask.' Here I enquire: In what way is that issuing of waters from the Pacific to the Atlantic to be understood? for if it does occur, [the waters] will, I think, pass from the one ocean into the other through the Strait of Magellan, i.e. from west to east. But the surface waters, which alone are visible, are [generally] believed to move in the opposite direction, [i.e.] from east to west, being activated by the wind that moves with the sun; but the deeper waters are not open to the view. In what sense, then, are we to understand (as is proposed for solution here, in place of the phenomenon [under scrutiny]), that such a rush of waters does in fact take place? Either there is no exodus of waters like this or, if there

FOL. 182v

is one, it does not square with the explanation given.

11. What he says next about the beginning of this motion[15] is open to criticism and is reprehensible in two ways: he pays too little attention to what he has laid down earlier, and he is too free in framing hypotheses, for he argues thus: 'As you are asking about the principle of this motion, you must allow that a principle existed and, as logically follows, that the sun and the earth are placed in a predetermined position [each to each]. We can set this position where we please; let us fix the sun at

---

[14] Ambiguous: Is oscillation meant, or a single movement only? Drake, pp. 151, 226.
[15] I.e. the alleged west-east motion of tides.

the equator so that it strikes perpendicularly either upon the Isle of San Lorenzo or on another of those Fortunate Isles [the Canaries] that stretch between it and Portugal. Let us at the same time study the effects of doing so'. But the question is: What would happen if we located the sun not at that part of the equator but at another part, or somewhere outside the equator? 'So', he will reply, 'you refuse to allow nature and her Parent [their] prudence in positioning the sun and earth where these may best serve their *raison-d'être* and functions?' This, I declare, is being far too free with philosophical postulates. It is not a hypothesis; for not only does he claim that, were the sun thus [advantageously] positioned, it would have to commence [its] daily motion,[16] but also he insists it *has* in fact been so positioned, as if in no other way would the Divine Judgment and the Works of God be served. Yet granted that matters are as he says, his opinions do not help; for, unless he had forgotten what he said about the cause of the wind that moves the earth (this is further grounds for censure), he would have seen that it is impossible for the motion of the earth to follow the position of the sun at the equator or be fixed and at rest outside it; for the vapours drawn up by the sun become (as he perceives in the First Problem of this Dialogue) a wind. This is because they are left behind when the sun or the earth moves ahead, and are condensed and sink again, but at an angle, i.e. as they follow the sun towards the west. Indeed, if the sun and the earth were both at rest, there would certainly be no reason why the wind should be easterly rather than westerly, or northerly, or from any other direction. So the motion of the earth (for he believes that the sun does not move) anteceded [that of] the wind, i.e. the earth's motion preceded its cause and its beginning! This argument cannot be sustained.

FOL. 183

FOL. 183v

12. Coming now to examine the reason for the [earth's] annual motion, he assumes that the initial motion of a wheel moving along a horizontal surface, if only it were driven by a plane touching it at the opposite side, is in a forward direction [*progressivus*]. This is true if the plane beneath (that is pressed backward) is hard and resists a part of the wheel that moves

---

[16] Presumably: 'It is evidence of divinity that, wherever the sun was placed or wherever its course arrested, it would still continue or resume its motion'.

forward in the reverse direction on the underside. Otherwise it is not true; for if the bottom point of the wheel is indeed moved in the opposite direction [to that of the top] a rotation will take place about the wheel's centre; but if [the lower point] is at rest, a rotation will take place about the point of contact. Again, the plane is hard; but when the point in contact with it cannot move backward owing to the roughness of the flat surface, the parts next moving up have to impose themselves one by one upon the plane. This is the only reason, [he says] why the wheel is rolled forward, with its whole body together with its centre, upon the flat surface. But reasoning of this kind can in no way apply to the progressive [forward] motion of the earth, where there is nothing corresponding to a resistant plane. Although, then, we grant him that the waters strike the earth [obliquely and] in the same way that the wheel is moved by the plane rested upon it, yet it will not follow that the centre of the earth is thence moved forward.

FOL. 184

Next he assumes that, when the water presses upon the earth from one direction tangentially, the part of the earth directly opposite is the while at rest. This he explains as follows: The daily motion beginning in one area of the earth has been propagated from part to part in turn; hence, when that part on which the wind has first rested was moved, another part, on the opposite side, remained still, until wholly overcome [by the need to move]. The result is, he says, that the earth is thrust forwards [circularly].

As regards this reasoning, [a] it can be true that one part [alone] of the earth is moved circularly, only if at the same time all the other parts are rotated—unless one part resists another. But if the latter applies, that proposition on which he based [his belief in] the daily motion of the earth, namely that the earth possesses no resistance to circular motion,[17] is untrue. [b] If we allow that [the earth's] motion is transmitted from part to part in succession, such motion will be none other than circular motion about a fixed centre; for until the motion reaches the opposite side there will indeed be a *conatus* of the earth and a movement of its parts, but no rotation of the whole. Such rotation will not commence until all the parts have absorbed the same *conatus*, as someone who shoves some heavy object with his finger along a flat surface does impart *conatus* for a space of

FOL. 184v

[17] Fols. 171v*ff*.

time [*mora*], and continuously [*successivè*], but not unless the body is a whole.[18] As soon as [he does so] it moves forward as the impressed force is applied to it. So it is impossible for the earth to be driven by the movement of the waters carried thereon in the manner of a plane surface in contact with it.

But if we are to grant him all this, where is the water that acts on the earth along a line or surface touching it [*contingens*]? 'Don't you see', he says, 'that the water cast forth from the Pacific swells up over the other water and so can move it? What kind of motion is this, [he asks]: circular or rectilinear?' And a little later, 'In what way do the separate waves strike the water thrust upon them? [Water and waves] seem to break one another, [either] as a plane [surface makes a section through] a sphere, or as the line [other than a tangent] touching a circle strikes it if the circle be moved [towards it]'. In the opinion of our author, therefore, the cause[19] of the annual motion is the upper waters that flow from the Pacific Ocean into the Atlantic. If so, then if this outflow takes place through the Strait of Magellan there will be a west-east movement of the upmost waters. Therefore according to his argument the deeper waters, which are the immediate cause of the earth's motion, are borne by a contrary motion [that operates] from east to west.[20] The earth, then, in its annual path moves in the same direction, [east-west]. This conflicts with the known facts; for the annual course, as the daily, is from west to east, except that it inclines a little to the north in one direction and to the south in the other.[21]

FOL. 185

Now, if he understands that the issue of waters occurs on the other side [of the Strait of Magellan], through the Indian Ocean, the same will be the case with the motion to which he attributes the earth's daily movement. Therefore either the annual and the daily motion will be [in] the same [direction] or at least will take place beneath the same circle, yet it is accepted that the annual motion differs from the daily [a] by an amount equal to the greatest declination of the ecliptic from the equator and finally [b] according to the speed required by the waters to

FOL. 185v

---

[18] Hobbes may mean: 'but only to the body as a whole'.
[19] Plural in the text. In the preceding sentence White seems to mean that either the rollers are pierced by the water approaching them, or else they deflect it.
[20] Or, 'A contrary motion carries them from east to west'.
[21] Is the reference to the obliquity of the earth's orbit?

bring about the annual motion. The latter is about three times faster than the daily, which, as said earlier, is ninety miles an hour.[22]

13. Lastly, the reason for the variation between the [earth's] annual and daily motion he ascribes to the situation of a certain strait beyond that of Magellan. Yet, because this strait seems to him to lie in a longitude[23] that is parallel to the ecliptic, I can scarcely believe that he puts this forward seriously, for who will credit that the greatest operations of nature are held in abeyance for the most trivial reasons? If causes so insignificant produce so important results, wouldn't the irregularity of the earth's motion—for the waters vary a great deal—be considerable? When the tide ebbed away from London Bridge, wouldn't the water under the bridge exert a great influence towards accelerating the [motion of the] earth, and again, towards retarding it when the water rushed up against the bridge? Wouldn't rivers and floods hurled from the mountains have a considerable effect in moving the earth forwards and back? Indeed, there are many places where the waters are swifter than in the narrows that he terms 'The Further Magellan Strait' (if any such strait exist). Moreover, in straits and enclosed stretches of water, and of dry land, many changes occur, so that we now find straits where once there were isthmuses or islands. On the other hand there are some people who believe, not without great reason and authority, that at the beginning the whole earth was under water. Therefore even if the force of the waters were sufficient to move the earth in the way described, we should not think he relies on this force as a cause [of the earth's motion], unless the straits were previously agreed to be of greater antiquity than the earth; for we may be certain that the latter is of the same age as its motion.

FOL. 186

FOL. 186v

14. Now in order briefly to explain my own views about the cause of the earth's motion, the annual as well as the daily (if it is indeed true that the earth is moved), and beginning with the annual, I think I must first examine that investigation of Galileo's. Here, when observing the movement of sunspots with the aid of his telescope, he noticed in the sun a certain

---

[22] Cf. fols 174, 177, 220.
[23] Perhaps merely: 'to extend in length'.

rotation about the axis of the ecliptic.[24] If this [phenomenon] is true, we must no longer doubt that of necessity the *earth* must be moved in the ecliptic.

It is agreed that there is present in the sun a certain motion which dilates it in every direction and by which the sun illuminates everything situated around it. That motion, combined with another motion, the rotary one, will compel the air or ether next to [the sun] to take on a circular action about the same axis, namely that of the ecliptic. Also the sun, through the rectilinear motion that causes it to dilate, will necessarily compress the adjacent air. At the same time, it will whirl round by its movement of rotation the air which it has compressed, and this at the same speed at which the sun itself is rotated. In this way, in the time that the first named air (that adjacent to the sun) completes its circuit, the air further away likewise completes a part of its own circuit equivalent to the whole circle of the first [named] air. Hence the times in which all the parts of the air are carried round are, each to the other, in the ratios of the straight lines indicating their distances from the surface of the rotating sun.[25] It is also clear that, being supported in air so moved, the earth will be borne in the same way in which the parts of the air that surround it are moved. Analogously, both the smallest and the largest ships in a river glide along with it, without [the use of] oars or sails, more quickly or more slowly according as the flow of the water they float on is swifter or slower.[26] If, then, the sun rotates about the axis of the ecliptic and at the same time moves as a luminary,[27] the earth will necessarily be carried through the ecliptic in a circular movement. As, however, a complete circle (either the earth's or the sun's) delimits or marks out the [terrestrial] year, the earth's moving in such a circle will constitute its annual motion.

For the same reason the remaining planets will also be

FOL. 187

FOL. 187v

---

[24] As regards sunspots, Galileo assumed the sun to turn not on an axis at right angles to the ecliptic but around a *tilted* axis. See Drake, p. 347; the general discussion occupies pp. 345–352.

[25] From the surface, rather than from the centre, of the sun. So much for the motion of the air round the sun; the similar distribution of light, however (Chapter IX), obeys a different rule.

[26] Cf. fols 139, 269v.

[27] By *habeatque motum illuminationis*, a phrase reminiscent of Kepler, Hobbes seems to mean a rectilinear movement of dilatation associated with the emission of light and with the compression of the adjacent air.

FOL. 188

moved through the ecliptic, except insofar as, by [their] interaction, one of them slightly turns another aside from this course. The sun's motion about the axis of the ecliptic or, at least, about one slightly oblique to it, constitutes a forceful argument, however,[28] because there is not a single planet whose path is not beneath the zodiac, i.e. [a planet moves] in a circle that never deviates from the ecliptic by more than six or eight degrees. Say he asks: 'Why has it not been known for any star to be borne through the Milky Way or from pole to pole, or from a south-westerly to a north-easterly position? Why does it go solely beneath the ecliptic?' I do not know whether any more fitting explanation can be given than for us to say: 'The cause of every circuit is the sun, but the latter has a clearly-marked circuit beneath the ecliptic'. If the earth is moved [as he surmises], therefore, I consider both that the sun rotates [as described] and that, if the sun indeed does so, the earth must be moved as set forth [above]. So much for what I think should be said concerning the annual motion.

FOL. 188v

15. As regards the daily, we must consider the nature of the earth itself: it is firm, consisting of parts tightly bound one to another so as not to be easily separated. Now our author directs his argument towards convincing us that, because the earth, considered as a whole, has no resistance, it is easy for daily motion to be imposed on it. I shall therefore with equal justice base an argument on the motion we may ascribe confidently (because the earth's parts *do* possess resistance) to the parts of the earth, but not to the whole.

FOL. 189

Resistance is nothing but contrary motion. Now if resistance is not motion, why then, if you prod any part of a stone with your finger, does it not give? You will answer: 'Another part next to it prevents it'. Very well, then; why does that adjacent part not yield? The only possible answer is either: 'The nearest part possesses some motion, inasfar as it affects the adjacent mover', or: 'Because the whole stone is homogeneous'. But a mass [*moles*] of water, air, mercury, or liquefied metal is just as much a unity as a mass of stone is; yet when you press an area of water, or liquefied metal, or mercury with your finger it yields easily, for the parts adjacent offer no resistance. Hence the unity of the whole body is not the reason for the coherence

---

[28] I.e. an argument that the planets move through the ecliptic.

CHAPTER XVIII: 15

of the parts (or hardness).[29] But what (to enable us, by using the term, to ascribe [to a body] the ability to do things, or force of action [*efficacia aut vis agendi*]) *is* unity? A unity, be it hard or soft, is that which we can delimit in a single sweep of the mind. We must conclude, therefore, that hard bodies –the earth is such a one–contain within their parts a motion by which the bodies cohere and through which we learn, as we would from their form [*tamquam formam*], that they are hard and also that, because of the diversity of the motions, different bodies appear hard, as do stone, metal, glass, etc.

Under these circumstances it is not difficult to believe that the earth, and each part of it, will, inasfar as it is not hindered, keep itself apart from any other mover by a distance at which that natural motion of its parts may enjoy the greatest freedom of action. An example of this occurs in bodies set in rotary motion. As soon as they knock against a wall or some other obstacle they immediately, by their very nature, recoil and continue spinning, for the rotary action unaffected [by the impact] having not yet been exhausted, a part [of the spinning body] renews the motion of the part that collides [with the obstacle]. Another parallel occurs with respect to creatures that, when exposed to a source of heat, of their own accord alter their position so that the vital motion may be kept at a constant temperature most suitable to its nature. If this behaviour [*conversio animalium*] is termed voluntary, there is no reason why it should not apply to the earth, for although this motion is not voluntary in the earth, yet in animals such an inclination, or the desire for warmth, is motion. So for this reason the earth seems to turn itself towards the sun,[30] i.e. to be borne in a daily motion, because the parts turned away from the sun approach it in order to preserve their nature or this essential motion. The parts nearest the sun, which bathe in its heat, are no longer at ease in such a position, however; hence, in order that the natural motion [of such bodies] may be the more freely set to work, they move away again. In this manner the circulation [*conversio*] takes place by which a day is measured out: it is therefore called diurnal.

FOL. 189v

---

[29] Cf. fol. 341.
[30] Again, surely borrowed from Kepler, as Hobbes concedes (*EW* I.434)?

CHAPTER NINETEEN

On Problems 6 and 7 of the Second Dialogue:
Why the earth keeps its axis parallel to
the axis of the [celestial] equator. The
cause of the precession of the equinoxes

1. A description of the parallelism of the earth's axis. 2. That the reason our author gives for this parallelism conflicts with what he said previously. 3. The same parallelism ascribed to the sun's motion that rotates the air in a spinning-motion around [the sun]. 4. What the precession of the equinoxes is, and how the Ptolemaeans explain it. 5. The reason the Copernicans give for the same is the more likely. 6. Our author's explanation is incorrect. 7. On the reason why the sun remains for more days in the constellations of the north than in those of the south; also (incidentally) on the reason for the daily motion of the moon and for the moon's approaching, and for her receding from, the earth.

1. To explain these problems we need a diagram so that the supposed motion of the earth round the sun—inasfar as this can be done—may be represented visually.

Let A be the centre of the sun, and B the centre of the earth initially; and let C, D, E in succession (and all other points on the circle BCDE) represent the path in which the earth is borne in its annual motion.[1] In addition, let a circle of diameter ZAY represent the ecliptic, [which is] in the same plane as the circle BCDE. This will ensure that the annual movement of

---
[1] Figure 30. Drake, p. 379.

the earth is shown as [lying] within the ecliptic. Now if we suppose that the axis of the earth, when the latter is at B, is the straight line PO, and that [the axis] is at right angles to a ray, AB, from the sun; then FG will be a straight line, i.e. ZY, representing the [celestial] equator in whose plane lies point A, which is the centre of the sun. So as long as the earth's axis is directly opposite the sun's rays, the sun is at the [earth's] equator. This should be noted. But if the earth moves forward to C then its axis, formerly PO, will be now RQ, parallel to PO, and hence the solar ray AC will not be parallel to RQ. Further, because the [earth's] equator is now taken to be HCI (the same equator as was previously at FG), and consequently the angle HCQ is taken to be a right angle, the angle QCA is not a right angle. Hence AC will not be in the same straight line as HCI, the equator. So when the earth is at C, then point A, i.e. the sun, is not at the [earth's] equator. By the same method we show that, when the earth is at D or at any other point on the circle BCDE except B and E, the sun is not at the [earth's] equator. That could happen only if the earth's axis moved in such a way that the position it occupied were always parallel to the one it had left. Because, therefore, the phenomena connected with the four periods of the year; with the changes of days and nights; and with other characteristics ascribed to this physical parallelism of the earth's axis are taken into account by those who maintain that the earth moves, it was relevant to our author's thesis to point to a physical cause that would bring about a parallelism of this kind.

FOL. 191v

2. He says on his page 194, therefore: 'Since, in the nature of things, the earth is freely suspended, untrammelled by external bonds, it seeks no reason *not* to move. What we must probe more deeply is this: Is there a cause strong enough to deflect from its path the earth's axis once the latter has been so poised?'[2] The general sense of these words is as follows: When the earth's axis is once proceeding in a given direction, say from O to P, from Q to R, or from S to T (all these have one and the same direction, from north to south), we need not ask why [the earth], which moves in an unchanged direction, is never diverted into another position such that, say, its axis

FOL. 192

---

[2] It is not clear whether by 'poised' (*libratus*) White means 'suspended and not moving' or 'oscillating, librating' in the air. Cf. fol. 199.

FOL. 192v

FOL. 193

FOL. 193v

would lie at any time on the straight line HI, KL, or MN. If it *were* thus turned aside, then we should have to seek the reason for the deflection.[3]

On occasion, it is true, man seems to accept as acts of nature those things that are merely botches or imperfections. This is because things we see *happening often* happen, we think, always in the same way.[4] When we whirl any body on a stretched cord round any centre, the same part [of the body] must touch the circle [of rotation] perpetually. For instance, if, in the diagram above, the earth, whose centre is B, were connected to the sun, A, by a rope AB and were thus rotated in the circle BCDE, the axis PO, which touches the circle at B, would touch it also at C, D and E. Therefore [PO] would never be parallel to itself. The particular explanation of this is not far to seek: The parts of the earth further removed from the centre A are moved more swiftly than those nearer to A, the ratio of the velocities being as that of the magnitudes of the circles [described]. However, some people who do not reason the matter out think that every circular motion without exception has, unless obstructed, the same characteristics. Hence they demand an efficient cause to explain the parallelism we have described. The latter is really due to the fact that, while some parts [of the earth] are moved and advance, others lack a mover, i.e. an efficient cause, and are therefore left behind. For although it is true that the parallelism of the earth's axis is not an effect [? of motion] but an absence [*defectus*] of motion, our author ought not to have said so here, because to do so would conflict with what he asserted about the earth's annual movement. If the motion of any part of the seas indeed drives the [earth's] centre B circularly through C, D and E in the ecliptic, then for the same reason the motion of another part of the waters drives point F through the circle FT, and point G through the circle GS. In consequence the parallelism of the axis will necessarily be violated. He has not, therefore, supplied any reason for this parallelism.

3. The true cause is to be sought in the reason for the earth's moving through the circle BCDE, i.e. in the efficient cause of the annual motion, namely the sun. If, therefore, we assume the

---

[3] Here the main clause is supplied by Mersenne.
[4] The italicisation (not represented by underlining in the text) removes, without syntactical complications, ambiguity in the sense. For 'things', perhaps 'the things'.

sun to rotate about the axis of the ecliptic and the sun's centre to rotate about itself, [the sun] will draw the adjacent air with it, and the latter in its turn will draw other air nearby, and so perpetually. Under the class of motion where one thing draws another by attraction [*adhaesio*], in a tenuous body, with very weak attractive power, it is impossible for an adjacent part of the air to follow at a speed greater than that which [the part] derives from the sun [when the latter is] contiguous [to it]. So the air (and with it any other hard and solid body floating in it) at any distance from the sun moves at a speed approximately equal to that at which the sun itself rotates.[5] The motion of the earth through [each of] the circles FT, BD and GS, therefore, is equally fast; and the times in which a circle is described in each of the several distances are, one to another, in the same ratio as the circles or their radii are. This being grasped, I add the following lemma:

FOL. 194

> If there are two concentric circles, and between their circumferences is drawn another circle, included and touching both circles, then the circumference of the exterior circle will be greater than that of the interior by a distance equal to double the circumference of the circle included.[6]

FOL. 194v

Referring again to the previous diagram [Figure 30], let us draw with centre A two circles, the exterior FT and the interior GS; also the included circle FPGO that touches the exterior at F and the interior at G. I say: The circumference of the exterior circle FTF is greater than the circumference GSG, the interior, by twice the circumference FPGO, because radius AF: radius AG:: circumference FTF: circumference GSG.

FOL. 195

But radius AF > radius AG by 2(radius GB),
Oce FTF > Oce GSG by 2(Oce FPGO). Q.E.D.
Hence Oce FTF > Oce BCDEB by (Oce FPGO).

It is clear, then, that point F cannot return to its own position in one journey of the centre B [of the circle on whose circumference F lies] around the circle BCDEB unless, through the

---

[5] The words 'to follow' are inadequate: to follow the part? But perhaps Hobbes means: "It is impossible for a tenuous body, with very weak attractive power, to pull an adjacent part of the air at a speed", etc. And by "any other hard and solid body" he seems to mean "as well as the sun".

[6] A lemma, yet 'proved' again four lines later?

earth's contrary motion, point F retrogresses and so slips into the inner circles, in which the circular motion is completed more quickly. [The latter occurs] because the time taken to traverse the [circumferences of] the separate concentric circles, one after the other, equals the time taken to move around the circumference of the middle circle BCDEB. The time taken for point F to rotate through the circle FTF will exceed that for point B to traverse the circle BCDEB by as long as is required for B to traverse a line equal to the circumference BCDEB. Finally, just as point F rotates, and with it the whole diameter FG, so the whole axis PO rotates in the antecedent. Here its whole axis must therefore always be parallel to itself. This we shall prove as follows:

FOL. 195v

Let the centre [of the circle] of the ecliptic be A and its circumference BCB.[7] Let B represent the centre of the earth initially, and let DE represent its axis, which touches the ecliptic at B. Let the centre of the earth move in the ecliptic according to the signs from B to C, describing the arc BC. Now unless point D moves backward, the axis of the earth when the latter is at C will be the straight line FG that touches the ecliptic at C. Next, let point F move backward to H so that the arc FH is to the arc BC as the whole circle of the earth is to the whole ecliptic, i.e. so that the angle FCH equals the angle BAC and that HCI, when drawn, is the earth's axis back again in the antecedent. I say that the straight lines HI and DE are parallel. Let HI be produced to meet AB at K, then ∠s ABD and ACG are rt∠s and are equal.

FOL. 196

But ∠ACG = ∠s ACK and GCI taken together, i.e. ∠ACK and ∠FCH, i.e. ∠ACK and ∠BAC.
But the external ∠BKC = [the sum of] the two internal ∠s ACK and KAC, i.e. = the same ∠ACK and ∠BAC.
∴ ∠BKC = ∠ACG.
∴ ∠BKC is a rt∠.
But ∠KBD is also a rt∠.
∴ KC, i.e. HI, is ∥ to DB, i.e. DE.   Q.E.D.

FOL. 196v

It will be clear, then, how the parallelism of the earth's axis can be ascribed to the same cause as the earth's motion, i.e. to a property of the sun; and that it cannot be due to the movement of the sea, as our author wished.

---

[7] Figure 31. The text is followed as it stands.

CHAPTER XIX: 5

4. So that we may realise what it is he calls the precession of the equinoxes, we must again refer to the diagram [to f. 190v] in Article 1. In this let the ecliptic of the eighth sphere be a circle. We may consider that its diameter ZY is the beginning of the sign of Libra at Z and that the first[8] star of Aries is at Y. Lastly, let the earth be at B.

FOL. 197

The sun, initially at the equator, has been rising together with that first star in Aries, but subsequently the sun at the equator begins to rise more and more before this star,[9] which begins to appear more remote every year from the equinoctial point about the time of the summer solstice.

The Ptolemaeans explain the phenomenon [of precession] as follows: They suppose that the sphere of the fixed stars moves in a slow, perpetual motion about the axis of the ecliptic through the signs of the Zodiac in order. So the first star of Aries, initially at Y, advances with that [slow, perpetual] motion, say to α. Subsequently, therefore, when the sun was at the equator it was necessary for that star to seem remote from the intersection of the equator and the ecliptic (i.e. from point Y) by a distance equal to the arc $Y\alpha$.

FOL. 197v

5. The Copernicans, however, give the following explanation for the same set of observed facts: They say that the earth's axis always stays parallel to itself, at any rate as we perceive it, but not exactly so, because a tilt about its centre enables it [and parts of the earth's surface] to complete a whole circuit round the centre of the ecliptic a little more quickly than the earth's centre completes its circuit. If this is true, the precession of the equinoxes results as follows:

Let the earth be at B, its axis FH touching the ecliptic.[10] Therefore FH and AB will be at right angles; and the sun will be at the equator, a fact we draw attention to in Article 1; so point D will be the intersection of the equator and the ecliptic. Initially the first star of Aries was at this spot. Then the earth moved from B through C, D and E. If its axis FH had, by tilting itself [so that point F or H described a circle within the

FOL. 19

---

[8] The first encountered, or the first in brightness? The phrase occurs twice on fol. 198.
[9] Doubly ambiguous: (a) 'more and more'–higher and higher, or faster and faster? (b) 'before' [*ante ipsam*] earlier than, or in front of (above)? See fol. 198.
[10] Figure 32.

ecliptic] indeed completed its circuit in exactly the same time as that in which its centre B went round the ecliptic, the axis FH would not have touched the circle of the ecliptic [tangentially] before the earth's centre had returned to B. But it is supposed that the axis, by tilting itself, describes its own complete circle a little faster than the centre of the earth is borne through the ecliptic. So the earth's axis will touch the ecliptic at a point beyond B, say at E. Here the earth's axis KL will be at right angles to the solar ray AE. Hence the sun will now be at the equator, whose intersection with the ecliptic will be at C. Therefore D, the first star in Aries, seems, though motionless in fact, to have moved forward through the arc CD. Such appears to be the true cause of the precession of the equinoxes, except that, so far, it gives no reason why the [earth's] axis keeps the same parallel position, not exactly but only apparently. But the difference from a true parallelism is so small, and the stars other than the sun that affect the earth are so many, that it is as rash to crave a reason for so small a discrepancy as it is difficult [to supply one].

6. The cause of the precession of the equinoxes our author explains on his page 197 as follows: 'It is the accepted practice', he says, 'to ascribe the reason to two factors: [a] that area of the earth which once brought us days of equal length may, by chance, have come to behave differently now; and [b] it may so happen that the earth's equator when immobile is perpendicularly facing the sun in a sign [of the Zodiac] different from what was formerly its usual one'. That is to say, it may be that the equator has moved, and hence that the terrestrial poles are different from before; or that, if the equator and the poles have not moved, a ray from the sun falls perpendicularly on the axis of the equator under a different sign from previously. The second of these proposals is tantamount to saying: 'It may be that the equinoxes do not match the same fixed star that they corresponded with previously', which is the same as saying, 'It may be that a precession of the equinoxes does take place' – thus the reason for the precession of the equinoxes will be the precession of the equinoxes!

He adopts, however, reason [a], declaring that the precession of the equinoxes is due to the fact that the equinox is always shifting from one part of the earth to another. This happens because the [Pacific] Ocean thrusts towards the north – through the Atlantic, I take it—a great force of water. The movement of such disturbances tilts the earth a little in a northerly direction. Thus the earth presents perpendicularly to the sun sometimes one axis of the daily motion, sometimes the other.[11]

He asserted earlier that the [earth's] daily motion (from west to east) was caused by the movement of the surface-waters from east to west; hence the deeper waters, driven by a contrary motion [exerted] from west to east, rotated the earth from west to east. So for the same reason, [it seems to me], the waters at mid-depth that flowed towards the north ought to have moved the lower waters southward; thus the earth would incline not to the north but to the south. This would occur not gradually but at a rate no less swift than that of the [earth's] daily motion, and this because the waters flow into the Atlantic no less quickly than they flow away across the extent of the [Pacific] Ocean. But even if all these points be granted him, he still does not show how the precession of the equinoxes can be ascribed to this inclination of the earth towards the [celestial] north, nor do I think that [this] can be shown.

FOL. 199v

7. It is known that there are more days from the spring equinox to the autumnal than from the autumnal to the spring. For that reason the Ptolemaeans say that the earth is not the centre of the sun's annual [circular] motion; and the Copernicans say that the sun is not the centre of the earth's, their case being very convincing. Our author, disregarding any reason for the eccentricity,[12] assigns the cause of this phenomenon to the wind *he* claims to exist beneath the sun.[13] In actual fact, in the period from the autumnal equinox to the vernal the sun stays above the [Atlantic] Ocean for a longer time than it does [in the period] from the spring equinox to the autumnal. This, he says, is the reason for the rise of a stronger wind [between autumn and spring than between spring and autumn], and consequently for a greater turbulence of water; and the

FOL. 200

---

[11] I.e. the same axis but at different times; it oscillates.
[12] Does Hobbes refer to the *rate* at which the orbit is completed, i.e. to heavenly motions, or does he refer to its form or *shape*? He is thinking, with Galileo, of the earth's orbit as being circular, not elliptical, but he may mean: '...reason, *based* on eccentricity'.
[13] It is not clear whether, by 'this phenomenon' [*res*], Hobbes means the eccentricity; the path of the earth in general; or the difference in number of days previously referred to.

latter makes the [earth's] daily motion swifter [between autumn and spring]. Therefore, [he infers,] the natural days are shorter in winter.

FOL. 200v

This goes against our experience of all timepieces, clocks as well as sundials; for the hours in summer are equal to those in winter.[14] It does not follow, either, that if the hours [in winter] were shorter, i.e. if a natural winter day were shorter than a spring one, there were therefore fewer days from autumn to spring; [it follows] rather that there are more [days from autumn to spring] than from spring to autumn; for the velocity of the earth's daily motion about its axis does not increase the rate of its annual motion through the ecliptic. The true scientific reason for the earth's eccentricity has not yet been explained by anyone, nor am I at all certain what this reason may be. One thing I am sure of: the earth always makes for that spot in which its essential motion, namely the internal motion which identifies it as the earth, is best conserved; and hence that it is always at the distance, both from the sun and from the rest of the stars, best fitting its own nature. Indeed, if there existed no heavenly bodies other than the sun, I should imagine [the earth] would lie at a distance from it that never

FOL. 201

varies, but for the eccentricity.[15] But because other stars influence the earth, its distance from the sun does vary, just as at full- and new-moon the moon changes both her distance from the earth and her distance from the sun: when, at new-moon, the moon is equally closer to the sun, she moves further back towards the earth, taking the position vacated by them both [i.e. the earth and the sun]; likewise at full-moon, when the moon is equally further from the sun, she draws closer to the earth, each adapting itself to the other so that it too may be nearer the sun.[16] Hence it has been noticed that when new and at the full the moon is nearer to the earth but that in the intervening periods she is farther away from it.[17] Likewise when one part

---

[14] Hobbes is pointing to the confusion between *number* of days and *length* of day, in hours. He seems to mean: An hour is absolutely and precisely the same measure of time, whatever the season i.e. there are always 24 hours in any day.

[15] Presumably, 'the eccentricity of the earth's orbit relative to the sun'. Perhaps we should translate: '...never varies without [there being] eccentricity'.

[16] Drake, p. 453. See also fol. 279.

[17] In the diagram to fol. 156 the distance of the moon from the sun at new-moon is AB−BE; at full-moon, AB+BE. But in the diagram Hobbes

CHAPTER XIX: 7

of the earth is exposed to the sun's rays at very close range, another part, one benefiting less from its warmth, strives to press forward into the same position, and the part that was there first is now sated and moves away.

This explanation of the [earth's] daily motion is, I think, not unlike the reason why living creatures are moved in accordance with the actions of those who treat them kindly, i.e. of those whose behaviour aids and strengthens the creatures' inner motion, which they have from birth.[18] For we draw up to the fire, and the parts of the body that are growing cold we turn towards it, one after another, even if we are doing something else.[19] I do not, however, think that the earth is animate; yet all consistent bodies, or those possessing coherent parts, have this in common: they maintain habitual motion inasfar as other motions, gravity especially, do not hinder [them]. But gravity cannot prevent parts of the earth from being moved in a daily rotation. As I turn the matter over, this thought comes to me: the reason why the earth is further from the sun when the latter is in [the sign] Cancer than when in Capricorn is the joint action of the northern stars, which are more numerous than the southern. The earth, therefore, moves farther away from [the northern stars] in a southerly direction when the sun is in Cancer than it does in a northerly when the sun is in Capricorn.

FOL. 201v

---

supposes that the orbit of the moon round the earth is a circle, whereas here he supposes her distance from earth to be variable. Hence 'equally further' may mean 'nearer the sun' and 'further away from the sun' equally, i.e. by the same margin.

[18] Or, 'which they find pleasant or comforting' [*genialis*].
[19] Or, 'they'. But in either case the sense seems curious.

221

FOL. 202

CHAPTER TWENTY

❧

On Problem 8 of the Second Dialogue:
That the [earth's] atmosphere is carried
round by the motion of the earth

1. The sun's rotation causes the air round the sun to spin because of the latter's tumescent motion or its [self-]dilatation. 2. A refutation of certain arguments customarily brought forward as suggesting that the air has weight.[1] 3. Why the water in the sea has no perceptible weight. 4. That although the air is an effluvium of the earth, it cannot have any weight. 5. That even though the air has weight, it is not necessarily borne round by the earth. 6. That though the air likes to cling to something dry, and resists the earth's motion hardly at all, it is not necessarily whirled round by the earth. 7. That our author's argument to prove that the air rotates is unsound; it is to do with clouds and a [particular] comet. 8. Motion does not have to cease when the mover leaves off.

FOL. 202v

1. We may accept, as we said in the chapter immediately above, that the rotation of the sun about its centre carries the air round with it. Now, apart from such spin, the sun has another motion, which causes it to spread out and swell in every direction and by which it illuminates everything. When, by means of this motion, the sun thrusts forward and compresses the air on all sides, that other motion, one of rotation, is added, which cannot but draw the air round. Now if the sun possessed, not the motion by which it drives the air forward, but only the

---

[1] Perhaps throughout the chapter 'has weight' should read 'gravitates'.

rotation about its own centre, there would be no reason why the air should follow the motion of the sun. For if the air does not move strongly towards the sun, or the sun towards the air, but if they merely touch each other, with no pressure or *conatus*, it is not at all evident how the sun can in any way impart that rotary motion to the air. I think our author has realised this; so, in order to make it feasible for the air to be carried circularly by the earth's motion, he assigns to the air weight, which, when[2] [the air] seeks the earth's centre, makes it press upon the [earth's] surface. Owing to this pressure, as if in an embrace, it clings so fast to the earth that it is carried round by the latter's daily motion.

FOL. 203

2. Even until the present day it has been a matter of dispute whether the air has any weight, or whether it has none. Of those who think that it has weight, some consider it reasonable to argue that as soon as you dig a hole in the soil the air rushes down into it. But by the same reasoning they could equally [well] prove that air rises; for if a hole is bored in the ceiling of a bedroom that is sealed up, the air will ascend and fill it. The reason for both these facts is that the borings from the hole are removed into a different place [from where they were], and that, from the place they are moved into, they drive out the air that was there previously; the air so driven propels other air, until at length the air nearest the hole is driven towards it and enters it. In this process nothing takes place that is caused either by gravity or by levity.

FOL. 203v

Others seem to have settled the [question of the] air's weight for themselves by means of weighing; they have found that an inflated bladder weighs more on the scales than the same bladder uninflated. Now, first of all, it is possible that the air blown in, i.e. forcibly introduced into the bladder and not only enclosed in it but also whirled round in a ceaseless circulation brought about by the blowing-action, seeks the centre of the earth not because of gravity but because of [its very] turbulence. It is not yet clear what constitutes gravity, i.e. what gravity derives from, for stones and the other bodies we call 'heavy' when thrown upwards fall again as though of their own accord and for no apparent reason. Unless, therefore, we know what gravity is, we cannot know whether that extra weight of air

FOL. 204

FOL. 204v

---

[2] Perhaps, 'because' (text has only the participle).

enclosed in a bladder is due to gravity or to some other cause.

If the air is understood as being contained and enclosed within a bladder that has not been inflated but is pressed upon gently [by the air] all round it, there is no reason why anything should add to the bladder's weight.[3] Even if a bladder is filled with water this adds nothing to the weight if the bladder is weighed on a balance immersed in water in the same manner as we weigh an air-filled bladder in air. From such arguments, therefore, it is not adequately made out, nor do I believe it to be at all true, that it is the nature of air to press towards the centre of the earth for the same reasons that water, earth and the other heavy things descend.[4]

FOL. 205

3. Since it is commonly said that the [four] elements have no weight when in their [proper] place[s], let us see how far and for what reason this may be true.[5] It is well enough known from experience that the water does not tend downward in the sea, for those who have been plunged beneath the water, even as far as the sea-bed, feel no weight of water resting on them. But this is not due to 'the [sea's proper] place',[6] because the same thing happens at the bottom of standing-water that is not in the sea. The sea itself, that tends perpetually towards the centre, with its whole mass and any single part of it, shows by that very striving that it is not in its own place, where it would properly be at rest. Besides, if [the need of water to seek its] 'place' be the reason why the water does not press down upon a man held under it, how comes it that the earth, which is just as much in its [proper] place as the sea [is in *its* place], *does* squeeze and press upon a man who has become buried?

FOL. 205v

The true reason why the weight of the water is not felt

---

[3] How can the air press on a bladder, inflated or uninflated, without adding to the bladder's weight? If the bladder in uninflated, how can it be said to contain air?

[4] Perhaps, 'It is not for the same reasons for which water, etc., descend that I believe it to be true that the nature of air is such as to press towards the centre of the earth.'

[5] Properly, one translates the indirect question '...is true'; but perusal of the argument that follows shows that Hobbes believes the 'reason' not to be true.

[6] Such seems the natural interpretation; clearly the reference is to the four elements' ceasing to sink or rise when they have reached their 'proper' place. Hobbes may mean, however, for 'the sea's "proper place"', 'their position under the sea', or 'the situation of the sea-bottom'.

underwater is this: a man, or any object other [than a man] which is immersed in water, displaces a quantity of water equal to his [or its] own volume [*moles*]. The displaced water, in order to ensure that its distance from the centre remains equal with [the distance of] the rest of the water [from the centre, then] presses towards the bottom, and the water driven from there pushes upwards and repels the thrust of the water resting upon it.

So that this may be more intelligible, let A, B be [points on] the circumference of a great circle on the surface of the sea and CD be on the surface of the earth where the sea-bed is.[7] Let the centre of both circles be O, namely the centre of the earth, and let E be a submerged body [between AC and BD]. Now if the water pressing upon E, i.e. that water contained within the straight lines AC and BD, exerts pressure on E, then the water on the [curved] line AB must sink towards E. If the water does sink from AB, then a like quantity of water must rise from CD, and with the same force. Therefore the water that was sinking will be thrust back a distance equal to the distance it was sinking through. Hence the water at AB exerts no force on E and therefore does not press upon it. If ABCD is any container, and is not in the sea, the same thing happens, unless the body E exactly covers the entire bottom of the vessel, making it inaccessible to the water; for in this instance the body E will be the equivalent of the base the water rests on. Therefore, just as the base of a vessel may happen to be split open, through an excessive weight of water, so also a body which exactly covers the whole base can be crumbled to pieces [by the weight of water]. With the earth, however, things are otherwise, for a higher area does not cause a lower one to swell up against it with a pressure contrary [to its own].

4. The reason our author adduces for the air's having some weight–and the more so, the nearer the earth it is–depends on his hypothesis that the air is nothing but a kind of vapour from the earth, or is a gaseous fluid. Even if we grant this hypothesis, though it is not a very convincing one, I still do not see why it is not more correct to say 'A vapour of a heavy body is heavy' than to say 'The vapour of earth is earth'. Indeed, even though we must call 'earth' the whole of the

---

[7] Figure 33.

sphere whose radius stretches from the centre of the [planet] earth to the highest vapours, then [under these circumstances] air is correctly termed 'heavy', for air is [then] earth, and earth is heavy. Conversely, if all air is a vapour of earth, and no vacuum exists anywhere,[8] then there must have issued from the [planet] earth a quantity of [the element] earth in excess of what the [planet] earth itself contains; this difference between the two amounts being that by which the sphere whose radius stretches from the centre of the earth as far as the moon exceeds the [size of the] earth itself. This is impossible. But if the effluence consists of vapours and earthy particles exhaled but scattered, it is not surprising if such a vapour is heavy; but the latter is no longer the air under discussion, for the investigation is about the weight of air, not about the weight of vapours or of soot.

FOL. 207v

Again, if it be postulated that the whole of that body, whatever it be, that we call air–be it vapour or pure ether or some mixture of the two–nevertheless will have weight even though there is no vacuum, then the consequence will be that such air itself has no weight.[9] Since the whole of that [mass of] air is of a spherical shape, there must be created, when the air's outermost parts thrust downward in spherical formations that grow ever smaller,[10] a kind of enclosing vault. Hence if one imagined the earth as being removed, then because of this vault the air would descend only if what was contained within the outer (and larger) sphere could also be included within the smaller. Without a vacuum this is impossible.[11]

---

[8] I.e. the vapour is solid, as contrasted with what follows concerning vacua among the vapour's particles.

[9] Partly because of the punctuation of the original the meaning is not fully clear. (Removal of the comma after 'it be' alters the sense, and the exact placing of the parenthesis is dubious.) The Latin of 'even though... vacuum' is ambiguous: *nullo existente vacuo*: (1) '[though (or when) in the situation described] no vacuum is present', or (2) 'even if [we concede that] a vacuum does not exist'. Perhaps Hobbes means: 'One postulates that air has weight and that vacua do not exist; but these claims are incompatible, as we shall show'.

[10] 'Downward', i.e. towards the earth, 'inwards'. What is the cause of this thrust? Weight? Gravity? And what is the nature of the 'enclosing vault'? It can only be air?

[11] I.e. if the earth was withdrawn, the air would still not descend to fill the gap left. 'Without a vacuum': elliptical: (1) 'unless there was a vacuum present', or (2) 'unless one grants the existence of vacua' (whether contained by boundaries and absolute, or dispersed among the particles of bodies). Drake, p. 244.

5. Very well, then; say the air does have some weight and rests on the earth—it follows, not that all the air as far as the moon, but only that the air contained in the irregularities of the earth's surface, is carried round by the earth's spin together with the earth. If the surface of the earth were indeed smooth and mathematically globular it would be inconceivable that any part of the earth's surface could be on the same circumference of the circle as any part of the air,[12] and hence that a part of the earth should drive a part of the air through the circumference of a circle.

Certainly I see that men on board ship are carried forwards by the motion of the vessel.[13] That, however, is not caused by the craft's motion alone: there is needed partly [i] a hardness of body[14] whereby part clings to part, for it is thus that stones and other hard and heavy lifeless objects stowed in a ship are carried forward together with the craft, and also [ii] (if a man is to be carried thus) that motion termed 'animal motion'. Otherwise, while his feet are carried onwards, his head will still resist the motion, and he will fall flat. We see that this happens to those who have not yet mastered, through experience of seafaring, the art of adapting themselves to the motion of a ship.

6. Our author, however, on his page 206 derives the motion of the entire air from that of the earth, thus: 'The following, I see, constitute the reason why the air is not forcibly separated from the earth: gravity and that [property of] adhesion by which damp objects tend to stick to something dry; but I see no reason why [the air] resists motion'. A little further on he concludes that 'all this air, therefore, will be carried round with the earth, to the same degree, in fact, as the earth is carried'.

As to why the air is not forcibly separated from the earth, the first of the [two] reasons, namely gravity, we discount from what we have said already, because if the air does gravitate [initially], this will set up a vault-formation; hence the air cannot gravitate. But the second reason, i.e. the things that

---

[12] The innermost circle of air lies outside the outermost circle of earth, nowhere overlapping.

[13] The transition is abrupt; Hobbes is taking up a point made by DM, p. 207.

[14] Or, perhaps better, amending the text: '...needed a hardness of the parts of a body'.

FOL. 209v

adhere to something dry, has no relevance save as regards that part of the air which touches the earth. For the upper air is in contact not with the earth or with [anything] dry, but with the [lower] air and with dampness.[15] Hence there will be no adhesion of air to air except [that of] contiguity, which does not seem to conflict with the [concept of] separability. But in saying 'I see no reason why the air should resist the earth's motion'[16] he acts rightly. From this it may be inferred that the air below the mountains is necessarily borne round by the earth, because the parts of a mountain and those of the air that attaches itself to it are in the same circle [of rotation]. Hence, as the mountain moves through a circle the air will be driven forward through the same circle. But this has no connection with the air that rests above the highest mountains: for air to move, it is not enough that it does not resist; a mover must be present.[17] Again, beyond the highest mountains no part of the earth is on the same circumference of the daily circle [of rotation] as any part of the air is. Therefore no part of the earth will move any part of the air through the circumference of the daily circle, whether the air resists motion or not.

FOL. 210

7. To prove the motion of the air to be circular he brings forward on his page 208 two ideas, one connected with clouds, the other with the comet of 1618. He says it is arguable that the air is borne circularly with the earth because the clouds are not seen to move swiftly westward. [I say:] Consider, however, the clouds that are lower than the highest mountains (that is, as many people believe, *all* clouds). Undeniably, the clouds are not seen to move westward, because the air in which they rest is incontestably moved eastward with the earth in its rotation. The impulse from the mountains [in motion] (they being on the same circumference) supplies [the clouds] with a point of contact [*labé*, with the mountains]. But as for the clouds–if such exist–at a greater altitude, no reliable evidence is to hand: the winds vary, and these clouds are themselves constantly melting away. Yet whenever any particle of a cloud had an eastward motion impressed on it before it rose above [*egredi*] the

FOL. 210v

---

[15] Or, 'with damp air'.
[16] Properly, 'he says he sees no reason why the air *does* resist the earth's motion'.
[17] Perhaps, 'That the air offers no resistance does not put air into motion; it must be [already] in motion as it approaches'.

height of the mountains, however, such impression [*impressio*] must not be construed as being lost at once. By virtue of the motion received from the earth, then, a cloud may follow the path of the earth to some extent, even if the air itself in which the cloud exists is caught up only slightly in this same motion.

From the 1618 comet he adduces evidence so puzzling that I do not know what he is saying. 'At length', he writes on his page 209, '[the results of] an incontrovertible experiment of his [Fromond's] convince us that the [comet's] tail, which was very close to the moon, was carried circularly with the earth. It must be admitted of necessity that it moves away from the sun and returns to it.'[18] What is this [talk of the] 'comet's tail that was very close to the moon'? [a] [Was the tail] very close to the body of the moon? Or [b] was the altitude, or the distance of the comet's tail from the earth, very nearly equal, i.e. almost equivalent, to the distance of the moon from the earth? In which of these two senses he intends to be understood I do not know. But far from our having to concede that a comet has moved away from, and has returned again to, the sun, the contrary is apparent from this very account of the comet under discussion. Just as, [to an observer on earth,] the sun changes its position daily, so a comet[19] does likewise, not because it is rotating, but because the earth is. As in reality the earth is moved from west to east [in its daily rotation], then the comet, too, like the sun, will appear [to us on earth] to be moved from east to west, except inasfar as the comet is said to have previously veered a little towards the west by reason of its own proper motion.[20]

FOL. 211

8. After satisfying himself that the air is borne round with the earth, he puts the question: If the earth were suddenly to stop moving, would the result not be that the air in this rapid spinning-motion would hurl off all the bodies resting on the

FOL. 211v

---

[18] Even though the words 'at length' are in the *DM*, it is not clear whether Hobbes intends them as part of his quotation. A seventeenth-century reader of the copy of *DM* now in the British Library alters White's words–perhaps rightly–to the effect that the fact here given confirms an experiment of Fromond's, not the experiment the fact. And 'of necessity' may go with 'returns'.

[19] Apparently any comet, not necessarily that under discussion. See also fol. 54v.

[20] I.e. the rate of the comet's observed east-west progress will be reduced by the rate of the comet's own west-east motion relative to the sun. But the relative motion of the comet will still seem to be from east to west.

earth? This, he says, will not happen, for the following reason: When the earth's rotation ceases (which is the cause), the spinning of the air (which is the effect of the cause) must cease at the same time. To see whether this is correctly argued we must revert to [the question of] the nature of motion discussed a little earlier.[21]

First, it is agreed that nothing begins to move of itself, but that everything moves for a reason, and has as the initiator of its own motion the movement of some body that touches it. This is so because–to put it briefly–a body, once it starts to move, has within itself everything necessary for motion. It follows from this that, if it has been moved of itself alone, [the body] possessed, before it was moved, everything necessary for motion. So it ought to have been moved before [it actually was]. It does not, therefore, begin to move, as was imagined [by White], now for the first time. For the same reason one can prove that nothing in motion can acquire rest from itself. But just as for [generating] motion in a body at rest, so for [generating] rest in a moved body, some agent is necessary that is external to, and in contact with, the body. Next, suppose that in its daily rotation the earth was drawing the air along with it, and an external agent presented itself that resisted the earth's motion and suddenly stayed the earth, bringing it to a halt. Then if the air is brought to rest together with [the earth], the same (or another) agent must also be applied to the air in order to prevent it from moving.[22] The reason is that the earth's coming to rest or being at rest, [x], is not an action (for all action is motion). Nor is x applicable to the earth [when the latter is] in a place facing the air so as to resist [the air]. Nor, if (as our author tells us) x were put to the scrutiny [apponeretur], would x be capable of arresting the air; but [the earth] itself, being (according to Problem 5 of the Second Dialogue) something that has no resistance to motion, would be carried away by the same motion of the air.[23]

---

[21] Perhaps the middle of the sentence should be: '...we must consider again, in rather greater depth': Hobbes's *altius* is ambiguous. But either interpretation is possible here.

[22] Either, 'to hold it at rest' (*sistere*) or, 'to prevent it advancing [when the earth is stationary]'.

[23] The last two sentences (part of a single sentence in the text) are carelessly composed by Hobbes. It is unclear what the subjects of several of the verbs are. One assumes that the subjects of consecutive verbs are the same: hence the device of x. For x read, perhaps, 'agent' in the two sentences.

Moreover, frequent experiments show that when the cause of a movement ceases, the effect [of the movement] does not. I say 'effect', not only 'work', for no-one can doubt that, though the painter dies, the picture remains; when a bow ceases [vibrating], the arrow continues in flight; when a ship runs aground, those standing on the decks are straightway flung on to their faces; and after gunpowder is consumed, the discharged ball flies onward. Why, then, when the earth stops moving, should the air not continue its movement if nothing else hinders it?

Our author had taken note of two experiments which seemed to contradict him. One was that a stone falling into water moves it to such a degree that on its surface appear little circles which grow ever wider, such movement being continued even when the stone has lain on the bottom for a long time afterwards. The other [observation] was that, when the mover ceases, the motion of pendulums returns to rest only gradually and after a long time. To these [phenomena], therefore, he assigns a particular cause, namely that upon the mover lies its own gravity.[24] [I answer:] But what if [for a thing] 'to be moved by its own gravity' does not mean 'to be moved of itself'? Furthermore, as substance is 'whatever is moved', so substance is also 'that which moves [something else]'. Thus gravity, which is *accidens*, does not [itself] move; indeed, gravity is the commencement of motion and therefore *is* motion but cannot be *in motion*. But as regards the case of the arrow or the cannonball projected perpendicularly upwards and continually striving to rise, even when the bow or the charge has ceased to act, will he ascribe this also to gravity? Will he say that stones, iron and lead are carried upwards by gravity? It seems he will. 'Perhaps', he says, 'gravity or some cause like it will be found, if we look more closely, [to be] the common cause of every motion that does not cease when the mover ceases.' Yet what he adds—'some cause like it'— is of no assistance to him, because gravity alone resembles gravity. The reason why the motion of pendulums is sustained because of their contrary movements, the rise and the fall, in the familiar manner, i.e. through oscillation, is twofold: The descent is due to the same cause as gravity is, but the rise is due to the fact that the fall continues obliquely. Thus it seems that, of all experiments, this motion of

FOL. 213

FOL. 213v

---

[24] Perhaps White means: '...a particular cause, namely a mover exerting its own gravity (or weight) upon the pendulum'.

pendulums is the one that conflicts most with our author's opinion when he says that 'withdraw the mover, and the motion ceases'. For the pendulum, when it falls through the force either of something that drives it or of something that pulls it towards the earth, strives continually to move even against the force of the same propellent or attracting-agent. That is, when the mover retires the motion persists, against the force of the mover.

CHAPTER TWENTY-ONE

FOL. 214

On Problem 9 of the Second Dialogue:
That a body is moved in position either
of itself or *per accidens*[1]

1. The circular motion of the entire air with the earth discounts our author's teaching about the earth's motion. 2. What he reverts to, concerning the nature of 'place', conflicts with what he said about place in Problem 3 of the First Dialogue. 3. That place is not the outermost surface of a body itself. 4. A distinction proposed by our author between 'proper motion' and 'motion *per accidens*'. 5. [Some] absurdities necessarily arising from this. 6. Why, according to our author, the earth's daily motion is to be called 'true and real motion'; and the nonsensicality of his reasoning. 7. The distance from the centre is insufficient to cause the outer limits of a sphere to move together with, and faster than, the middle. 8. That the motion of the parts close to the centre of a sphere or of a wheel is not an efficient cause of the motion of the parts on the circumference. The efficient cause, both immediate and mediate [at one remove], of (9) straight motion, and (10) rotary motion, in hard bodies. 11. The earth causes to rotate, not all the air as far as the moon, but only that portion contained within the irregularities of the [surface of the] earth. 12. An inspection of the comparison our author draws between his own teachings about motion and those of Galileo. 13. A scrutiny of [our author's] unfair criticism of certain experiments that are cited by Galileo. 14. That the velocity of a body in motion is fixed by its own mover,

FOL. 214v

[1] The main parallels here are fols. 109, 292v and 316.

not by a resisting body only. 15. That our author's account of a Turk's falling from the top of a tower is incorrect.

FOL. 215

1. In the Problem immediately preceding, our author had contended that all the air as far as the moon is carried round by the earth's motion with the earth, namely from west to east; he now endeavours to show how this can occur. If he does so, [however,] it seems to me he will vitiate his own earlier explanation of the earth's movement. He derived the latter from the motion of the waters on the sea-bed that surge from west to east. The motion of these waters he inferred from the motion of the waters at the surface of the same sea that flow from east to west; and the movement of these surface-waters is brought about, he declares, by a wind that always accompanies the sun from east to west. Now, [I say,] either that wind from east to west *is* the air itself, so moved; or at least it moves the air, as described. Hence [in either event] the air is carried from east to west. Again, if what he says is true–that all the air is carried

FOL. 215v

with the earth from west to east–then the same air will be carried simultaneously in two diametrically opposed movements, which is ridiculous.

2. Before showing how the air can be whirled round by the earth, he distinguishes, on his page 213, motion into 'motion, properly termed' and 'motion *per accidens*'. To make this comprehensible, he gathers up certain of the points about the nature of place that were confuted [by me] above. 'You are aware', he says, 'that place is defined as: the outmost surface of a thing, which puts the thing into contact with the edge of the medium that surrounds it'. (Let us omit the fact that these words contradict what he said on his page 28 [fol. 17 above], where he writes: 'For "place" everyone understands "that which encloses and (as Aristotle calls it) is some kind of unmoving vessel"'.) Now, upon this definition of 'place' depends the whole reasoning by which White has wished to prove that there cannot be more worlds than one; for 'Place is the concave sur-

FOL. 216

face of the enclosing body'. But by no means does he prove there cannot be another world; since [if there were] it would lack an enclosing medium within whose surface it could be contained. Here, however, he no longer calls 'place', 'the concave surface of the ambient medium'; it is now 'the convex surface of the thing itself that has been placed'. Let us leave

CHAPTER XXI: 4

this contradiction out, be it as it may, and see what comes next.

3. He adds that 'place is relative, as are something "similar" and something "equal", etc.' Now, all things other [than place] that are interrelated have some basis of connection, e.g. when two white things are similar, whiteness is the basis of that resemblance, and when two bodies are both a cubit high, that cubit is the basis of their being equal. Similarly (he declares)[2] it is also necessary for 'place', being something relative, to have some basis for its relationship [with other things]. But what this basis is he does not know. 'It is hard to assess', he says, 'what the fact of *being in a place* has to offer' (i.e., as he puts it, what basis of relationship it has [with something else]). Suppose that space is the surface of a containing medium: it is [still] difficult to acknowledge that the basis of a relationship is the constitution of the body contained in a place; for as the surface of the container is termed the body's 'place' because the body is tightly enclosed [by it], so this constitution [of the body] will be the basis of the relationship of it [the place] to the object contained. Likewise if something is said to resemble something [else], in that it is white, white will be the basis of that similarity. Since, however, he has altered his definition of 'place' (it is no longer 'the surface of the enclosing medium' but is 'the surface of the thing itself placed'), he strives in vain to prove that place possesses a basis of relationship [with anything]; no longer will it be relative. We have shown above, in the said Second Problem of the First Dialogue, that 'place' is not the surface of an ambient [medium]. It is accepted that ['place'] is not 'the surface of the body itself that is placed', in that if a body were moved, its place would be moved at the same time. From this it would follow that all moved things are always in the same place, i.e. that moved bodies are not moved; for things that are always in the same place are not moved. I still do not see why our author has wished to alter his former definition of 'place' to this latter one.

FOL. 216v

FOL. 217

4. Having propounded these views on the nature of place, 'That body', he says, 'in which a change of the joining of surface to surface occurs, *verè et realiter*, has *verè et realiter* that pro-

---

[2] But the preceding statement is not in *DM*; it is Hobbes's own.

FOL. 217v

perty³ from which it follows self-evidently that the body has changed its place (though it is contingently possible that this same body has not changed its place). That thing whose surface does not alter its contact with [the ambient medium] does not produce this effect'. This means: 'When the surface of a body is not in contact with the same surface of the surroundings it was in contact with previously, then the body is said to have been moved, *verè et realiter*; when it is in such contact it is moved, not *verè et realiter* but *per accidens*'.

FOL. 218

5. The first absurdity we may note in these words is that enclosed in the parenthesis, namely '(though this body may not have been changed in respect of its place)'. It is most glaringly inconsistent to say 'a body has such a property as will subject it to a change of position' and yet that 'the body may not have experienced a change of place'. The second absurdity is this distinction between two [kinds of] motion: [*a*], motion 'truly [defined]', and [*b*], motion 'not correctly [defined] and not in accordance with the truth of the matter'. This sounds the same as distinguishing motion into 'motion' and 'rest'; that which is truly not in motion is in fact at rest, and as often as we say 'It is moved' we do so erroneously and falsely.⁴ The third absurdity is that this distinction [of his] would imply that whenever a tower, or any other stable object, was faced by a different front of air as the wind drove [the air] about, we could say, *propriè et verè*: 'That is the cause that moves [the tower, etc.], for in this instance the juncture of surface to surface is altered'. The fourth absurdity is that a result of the same distinction would be that anything can be transported as far as the Indies without having been moved; for wine sealed up in a flask can be sent there without the [wine's] surface's altering its contact with the surface of the flask. Here is the final absurdity: A consequent [of his distinction] is that the earth, with which all the air is carried round and which is postulated as being the reason for such a rotation, is itself unmoved. For if, [I say,] all the air is borne circularly with the earth, then the same surface of the earth must always be attached to the same front of air; therefore the earth will not be moved—yet the earth itself is supposed to bring about the actual rota-

FOL. 218v

---

³ See the Introduction, Section VII, on technical terms.
⁴ I.e. we are wrong both as to fact and as to terminology.

CHAPTER XXI: 6

tion of the air! Nothing is more common than for nonsensical things to be deduced from nonsensical things; no wonder, then, that a false conception of 'place' generates preposterous dogmas concerning motion.

6. This last absurdity he attempts to answer. On his page 216, 'In our enquiry', he writes, '[we have seen that] the earth has a motion due to a friction with the waves. In part, the earth moves away from them,[5] in part it draws them along with it. And to the extent that it moves away, so it begins to experience the correct [*proprius*] condition for motion. Inasfar as it draws them with it, it is held together [*contentus*] by the common nature of the same [motion]. For you see that the waves have two motions: [a] a proper one, by which they always wash against the earth, some parts against some waves, some parts against others, and [b] a common motion, by which they [the waves] are carried in a circular motion on the back of the land. Nothing arises here that represents any need for you to untangle this problem more broadly or more clearly.'

FOL. 219

He wishes, therefore, [a] the motion of the earth to be 'proper' motion, for the following reason alone: When the wind is driving it, the water is not always in contact with parts of the earth along the same surface; and [b] that that varying of surface arising from the ebb and flow of the water is, *propriè verè et realiter*, the earth's motion.

He falls, therefore, into the third absurdity of those I have listed in the preceding article: Owing to the ebb and flow, different surfaces of the water are in contact with other parts of the earth. So, [I say,] for the same reason one surface and another of the air will touch some parts or others of a firmly-fixed tower owing to buffeting by the wind. Just as, in the latter instance, it is not the tower, but the air, that is truly moved, so in the former it is not the earth but the water. Since the surface of the same particle of water, although not always directly above the same particle of the earth, does not quit it for some time,[6] he ought to have given a reason [to explain] not

FOL. 219v

---

[5] I.e. the earth moves; the seas move a different distance in the same time.
[6] The translation aims to preserve Hobbes's ambiguity here: (1) 'A long time passes before the particle of water abandons the particle of land beneath it', or (2) 'The particle of water is distant from the particle of earth for a long time'. But he may mean: 'One particle does not stray far from the other particle'.

FOL. 220

only how this pulling-apart of surfaces constitutes motion but also how the motion would be so strong that any individual particle of the earth close to the equator covers fifteen miles every minute,[7] whereas the actual water is moved forward only slowly by the wind which, he says, exists beneath the sun.

Now, I see that a horse is roused to swift motion by a slow cut of the whip; no matter what our author says about the earth's being lashed by the seas, however, a reason applicable to animals is not valid for the earth.[8] He ought to have shown also how what he laid down previously—that that variability in the connection between the surfaces of the water and of the earth is the cause of the earth's motion—agrees with what he says now. Earlier, on his page 178, he told us that the cause of the earth's daily movement is the waters whose depths, being in contact with the earth, would be driven from west to east.

FOL. 220v

Now he asserts that that forward thrust of the waters is not the cause of the earth's motion, but that its [earth's] own motion, real and actual, is.[9] He ought, I say, to have adduced some reason for, or given some account of, these matters, but he declares flatly that he is not going to do so, stating merely: 'There is nothing here that represents any need for you to untangle this problem at greater length or more clearly', i.e. that the difficulties he has raised for himself through his ridiculous definition of 'place' quite inhibit his further examining them.

FOL. 221

7. Next, having accepted all the preceding [as being true], he ought to have shown how, by its rotation, the earth can generate a circular motion in the upper air. Here the motion should be faster than the earth's in accordance with the ratio of [a], the radius of the whole sphere comprising the earth and the air as far as the moon, to [b], the radius of the earth alone, i.e. about sixty times greater, for it seems impossible for a moving body to impart a velocity greater than its own to a moved body. He therefore brings forward the instance of a rotating grindstone on which iron tools are being sharpened, the parts near the centre being moved slowly, and those near the circumference quickly, according to their distance from the axis.

---

[7] Cf. fols 174, 177, 185v.
[8] Cf. fol. 201v.
[9] Or, 'but is itself its (the earth's) motion'.

From this example, indeed, it is clear that such is the case; but whether the parts close to the circumference derive their motion from that of the parts around the centre cannot be proved from examples. Nor does he give a single reason to explain why, in the rotation of hard bodies, the motion of the parts further from the axis is swifter than that of the parts closer in; but, he says, 'No reason need be supplied other than the subject's condition itself,[10] which compels the more remote objects to be interchanged with the nearer ones'. We query, however, in what that 'state of the subject' consists, if the parts further from the axis have to be rotated at the same time as those nearer the centre. We see that this happens with hard bodies, but not in liquid ones, for if someone were to whirl water round in the middle of the Caspian Sea, I would not imagine that this would cause the whole of that sea to spin round sympathetically, even as far as its shores! That condition, then, is not of any subject–only a solid one. We were expecting, therefore, and rightly, from our author an explanation why the rotation that he ascribes to the sphere made up of the air and the earth (i.e. the rotation of the furthest parts together with the centre) would be more in accordance with the rotation of a grindstone than with that of the water in any great sea; for it appears that the nature of air is closer to that of water, a liquid, than to that of a grindstone.

FOL. 221v

FOL. 222

8. Our purpose, then, is to look into the means by which, in the rotation of a wheel or a sphere constituted of hard material and having its parts firmly cohering one to another, the motion which, near the centre, is slower than any given motion, becomes even swifter with the increase in distance from the centre, finally becoming faster than any given motion. We shall first note the equivalent: that a slowness greater than every given slowness, as though from an efficient cause, cannot produce a velocity greater than any given velocity. On the contrary, it is universally true that no body can produce, by acting with its own motion upon another body equal to itself, a motion more swift [within the second body] than the agent itself has produced. Far less, then, can a corpuscle located at the centre of a very large sphere, and there rotated very slowly, set [the sphere] rolling at that very fast rate necessitated by the [sphere's]

FOL. 222v

---

[10] What White means by *conditio subjecti* is unclear.

surface's being very far removed from its centre–set it rolling, that is, by itself, by its own force alone, and without the aid of any agent.

Imagine a sphere, of centre A and radius AB, that is supposed to turn about the centre-point A at any given rate.[11] Close round this ball let there be placed a spherical shell whose thickness BC equals the radius AB. So if the set [rate of the] motion of the ball AB is, say, 1 [cm/sec], that of the shell AC will be 2 [cm/sec]. (I refer to the outmost circumferences of both spheres.) The single motion at B, i.e. 1, cannot, I say, generate the double motion, i.e. 2, at C. Now suppose that while B is moved at its own pace, i.e. 1, the shell BC is placed round the ball AB so that BC suddenly becomes affixed to it and they constitute a single hard body. Since that shell is not yet in rotation, but has resistance to motion, it will resist the rotation of the ball AB. From now on, therefore, B will be moved more slowly than before, i.e. than rate (1). Hence C will also be moved more slowly than rate (2). So motion at B does not generate motion at C in the ratio of their [respective] distances from the centre.

9. We must note, furthermore, how a straight motion is propagated in a hard body. Imagine a straight wand of wood, iron, or any hard material, lying on the ground in a rather slippery spot, so that the wand is pushed by prodding with the finger at one end. It is through the pressure of the finger, therefore, that motion is imparted to that end. Its material, however hard, yields laterally to this pressure by a very small amount and hardly perceptibly, until, compressed more than its nature will endure, [the rod] springs straight again in order to recover its former state. Hence the further tip begins to move in a straight motion at the same instant that the tip being pushed commences to move, and both ends at the same pace. Now if someone asks: 'What is the efficient cause of the motion at the far end of the rod?' I reply: 'The immediate cause is indeed that motion of resilience which the parts have collectively, or [the motion] because of which the rod is solid. The mediate cause is the motion of the finger that exerts the pressure'. Therefore the motion of the rod as a whole, which is the same as

---

[11] Text: 'at any given motion'. Figure 34. In what plane is a section through the sphere's centre supposed to rotate?

that of both ends, is generated in the same instant at which the finger's *conatus* reached a degree of motion sufficient to move the wand; and that degree of velocity with which the finger was pressing was equal to the velocity with which the wand was recoiling.[12] It does not matter if the wand is thicker at one end than at the other, for this reason: Before the wand springs back in straightening itself the compression must always be so strong that [the wand] as a whole springs back; for if the far end is fixed, the other end cannot give.

FOL. 224v

10. In the same manner we may set about discussing rotary motion in a sphere or a wheel composed of hard material. For the hand (or a handle) works on the shaft, not with a *conatus* that is rectilinear, but by twisting circularly. By this pressure the parts on any given radius are forcibly, if imperceptibly, displaced. That is, they resist [the force] as long as their nature allows. [There comes a time when,] owing to the innate motion by which they cohere to one another and in which the nature of hardness consists, they can no longer be twisted round. Then suddenly, all together, and as a single mass they submit to the pull exerted by the handle. The parts near the centre cannot be moved at a pace of their own, however slow, until the outmost parts have been subjected to a twisting-motion so great that they can free themselves only by springing back [from it]–and this at a speed proportional to their distance from the centre. When this twisting has taken place the whole sphere turns, all of it at the same time, at a rate that varies as the distances [of its constituent particles] from the centre in every direction. So the direct cause of the rotation of the parts further [from the centre] is the motion by which the sphere frees itself from the torsion and from the forcible separation of its parts; but the indirect [*mediatus*] cause is the hand or the crank; and the larger the sphere to be rotated, the greater the application of effort needed to turn it.

FOL. 225

11. On the other hand, if a globe is hard near the centre but fluid in the other parts, as is the sphere comprising the earth (i.e. a hard substance) and the air (a most fluid body with very easily separable parts), there is no necessity for the fluid parts to be rotated also when a hard part moves. For as they have no

FOL. 225v

---

[12] Tenses are as in the original.

resistance, or an imperceptible one, to being separated, there can be no reason why they should resume their previous form, [as would solids theirs]. Indeed, where there is no restoration of the parts to their previous position there is no direct efficient-cause of rotation. Therefore all the air from the earth as far as the moon is not rotated together with the earth. It may be that the air caught up between two mountains is rotated with the earth–indeed, this must be so if the earth revolves; such is Galileo's view.[13] Furthermore, Galileo states that bodies projected upward follow the earth's motion. Correctly so, unless they are shot too high for the air to be carried with them.[14] We must not overlook the fact that missiles retain that impetus which they had received from the earth's motion before they were shot off; hence they also follow the earth's rotation,[15] but not closely, because the velocity they attain [when fired off] on the earth, [i.e.] in a lesser circle, does not agree closely with [the velocity of] a rotation that takes place in the air within [sub] a greater circle.[16] If, however, projectiles are shot above air that is hemmed in by mountains, then they are obstructed by that air, and they gradually lose their impetus towards the east, that is, when the air that is not borne circularly with them resists [them].

FOL. 226

12. From this, [our author] proceeds to compare his own tenets, under certain headings, with those of Galileo on motion. White says on his pages 218 and 219 that, in Galileo's contention, common motion, even with respect to particular things, is in fact motion, but he [himself] contends that it is not true motion. To understand the circumstances of the problem, imagine that a man driving in a carriage is setting out from here [Paris] to go as far as Lyons. The motion of the carriage and of the man is the same and is common to both. But if that man

FOL. 226v

---

[13] Drake, p. 438.
[14] Sc. into a vacuum of space beyond the earth?
[15] The context does not make clear in what sense we are to take 'follow' (sequuntur). But the general sense seems to be that to the velocity imparted by the propellent must be added the motion already present (relative to, say, the sun) in the gun before firing. Hence part of the missiles' velocity is due to, and matches, that of the earth.
[16] Presumably the 'lesser circle' is that of the earth alone, and the 'greater circle' that of the earth together with the vapours and clouds immediately encircling it, where in unit time the missiles' velocity along a larger circumference would be slower than when closer to earth: cf. fols 36v, 195v.

happens to cross over from one part of the coach to the other, this motion is particular to the man. Now, following Galileo's opinion, we say that, even when that man does not shift his position from one side of the coach to the other, he is, properly and correctly speaking, moved. Rightly so; for the person who was at Paris is now at Lyons; he is not in the same place he was in; therefore he has, indeed, been moved. Our author denies that he has, because the man always stays in the same part of the coach. Likewise if a ball is shot from a piece of ordnance, the motion of that ball is common to all its parts–as much those near its centre as those close to the circumference. Now, Galileo would also say that those parts near the centre are moved, *propriè et verè*, because they are not where they were before. Our author denies this, on the grounds that the parts not in contact with the air are continually inside the surface of the same enclosing medium, i.e. in the same place. In my opinion, two points escape him. [i] According to the opinion just stated, nothing whatever of a cannon-ball will move save the mathematical surface or, at least, the ball's surface-covering, which is thinner than any measurable quantity. [ii] Galileo asserts that impetus is present in every part of a moved body, but White says that this is true of the outermost parts alone.

FOL. 227

It must be noted, too, that motion and impetus are the same thing, and are called 'impetus', and also 'force' (for 'impetus' and 'force' are synonymous, inasfar as [a force] brings about another motion). Galileo rightly took this point of view; for when, as a ship strikes a reef, the voyagers fall flat with the impact or when, if a horse is startled, the rider is thrown from the saddle by the impetus, who does not see that that impetus (though derived from the motion of the ship [or of the horse]) is particular to the voyagers and to the rider? We see that, of two bodies of the same material, the greater thrust belongs to the greater and the lesser thrust to the smaller. Can we doubt, then, that that difference in impetus is due to there being more material in the larger than in the smaller? It follows from this that everything that contributes anything to the magnitude of a movable body also contributes to the impetus of the same. Hence every part of a body that is moved must possess its own impetus, as Galileo was disposed to believe.

FOL. 227v

FOL. 228

But as for the startled horse and the ship that runs aground, our author thinks he has made sufficient reply when he says on his page 226: 'That there is present [*inesse*] a motion as a result of the riding is one thing; that there is impetus sufficient to generate motion after the source is spent is another'. Who can understand or reason out these questions? Let us imagine that a ship is being carried forward, and a passenger with it. A motion is therefore present in the passenger because of the [ship's] forward movement. Suddenly the ship halts, having run on to land; but the traveller is not cast ashore. The forward motion therefore ceases as regards the ship but not as regards the voyager; it stays with him; hence he falls flat on his face. So the motion forward is the impetus that throws him forward. The same applies to the impetus of the rider whose horse is startled; therefore the motion due to the riding does not differ from impetus or from motion when the cause has died away. What, then, does our author mean when he speaks of 'the impetus sufficient to generate motion after the source is spent'? Does he think that Galileo states or believes that the voyager acquires impetus from the ship or the horseman from his mount, and that this impetus then imparts a motion which causes the man to fall when the ship strikes land or the horse rears startled? Indeed he does, but he is fooling himself: Galileo holds that the man's movement (which derives from his being carried along, and is continued after this latter motion ceases) and the fall itself are, in number, one and the same motion, [and this] no less [so] than that the man who was moved not only with the ship but afterwards also (owing to the ship's running aground and being wrecked) was the same man in number.

FOL. 228v

FOL. 229

Again, concerning the increase in impetus stemming from an increase in the magnitude [of a body], our author thinks he is answering [the problem] when he says on his page 228: 'that a greater impulse is due, not to impetus alone' but also, in part, to the portions [of a body] that are other than in a state of motion, i.e. that lack impetus. So now impetus *is* motion (or velocity); impetus is not the cause or the effect of motion!

Third, he says that, in Galileo's contention, impetus is the quality first implanted in the body itself[17] before motion [takes place], but that he himself believes impetus to be nothing but the swift motion of the body itself. Both these statements are wrong: [a] Galileo says that initial impetus does not precede motion or is anything other than motion itself, and [b] in the

---

[17] Or, 'actually in the body'.

words quoted earlier in this very article our author denies that impetus is the same thing as motion, for he says:[18] 'That a motion is present as a result of the riding is one thing; that the impetus is sufficient to generate motion after the source is spent is another. The latter motion I eschew; the former I accept gladly, as being present in almost everything'.

FOL. 229v

Fourth, he asserts that Galileo postulates two motions, each contrary to the other, in the same body, but that he himself postulates one true motion only, the other being *per accidens* and [therefore] not true. I do not recall Galileo's having written anywhere that there can be two opposed motions in the same body at the same time; and absurdities like this could not have occurred to him. He did hold it possible, I believe, for something that is carried about by something else, or is moved as part of it, to possess also a motion restricted to itself; and he thought that two such motions can have opposing limits [*termini*] and yet be, not two contrary motions, but a single [compounded] motion generated by two things themselves in motion.[19] For instance, if a ship sails onward and if a man meanwhile walks from bow to stern we have here contrary ends,[20] and also two movers, namely the wind that moves the ship (and the man on board) and the mind that directs the man towards the opposite end.[21] But these, [says Galileo,] combine to produce one motion only, or mere rest; for if the rate of [motion of] the ship equals the speed at which the man paces, then he gains no ground by walking thus: he has, indeed, placed one foot forward after the other, but the man as a whole has remained at the same spot. Therefore he was not moved in a direction opposite [that of the ship]; but if the latter's motion is slower than the pace he walks at, then he was moved, in the same direction as that of the ship, a distance that anyone can easily work out for himself.[22] Galileo's view was precisely that just sketched. Our author's claim that he himself is demonstrating [the existence of] one true motion, and of another

FOL. 230

---

[18] Cf. fol. 228v.
[19] Drake, pp. 149, 154, 175–177.
[20] 'Ends': *termini*. Here, and at the close of the sentence, Hobbes seems to mean 'limits reached'. At fols 455vff he distinguishes *terminus* from *finis* ('objective not necessarily, or yet, reached').
[21] Presumably, a limit lying opposite that towards which the wind drives the ship.
[22] The man's movement or non-movement is of course seen relative to a fixed point, e.g. on the land. Hobbes's tenses are here retained.

FOL. 230v motion *per accidens* is tantamount to his saying, 'I postulate two contrary motions in the same body. Of these, one is motion, the other is rest'—which is absurd.

13. The remainder of this chapter contains a censure of Galileo and a disparagement of the [results of] experiments he has cited as being consistent with the earth's moving. First of all, says White, Galileo cannot explain why we do not perceive so swift a motion of the earth [as there is], even though Galileo has shown more than adequately not only that common motion is unperceived, but also that we undoubtedly perceive, as if it were a wind,[23] the motion of resistant air in unenclosed areas and near the equator. Second, the instance of the stone that is released and falls close to the mast of a ship in motion White alleges to be unreliable and contrary to the belief of all

FOL. 231 those who, so far, claim to have made such an experiment. Third, he chides Galileo for holding that the movement of a sphere over a flat surface will, once commenced, continue indefinitely. Indeed, Galileo does believe that any motion, once imparted, will persist for ever unless obstructed by another mover proceeding in the opposite direction. This opinion, as I say, our author criticises, but he offers no reason why matters should be otherwise; he says merely: 'An impressed quality of this kind I know no more than Simplicio does'.[24] Yet (as it seems to me) I believe it no more difficult to grasp [a]: that a thing which has been moved is moved continuously until some body, pressing forward in the reverse direction, itself halts it, than to understand [b]: that a thing at rest is always at rest,

FOL. 231v until some [other] body drives it from its place.

14. Now, Galileo assumes that when a sphere rolls across a plane all external hindrances to motion have been removed.[25] [Our author] construes this as suggesting [that Galileo proposes] the removal of the medium from the resistance of which every limitation of motion or velocity is assessed.[26] This view [of

---

[23] Perhaps, 'as we would perceive a wind'.
[24] In Galileo's *Two chief world-systems* Simplicio represents the Aristotelians.
[25] This is the translation, but Hobbes probably means: 'Galileo asks us to assume the existence of a resistanceless plane, over which a ball is rolling'. Galileo's earliest statement concerning inertia seems to be that in his *Letters on sunspots* (1613).
[26] For 'assessed', perhaps 'differentiated'. White is surmising that Galileo asks us to discount the resistance of the medium traversed. The text

White's] is erroneous; for even were the whole world moved (and it is not situated in the midst of any corporeal medium) and we could not tell whether it was moved or not, its motion would still be of a determined velocity–otherwise it would not be moved, for there is no such thing as unlimited speed.[27] Therefore the velocity of bodies is limited not by the resistance [of the medium traversed], but by the mover. Now, the medium may, by resisting, reduce the velocity (which is limited by the mover); but this does not prevent our assigning limits to motion, even within a vacuum. In supposing the absence of external impediments [to the motion of the earth,] then, Galileo made no unreasonable assumption.

FOL. 232

15. Lastly, White wishes to produce some experiment that will appear to support this opinion of his, that 'the parts of a body in motion–not all of them, but only those nearest to the outmost surface–have impetus'. He relates on his page 228 that, on the word of a man of integrity whom he knew, a certain Turk caused a half-sphere (a kind of container, as the name tells us)[28] to be hung from the top of the tower of St Mark's at Venice. He placed himself upon the vessel itself [*sese imposuisse in ipsum vas*]; then the rope was cut and the hemicycle was smashed into a thousand pieces as it crashed to earth. A little before it reached the ground, however, the Turk, springing out [*ex hemicyclio*], landed lightly and without injury on the pavement.[29]

FOL. 232v

A piece of evidence, this, very well worth relating, and one on which a philosophy such as our author's might well de-

---

implies that Galileo believes that not necessarily all such media offer resistance; if he means all, then 'from the resistance...assessed' should have brackets added.

[27] Perhaps by 'unlimited speed' (*indeterminata velocitas*) Hobbes means indefinite, 'indeterminate' velocity rather than infinite speed, and for 'otherwise it would not be moved' perhaps he intends: 'for in no other way would it have been moved', or 'otherwise, it (i.e. the progress of the world through the skies) would not be (truly) "motion"'.

[28] Perhaps, for the last clause, 'as appears from its shape'. According to Daremberg and Saglio, *Dictionnaire des antiquités grecques*..., a *hemicyclium* was a semicircular structure resembling a summer-house or arbour; here it was apparently a large hemispherical vessel, perhaps of glass.

[29] White, though he does not remember whether the Turk was standing or seated, places him *inside* the vessel (*in hemicyclio*). It is perhaps not surprising–see next note–that it has proved impossible to document the account: even were some kind of parachute used, could one jump nine feet on to some flags and be uninjured?

pend!³⁰ We notice that tumblers need a firm place to leap from. Supported by this, they can draw from its resistance the force required for a leap upward. Or, at the least, they need a rope of some kind, or some other support, from which, when they hang aloft, they can raise themselves up by animal motion. But in the instance under discussion, when the hemicycle fell after the rope had been cut, there was nothing for the Turk to support himself by; hence it was impossible for him to leap out [in flight].

---

³⁰ Hobbes's heavy irony ('Such is the sort of flimsy belief that White can be trusted to rely on') shows that he personally is in no doubt over whether to accept or reject the tale: see the close of the previous note.

CHAPTER TWENTY-TWO

FOL. 234[1]

On Problem 10 of the Second Dialogue:
That circular motion does not, through
its own force, hurl off bodies lying
on [one of its radii][2]

1. Two experiments by which it may be seen that circular motion hurls off the bodies lying on [a radius]. 2. That this cannot be gathered from the said experiments. 3. That bodies driven outwards by things so moved do not fly at a tangent [to the circle of rotation]. 4. That, as shown by Galileo, although heavy bodies are hurled tangentially [on their release from circular motion], heavy bodies situated on the earth are not necessarily thrown off it. 5. Other experiments making it clear that circular motion does, through its force, drive off the bodies lying on [a radius]. 6. That mountains are not necessarily torn apart as a result of such [circular] motion of the earth.

1. It was Ptolemy's opinion that if the earth were rotated in a daily motion the consequence would be that fortresses, towers, trees, mountains and everything else raised above the surface of the earth would all be flung off by that motion. He seems to have arrived at such a viewpoint through his belief that all circular motion by its own nature hurls outward all heavy

FOL. 234v

---

[1] Fol. 233rv is blank.
[2] It might be thought more correct to say '[its circumference]'; but throughout the chapter 'radius/radii' is retained, because what Hobbes says of expulsion applies to any point on a rotated disc (or within a partially hollow sphere), not merely to the outermost or to a body whirled on a string.

THOMAS WHITE'S *DE MUNDO* EXAMINED

bodies lying on [a radius of] itself. Galileo brings two experiments to bear on the question, one being as follows:

FOL. 235

Take a vessel AB and fill it with water. Then with the aid of a cord EF, [F being fastened to the open end, A, of the vessel,] let your hand, at E, whirl the vessel ACB round within the circles B and A.[3] Now whether the slinging-action is done in a horizontal plane, or whether in a vertical, the water contained in the vessel will not trickle out [at A], for it will be drawn back towards the bottom of the vessel. If a hole is drilled here at C, the water will gush out in the direction CD. It seems that this can happen only if the whirling or the circular motion of the vessel exerts on the water [sufficient] force to move it away from the centre [of rotation]. It is also consistent with this experiment that those who quickly swing round themselves some heavy object tied to a string find that they are drawn out of their position [and] towards the circumference of that circle described by the axial motion.

The other experiment is based on the firing-off of a pebble placed in the hollow[4] of a split cane: the motion of the arm that fires the stone is seen to be circular, yet the motion of the stone is most obviously rectilinear. On this evidence, therefore, people infer that every circular movement contains the force to hurl forth heavy bodies placed on [a radius of] it.

FOL. 235v

2. To be honest, I do not see how such a conclusion can be drawn from the said experiments. For if the vessel AB, or the water contained in it, indeed strives to move away from the centre [of rotation], namely from C towards D, as agreed from the experiment, then the hand placed at the centre E must also try to stray from the centre towards the circumference. (Experience also proves this.) The hand, therefore, has a motion compounded of the straight and the circular, so that the path described by the vessel is not a perfect circle but is a kind of spiral shape (in which the circumference is constantly augmented). Hence the water is driven outward not only by the force through which the hand describes a circle, but also by the force that continually enlarges and increases the radius of the circle described; i.e. [the water is driven outward] by a straight motion.

[3] Figure 35. Cf. fol. 121.
[4] Bold and clear in the manuscript, and not queried by Mersenne, the word is *frena* or *crena* (sic); *cavea* seems required; cf. *viscitus*, 102v.

CHAPTER XXII: 2

If the hand does not [initially] travel in a straight movement it never stretches the cord to such a degree as to carry the vessel from the hand as far as the cord will allow. The reason is: To the hand, moving circularly, is added a straight motion arising from the weight of the flask. This last, when it arrives at a position very close to the ground, at GH, by its own weight drags the hand downward. Hence the hand describes a larger circle than before, and when the flask again comes round to GH the hand is again drawn downward. Thus little by little the circular motion grows larger, but of irregular shape, approaching a spiral line. So the instance of the vessel whirled in a circle is not sufficient to prove that circular motion tends by its nature to drive outwards heavy bodies on [the radius of rotation].

FOL. 236

As for the example of the pebble shot out of the split head of a cane, the pebble's motion is indeed straight, but the motion of that part of the cane in which the stone is inserted seems also straight. That propulsion is no different [from what it would be] if the stone were thrown by hand. Even with pebbles that have to be thrown manually the hand is drawn back as far as the shoulder in order to initiate motion in a straight line. Also, if anyone considers carefully the motion of the base [of the slot in the cane] before [the pebble] flies out, he will find that such motion is not circular when close to the point of release.

FOL. 236v

Furthermore, suppose that someone holds in his hand a cord from which some heavy object is hanging. Let him try whether, if he whirls his hand round in a very small circle, however fast he turns it[5] he can make the heavy body whirl round [with his hand] and describe a [perfect] circle with the length of the string as radius, i.e. preserving the ratio of the radius [to the circumference] that must govern the shape of a circle. He will find that it cannot be done. Before his hand begins [to describe] its circle he must impart a straight motion [to the object], so as to remove it from its perpendicular position. Hence [i], [initially] the string is kept taut by the downward pressure [*nisus*] of the heavy body towards the centre of the earth, and also [ii], the whirling-round [imparted] by the hand's movement is constantly increased. Without straight motion this cannot be done. The said experiments therefore carry no conviction that circular motion drives outwards by its own force bodies lying on [a radius of rotation].

FOL. 237

---

[5] Or, '...circle, and at any velocity he pleases'. (The manuscript's *uoluerit* may be read as *volverit* or *voluerit*.)

3. The motion of a heavy body cast outward by circular motion in the manner described above takes place, it is said, along the tangent at the point from which the body is flung off. Whether this is so, or can be proved by any experiment, I do not know, for it is hard to observe at what point on the circle the body flies clear. Suppose we believe that someone who spins some heavy object round him [in a circle] with himself as centre is drawn, by this motion, from the centre to the circumference. Then we must [also] believe that, if the cord to which the object is attached happens to be snapped by the rotation, [the object's] movement must also be towards the circumference so as to cross it at right angles [at that instant].[6] If to this last motion is added the motion caused by the rotation, i.e. the tangential motion, then the true motion of that heavy body will be compounded of two motions, i.e. tangential motion and motion rectilinearly from the centre. The [body's] path will therefore lie between these two lines, and will make an angle with the circumference in accordance with the ratio of the circular velocity to the straight force [*conatus*] from the centre.

FOL. 237v

4. Galileo has shown, in my opinion soundly enough, that even if we concede the motion of heavy bodies driven outward [from a circle] to be tangential, they still cannot, by such motion, be flung off the earth (to whose centre they tend). For that demonstration it was not necessary to assume, as our author thinks it is, that gravity is not supposed to be *ab externo*. Galileo made no use of such a supposition in his argument, and yet his reasoning is legitimate–at least, our author does not show where in the argument the flaw, if any, lies. What he states next, however, is irrational: 'Besides, Ptolemy would say that, if we suppose the earth to move, there will be no gravity'. The explanation that White adds is that the earth's motion would throw the order of nature into such confusion that there would be no further reason left why gravity should exist. [It is irrational] because, no matter whether the cause of gravity were internal or external, the effect of gravity would be equally disordered by the earth's rotation.

FOL. 238

---

[6] Presumably because of the spiral movement of the man as described above. Unfortunately the Latin is ambiguous, it not being clear whether for 'the object's movement' we should say 'the man's movement'. In the sentence following, for 'If to this last motion...' the text has: 'If to his [or its] own motion'.

5. There are other experiments by which one may prove very clearly that motion–even purely circular motion–hurls away by its own force bodies that rest on [something]. First, if a spinning-top is made to turn fast and is whipped strongly for a long time so that it begins to spin very quickly indeed, it will throw off whatever you place on it. Second, let us think of a [thin] iron strip with one end fixed firm at a centre A and the other [able to describe an arc of] the circumference BC.[7] Let us suppose, further, that the end BC has been forcibly drawn back to DE, from which, if released, it will return to BC. Meanwhile, say there is placed at BC a heavy body that the iron strikes as it springs back. From [this] experiment it is clear that the iron strip, though it moves from DE to BC in a purely circular motion, will not, by its impact with BC, cause the heavy body at BC to describe the circle DB or EC.

FOL. 238v

An explanation confirming [what] the experiment [demonstrates] is not wanting. Bodies that are [approached and] struck recoil from the striker by the shortest path possible, i.e. in a perpendicular line.[8] It will follow, then, that the motion of the heavy object placed at BC and driven thence by the movement of the iron strip will be in the straight line perpendicular to BC, i.e. through the tangent BF or CI. Every motion due to resilience is like this one; hence every blow that propagates rectilinear motion will serve as evidence to prove not only that circular motion casts off heavy bodies by its own force but also that it expels them at a tangent.

FOL. 239

6. It does not follow from this, however, that the earth's spin will hurl away mountains or trees or buildings. Let us say that a radius AD [on the last diagram] is moved to AB, and then successively in a circle. Hence if someone puts any heavy object [on the radius] at point B so that that object [when placed] is not yet in circular motion, it will be thrown away from B at a tangent because of the circular motion previously existing at B. If, however, the heavy body at B has the same circular motion that the earth has, and its spin is due to the same agency as that to which the earth's spin is due, the heavy body at B will strive to stay [where it is. This is] because its

FOL. 239v

---

[7] Figure 36.
[8] Surely Hobbes means 'in the striker's line of advance, but in the opposite direction'? So in the sentence following, and cf. fol. 246.

motion is equal to the motion of the part of the earth it rests on; hence the latter cannot propel it. Now, say that the reason why the part of the earth on which the heavy body rests is rotated within twenty-four hours is the same as that for which the heavy body resting on this part is rotated; and that the daily motion of any separate part of the earth is not due to any other part. There will be no reason for heavy bodies to be hurled off by the earth's spin; nevertheless it will remain true that circular motion does cast off, by its own force, bodies placed on [the radius of rotation]. But the daily motion of a mountain or of a building situated on the earth is not caused by the daily motion of the part of the earth it rests on. If it were, the entire cause of the daily motion would lie ever more inwards towards the centre, i.e. it would approach zero. So the agent would necessarily be in the centre itself and in magnitude would have to be less than any magnitude possible, which simply cannot be so.[9] Therefore the agent that rotates the earth in its daily motion lies outside the earth itself and it acts on the parts near the surface before it acts on those closer to the earth's centre. Hence it is more likely that the spin of a part lying near a mountain is due to the mountain than that the motion of the mountain is due to the part adjacent to it. But it is very probable that the parts of the earth, the outermost no less than the innermost, are all rotated, at the same time and in the same manner in which a living creature turns round, there being suffused throughout the whole earth a most subtle body, that is, a spirit working within it. Thus at every distance from the centre of the earth bodies within the earth's sphere of action have a circular motion proportional to their distances from the centre. The earth, therefore, governed as it is by its own circular motion, will not act on a mountain situated on it, nor will it therefore shake the mountain itself to pieces even though the whole earth is rotated in the daily motion.

---

[9] I.e. no force can be zero or less than zero. *EW* I. 212.

CHAPTER TWENTY-THREE

FOL. 242[1]

❧

On Problem 11 of the Second Dialogue:

That the angle of contingence[2] is

not less than every right angle

1. Why geometrical demonstrations are more certain than those of physics and of moral [philosophy]. 2. The controversy over the nature of the angle of contact has arisen from a shortcoming in the definition of 'angle'. 3. The argument by which Clavius claims to show that an angle is quantity[3] at the point of contact. 4. Peletier's [Peletarius's][4] arguments [to show] why such an angle is not quantity. 5. Three definitions of 'inclination' that are at variance with Euclid's definition of an angle.[5] 6. Measure, and the quantities [dimensions] of that which is measured are of the same kind. 7. That Clavius's argument is a paralogism. 8. Certain pronouncements taken from the same Clavius contradict his opinion about the quantity of the

---

[1] Fol. 241rv is blank.
[2] Drake, p. 203. The angle of contingence or of contact (Hobbes uses either term indifferently) is the angle between the arc of a circle and a tangent to a point on the arc. Subsequent references to Euclid in the notes are to the edition of Sir Thomas L. Heath (1908, 1925, etc.). Attempts to trace the Latin text of Euclid used by Hobbes have failed; the wording is not that of the editions of Clavius and Peletier referred to below.
[3] Christopher Clavius's edition of Euclid appeared in 1574. Though perhaps less natural in comparable English contexts than its adjective, the noun 'quantity' is preferred in the narrative, because arguments are to follow that only a (finite) quantity or sum is measurable in terms of number and, conversely, that anything to which such measures alone are applicable must itself be number or quantity.
[4] Jacques Peletier, 1517–1582. His edition of Euclid was pulished in 1557. See Heath, I. 104.
[5] Hobbes distinguishes between 'inclination towards' and 'declination from'. See fols 74v, 257v.

angle of contact. 9. That Peletier's argument, though more or less true, is yet not an irrefutable one. 10. The absurdities resulting from the opinion of Clavius and of those who say that the angle of contact is quantity. 11. The absurdities which, according to Clavius, stem from Peletier's opinion, are not absurdities in the sense which Peletier takes them. 12. Two inconsistencies resulting from the definition of a plane angle as proposed by Euclid. 13. That the definitions of a straight line and of a plane surface have not been correctly given by Euclid. 14. What 'to coincide' and 'to be contiguous' are. 15. That two points taken in the mathematical sense cannot be considered as touching one another, each to each. 16. That two lines, interpreted mathematically, cannot touch one another laterally. 17. Lines said to touch one another are considered as possessing width. 18. In what manner adjacent lines may incline one to the other so that the inclination is measurable and quantitative. 19. That the inclination [of two lines] brought about by circular motion is at the point of contact itself. 20. That the drawing-away of a line from another one touching it, not through circular motion but in a curve, is neither measurable nor quantitative. 21. A definition of the plane angle. 22. How two straight lines laid end to end rectilinearly constitute an angle; and how curvature and the complete circumference of a circle constitute one angle. 23. Our author's interpretation of 'the nature of an angle'.

1. Over all other branches of philosophy one outstanding and special discipline, geometry, has always held sway,[6] for rarely has this discipline been laid open to controversy, and if by chance any contention has arisen, this has[7] soon been quelled. The reason is that geometers, even before Euclid, never examined the rationale of lines, surfaces and bodies or argued about the properties of figures, save in one respect: the meaning of the terms used when we describe things as 'figures' and 'quantities'. Thus definitions were fixed at the start, with the aim of excluding wholly all ambiguous and meta-

---

[6] This rendering, based upon the alteration of a single letter in the Latin, seems to be far nearer Hobbes's intention and more in keeping with the clauses following than would be the translation of the original as it stands: 'One outstanding and especial feature of geometry has always held sway over all other branches of philosophy'.

[7] Perhaps two conditionals are required here.

phorical language. Two comments may be made on the definitions required by geometers: first, they should be free of difficulty; and second, they should not obstruct our purpose (for we seek only the truth); hence we easily understand them and readily keep them in mind. Yet natural philosophers are concerned with terms very difficult, if not impossible, to define[8] (e.g. damp, dry, hot, cold, rare, dense, opaque, transparent, liquid, and so on). In consequence, the uncertainty of [these] notions leads them to hold dogmas that contradict one another. Moral philosophers, on the other hand, although they are concerned with terms easy to explain and understand (such as 'just', 'unjust', 'good', 'bad', 'honourable', 'shameful' and the like), reject their own definitions because the conclusions [based on them] are, from time to time, disadvantageous [to the philosophers] as individuals, so they look for different, more convenient ones. Hence most of those people who have found certainty in the dogmas of geometry, but uncertainty in those of physics and of moral philosophy, have denied that there is any place at all, in the latter, as there is in the former, for demonstration, i.e. for right reasoning[9]–and the least so in ethics, because the will, the source of moral actions, is not immutable (as though unchangeable theorems concerning morals could not exist!).

FOL. 243v

Actually, all the branches of philosophy, were they founded on sure principles, i.e. on received definitions, would be equally certain,[10] for the intellectual powers of geometers have not surpassed those of persons engaged in other writings. That certainty, however, belongs not to the geometers but to their method: [indeed,] we see that, when the basis of the reasonings they employ has rested on an inadequate definition [of terms], the most famous of geometers make mistakes and quarrel among themselves quite as much as any ordinary people do.

2. Between Peletier and Clavius, therefore, though[11] they were leading geometers, a great controversy arose, namely

---

[8] Text: ...*quae definiri...difficile...est*. Read *definire* (though the general sense is clear)?
[9] I.e. natural science's lack of an unvarying, accepted system of nomenclature and of proved fundamentals leads to its degeneration.
[10] I.e. 'as certain as one another'?
[11] Particularly in the context of the sentence preceding, the concessive is curious; for in what ages other than the sixteenth and seventeenth centuries were the leading geometers less at issue one with another?

about the nature of the angle of contact. In this dispute the arguments put forward for either side are on record for perusal at the end of the 16th Proposition of the Third Book of Euclid's *Elements* in Clavius's edition.[12] Together with Euclid, Proclus[13] and other ancient geometers, Clavius had believed that there is included between the circumference of a circle and the straight line that touches it a true angle which they call the angle of contact, and that such an angle is quantity. Peletier, as opposed to all those named [by Clavius], says that it is not quantity and not an angle. The controversy remains unresolved today. Why so? Only because they have not agreed on the definition of 'angle'? Had it been clearly understood what an angle is, and, if what is called the angle of contact had been examined against the definition of 'angle', such great geometers could not but have realised at once whether or not [the angle of contact] was an angle and quantity.

3. Clavius really does believe it to be an angle, because it comprises two lines, a straight one and a curve, meeting in one and the same point. Now although in Euclid there is considered to be no definition of 'angle' at large, he does define a plane angle as follows [Bk. 1, Def. 8]: 'A plane angle is the inclination to one another of two lines in a plane which meet one another and do not lie in a straight line'.

With A as centre describe the arc BC which the straight line BD touches at B.[14] CB and DB are in the same plane, namely in the plane of the surface of this paper, and each touches the other at B because, as they are not parallel, each is inclined to the other (for it is thus that the term 'inclination' seems here to be understood). So from this definition it appears that CB and DB contain an angle. Next, on centre F, [on BA produced,] and with radius FB draw the circle BE. Its circumference, EB, will fall between the lines DB and CB. Therefore the angle DBC is seen to be divided into two parts, namely into the angles

---

[12] The allusion is to Clavius's scholium to Euclid's Book III, Theorem 15, Proposition 16. The last reads: 'The straight line drawn at right angles to the diameter of a circle from its extremity will fall outside the circle, and into the space between the straight line and the circumference another straight line cannot be interposed; further, the angle of the semicircle is greater, and the remaining angle less, than any acute rectilineal angle' (tr. Heath.) See also Proposition 18.

[13] AD 410–485. See Heath, I. 26, 31f.

[14] Figure 37, repeated on fol. 244v.

DBE and EBC; so it seems to Clavius that the angle of contact is quantity, for nothing in the nature of things can be divided except quantity.

4. Peletier, however, believed neither that the lines DB and CB constitute an angle at the point B nor that [DBC] is quantity, for the reasons following:

First, if such quantity be less than the quantity of any acute angle, this will be seen to contradict the First Proposition of the Tenth Book of Euclid's *Elements*. Here it is said: 'Given two unequal magnitudes. If from the greater there be subtracted [a magnitude] greater than half of it; and again, if from the remainder there be subtracted in turn more than half of this: if such a process be carried out continually there will remain some magnitude less than the lesser magnitude set out'.[15] Even though a given acute angle Z is reduced by over a half and the residue is in turn reduced by one half of itself, and so infinitely, [Z] will never become smaller than the angle DBC. This was proved by Euclid in the Sixteenth Proposition of the Third Book of the *Elements*.

FOL. 245

The second reason [why the lines DB and CB do not constitute an angle at B] is that a line is seen to be inclined to [another] line when one of them, if produced, will cut the other. This, [Peletier] thinks, cannot happen with the angle of contact.

5. To see which view carries the greater weight we must first examine the definition of a plane angle as derived from Euclid. Since he believes a plane angle to be quantity, and progressively divisible into lesser angles, it will follow from that Euclidean definition that inclination is quantity. We ought therefore to consider in what consists that inclination which Euclid wishes to call quantity, and under what kind of quantity it is to be classed.

FOL. 245v

Given two straight lines AC and AB, or AC and AD, forming an angle at the point A where they meet.[16] AC is therefore inclined to AB and to AD. But its being termed 'inclined' can be understood only if 'to be inclined' is defined at the

FOL. 246

---

[15] The last seven words are from Heath's translation. Hobbes has: 'less than the other lesser of the two magnitudes that were given initially'.
[16] Figure 38.

outset. Nowhere is this done by Euclid. So we must find out what 'to be inclined' is.

We usually say that a straight line is inclined to any other line when, through circular motion with their meeting-point as centre, the first line has been removed from its position perpendicular [to the second]. For example: Given any straight line C[A]D[17] which a straight line BA meets perpendicularly. Now if BA is seen to have been moved about the centre A so as to lie at AE, AE will be said to be inclined to AC, and the amount of such inclination will be the arc BE and will be the same quantity as that of the angle BAE. This, however, is not the meaning of the term 'inclined' in Euclid, for if it were, BAC would not be an angle, even though DAC were a straight line, because AB has no inclination towards AC. We must therefore seek another definition of 'inclination'.

FOL. 246v

A line may be said to be inclined towards another line when the former has been removed from the latter through circular motion. Suppose, for instance, that a straight line AD has been shifted, by circular motion, from its position as far as AB.[18] AB will be said to be inclined to AD; and the measure or quantity of such inclination will be the arc DB, and the size of the angle DAB will be the same [as that of the arc]. But according to the above definition EAB, the angle [now] made by the curve EA[19] and the straight line AB will not be an angle; for the straight line AB cannot be conceived as having moved away from AE, which in fact could never have been coincident with AB: so we must probe the definition of 'inclination' even more deeply.

Let us say that a line has been 'inclined' towards another line when one of them has been moved away from coincidence with the other in a circular motion with their meeting-point as centre. For example,

FOL. 247

(a) Let there be initially two straight lines AB and AC, and let AB be understood as having lain on AC, i.e. to have been in contact with it (for two straight lines can touch one another at every point), but then to have moved away towards AB in a circular motion about the centre A. Its inclination will

---

[17] In the accompanying figure in the original (Figure 39) CD is curved. For this and the five diagrams following, the manuscript is followed, which prefers minuscules to majuscules.
[18] Figure 40.
[19] A straight line in the original figure.

be the arc BC, and the same arc will be the size of the angle BAC.[20]

(b) Let one of two lines, DE, be straight, and the other, EF, curved.[21] Now, let there be a tangent [GE] drawn to the curve such that DE may be understood [initially] as lying along it. Because DE is straight the tangent will also be straight. Let EG be that straight line. ED is [now] understood to have moved away from this line in a circular motion about the centre E. Then the inclination of the lines DE and EF will be the arc DG, which is the same as the inclination of the two straight lines DE and GE.

(c) Given two curved lines HI and HK.[22] Now if these may be regarded as [initially] coincident, and if HK be supposed to have left its position [and passed] in a circular motion with centre H until HK lies at HI, the described arc IK will be the amount of their inclination. If they cannot be understood as coinciding [initially], then let HK be touched at the point H by another curve that *can* be coincident with HI, say by HL. Then the arc IL will be the measure of inclination of the said HI to HK.

FOL. 247v

How, save as above, anyone can define 'inclination' such that it is still understood as being quantity, I cannot imagine. Yet the definition of 'inclination' will not fit an angle in this way, because if 'inclination' were so accepted, two straight lines lying rectilinearly end to end would have an inclination [one to the other], and yet in the Euclidean sense they do not form an angle.

6. Now, to find out in which class of quantity the quantity of an angle is to be placed, we must realise that all bodily quantity is measurable in terms of a line or lines, of a surface or surfaces, and of a solid or solids; and that the measuring quantity is of the same kind as the quantity measured. The quantity measured by a line *is* a line; the quantity measured in several lines, i.e. by a number of lines, *is* a number of lines; the quantity measured by a surface is a surface; and the quantity measured by several surfaces, i.e. by a number of surfaces, is a number of surfaces. So the measure of line is line, of surface is surface, of solid is solid, and of number is number; the quantity

FOL. 248

[20] Figure 41.
[21] Figure 42.
[22] Figure 43.

of time is nothing else but the quantity of the line described by a body moved in that time: therefore it has a line as its measure.[23] [Again,] the quantity of motion i.e. the degree of velocity, is also that of the line which is [a] described by a body in motion, but is [b], relative to another line also described in a certain fixed time. Hence velocity is measured by line; and lines, times, and speeds have a relationship or ratio one to another. Surface is compared with surface, solid with solid, and they have a ratio to each other; but a line cannot have any ratio to a surface, nor can a line to a solid, nor can a surface to a solid. [The size of Z,] a plane angle composed of straight lines, is measured by the circumference of a circle; but this is a line. The quantity of a plane angle is therefore of the same species as the line; likewise such plane angles as are composed of two parts of circles or of conic sections that cut one another are measured in terms of the circumference of a circle, i.e. by a line. Hence an angle and a line are of the same species. Likewise, spherical angles are measured in terms of the circumference of a circle and hence are of the same kind of quantity as lines are; a solid angle is measured by the [portion of] spherical surface marked off by the planes originating in the centre of the sphere. The solid angle is therefore of the same kind of quantity as that to which surfaces belong. When a plane is cut by a plane, the inclination of these planes is measured in terms of the circumference of a circle and such inclination is a quantity of the same kind as a line is. There remains [the question of] that special angle called the angle of contact—and consequently [that of the] inclination of the surface of a cylinder and of a sphere to a plane or other surface touched by it. It is doubtful what measure should be assigned to the angle of contact. If, however, this is an angle and is quantity, then it must possess a measure of the same kind as itself; whence arose the argument between Clavius and Peletier. Here it seems to me that (for reasons to be given) the opinion of Peletier, though true, is [based on] ineffectual arguments; but Clavius, besides putting forward an erroneous view, has contrived further mistaken tenets in order to lend it support.

7. So as to prove the angle they call the angle of contact to be quantity, Clavius claims that it is divisible. Let there be described about centre A the arc of a circle BC, which is touched

---

[23] See fol. 34v. EW VII. 270.

CHAPTER XXIII: 7

by the straight line EC at C.[24] The angle of contact, therefore, is that made by the meeting of the two lines BC and EC. Now take any point F on the diameter CA produced, and with centre F and at a distance FC draw the circumference DC, which will necessarily lie between EC and BC. Hence the angle ECB is divided into two smaller angles, ECD and DCB. Thus Clavius thinks he has shown that the angle of contact has been divided and is therefore quantity; but he is wrong. He has shown, to be sure, that the surface enclosed between EC and BC has been divided, but not that the angle has; for according to Euclid 'angle' is 'inclination'. [Clavius] should therefore have shown either [a] that it is the inclination of the lines BC and EC that has been divided, not the surface they enclose–unless, perhaps, he believes their inclination to constitute surface itself enclosed in such a way that the nature of 'angle' consists in such a surface; or at any rate, [b] that from a division of this kind there necessarily results a division of the angle. Each of these alternatives is wrong; for if the nature of an angle consists in the latter's containing the surface enclosed between the 'legs' that meet one another, then of whatever kind the division of such a surface is, the division of the angle will be of the same kind.

FOL. 250

For example, if the surface of a triangle HGI were the quantity of the angle at G, the area of the triangle having been divided in any way you please by [the straight line] LM, the angle HGI would be greater than the angle LGM, which is absurd.[25] Also it would follow that the [sum of] the three angles of the triangle is not greater than any one of them; for the surface enclosed within the three sides, i.e. the area of the triangle, is the same as that included within any two sides that constitute an angle. The same would have to be said of the triangle (if only it were a triangle) EBC, where the angles EBC and ECB, together with the rectilinear angle BEC, would not be greater than the one angle ECB or EBC, [either of] which is [in fact] smaller not only than the angle BEC but also than every rectilinear angle.[26] Again, although a line that divides a surface

FOL. 250v

---

[24] Figure 44.
[25] Figure 45.
[26] The reference is to Figure 44. To specify the angles here given in conventional form the text speaks of 'the angle at B', 'the angle at C', etc. It calls them both angles of contact; this, certainly as the figure stands, is not necessarily true of EBC, for the text nowhere specifies that the line EB is a tangent to the arc BC or that BEC need be a right angle.

passes through the point of contact, it will not on this account be said to divide an angle because it divides a surface; for it ought also to divide the angle into the same parts as those fixed parts into which, for the same reason, it divides the surface. But it is untrue [that it does so].

Let GH and GI be the sides enclosing the angle HGI, these being equal to one another.[27] Let the surface enclosed, or the area of the triangle, be divided by the straight line KG. Let HK be one quarter of HI: [hence] the surface is indeed so divided that its [two] parts are, each to each, as 1:3. But the angle [HGI] is not similarly divided. Again [in the previous figure], because the circumference DC divides the space included between EC and BC, it does not follow that the angle ECB has been divided, as [Clavius] imagines. He has committed a paralogism; surely Joseph Scaliger would not have committed a greater?[28]

FOL. 251

8. Further, in refuting Peletier, Clavius explains what 'to have the same inclination' is.[29] 'If angles of the same kind are equal', he says, 'then the inclination of the lines [forming them] must also be the same, such that, if one angle be superimposed on the other, the lines of the one correspond in every detail with those of the other.' He shows clearly that he believes inclination to be nothing but the form of the lines, namely straightness and curvature, for it is in respect of their shapes that they cannot correspond: straight with curved, circular with hyperbolical, hyperbolical with elliptical. Therefore inclination is not quantity, but quality; hence in Clavius's own opinion (and he by no means rejects the Euclidean definition of 'angle') an angle will not be quantity. In the same context he has the following: 'Any angle consists in a single point and in the inclination of lines that do not lie end to end and straight, as is agreed from the definition of the plane angle'. If this be true, the angle of contact will, in his opinion, in no way be quantity, for (he says) 'it consists in a point'–but a point is not quantity–'and in the inclination of lines (i.e. of two lines that touch one another)'–but the inclina-

---

[27] Hobbes reverts to Figure 45.
[28] Joseph Justus Scaliger (1540–1609), classicist, son of Julius Caesar Scaliger, thought he had performed the quadrature of the circle, and committed other follies.
[29] Cf. Heath, I. 194 and II. 41. Clavius, *s.v.* Euclid's Bk. I, Defn. 8.

tion of things in contact is not quantity, for it[30] is the same as contiguity. Therefore the angle of contact is non-quantity; hence it is not quantity.

FOL. 251v

Again, we must not fail to observe that these words 'as is agreed from the definition of the plane angle' [Clavius] does not quote correctly, for Euclid says[31] that a plane angle consists, not in a point and in the inclination of lines, but in 'the inclination of lines at a point'. This differs considerably from [Clavius]; for quantity cannot consist in a point, but it can consist in the inclination of lines at a point.

9. Peletier's argument by which he proves that the angle of contact is not an angle we derive from his declaring the form of an angle to consist in the section or dividing of lines.[32] But he does not prove this; he only explains it by means of an example–and if he did prove it, this would not help him in any way.[33] If on centres D and E equal [arcs of] circles AB and BC are drawn that touch each other at B, and if the straight line FG be drawn that cuts the line ABC in the point B, then according to Peletier the angles FBA and CBG are not truly angles or quantities, and yet it is clear that the line ABC is cut by the straight line FBG.[34] Therefore a line can not only touch [another] line but can also cut it, yet not so as to produce an angle; so even though the proof is a weak one, Peletier's conclusion may yet be true.

FOL. 252

10. For the contrary opinion [i.e. that of Clavius] is attended by many absurdities that cannot be sustained, such as 'It is possible to pass from the lesser to the greater, and the converse, and this through all mean quantities, but not through equality', i.e. 'there is no equal mean between greater and lesser'. So says Clavius, as do those who agree with him, citing no argument other than what necessarily follows if one supposes the angle of contact to be quantity: [as one may phrase it,] 'the part of a quantity is not homogeneous with its whole'; that 'there can

---

[30] Text repeats '...inclination...contact...'
[31] *Elements*, Bk. I, Defn. 8; Heath, I. 177f.
[32] Peletier, Bk. I, Defn. 8. After calling the Euclidean definition '*pervulgata*' he says: '*Angulus planus est duarum linearum in plano sectio. Cessante enim sectione, cessat angulus*'.
[33] See *Elements*, Bk. III, Prop. 16 and Heath, II. 41.
[34] Figure 46.

FOL. 252v

exist a part that is in no ratio with its whole'; and that 'there are as many kinds of quantities as there can be differences in curved lines, as regards both magnitude and shape, i.e. that the kinds of quantities are infinitely infinite'. For the shapes of curved lines are beyond number: circular, elliptical, hyperbolical, parabolical, conchoids and others without limit; any shape may be represented by lines either greater or lesser, the diversity of which is infinite; and with reference to heterogeneous quantities we can speak of 'greater' and 'less' but not of 'equal'. Clavius makes all these observations; and they all depend on that one proposition: 'that the angle of contact is quantity'. This, as I have shown above, has not been demonstrated, for he has proved that the surface included within contingent [lines] can be divided, not the angle itself.

FOL. 253

To the same opinion we must ascribe the fact that the same [Clavius] has been forced to deny 'that two straight lines, one finite, the other infinite, are the same kind of quantity', though he does try to confirm it by means of a certain kind of argument from Euclid (who had declared that quantities having a ratio one to another are only those of which the lesser, if many times increased, can at last exceed the greater). But it is not feasible merely on that account; for, given two straight lines, [one] finite, and [one] infinite, I shall easily show that they are in the same proportion [each to each] as are two finite lines. Suppose someone objects: 'An infinite line does not exist'. I reply: 'An infinite line is not quantity'. Say this person asks: 'How big (sic) is a line?'[35] Anyone who correctly weighs up the matter determines a [line's] length by comparison with some other [line or length] already measured, and will answer: 'It is of a given size', for all given size is finite. But a person who replies: 'The line is infinite' says that it is not 'of such or such a size', i.e. he denies that it actually does have any quantity, but [says] that we may assume that potentially it has. Moreover, if an infinite straight line is not quantity of the same kind as [is] a finite straight line (this latter being assumed a quantity), [the infinite line] will nevertheless come under some kind of quantity.[36] But an infinite line does not fall into the category of surfaces or solids. Therefore, if it is to be called quantity in any

---

[35] Or, 'How big is [this] line?'
[36] Mersenne makes changes to this sentence, the effect of which is to remove the words 'Moreover, if' and to assume that the closing clause begins a new sentence.

sense, [an infinite straight line] belongs to the class *lines*. True, Clavius believed that there is no ratio of finite to infinite, and that this agrees with Euclid's definition, namely the Fifth of Book Five;[37] yet [according to Clavius] this does not prevent a finite line and any other line from being of the same species of quantity, whatever the line you wish to make infinite. Hence homogeneity in a finite line will be understood in the same way that quantity may be perceived in an infinite line! Not satisfied with allowing absurdities of this kind, [Clavius] even wished to honour some of them, calling them the miracles of geometry, though in truth they must be termed–and with the more reason–monsters begotten by fertile imaginations upon vain words.

FOL. 253v

11. But the absurdities that, according to Clavius, attend Peletier are these: First, if of three or more circles each touches the others, then the figure included, a three-sided or a many-sided one, will have no angle. Second, if two straight lines that are tangents to the same circle meet, the three-sided figure will have only one angle.[38] Peletier, as appears from the intention of the present pleading, has not[39] attempted to do away altogether with linguistic usage: [he intends to say] not that 'the angle of contact shall no longer be termed an angle' but that 'it ought not be called "angle" in the sense that "it contains quantity"'. Now if Peletier concedes that a three-sided or many-sided figure that is contained within limits, each of which touches the others, is quantity; but if he denies that such angles are quantity although called angles, I do not understand why this is absurd. Nor would I consider absurd his saying that the three-sided figure contained within [the arc of] the circumference of a circle and two lines touching the circle is truly quantity; that its rectilinear angle is quantity; but that the two angles at the points of contact are not quantity.

FOL. 254

12. Again, as we have said that the flaw in the Euclidean definition of an angle has brought about the controversy described, we shall also show two anomalies[40] that result from this definition.

[37] Heath, II. 114.
[38] The third side is the included arc.
[39] The negative is added by Mersenne.
[40] By *inconvenientia* Hobbes may mean 'inconsistencies'.

(a) Let the segment of a circle be AB, which is divided anywhere by point C.[41] The two lines AC and CB do not, therefore, form an uninterrupted line, but touch each other at C, and according to Clavius's interpretation [of Euclid] (in this matter Clavius is not opposed by those others who explain inclination in terms of the meeting [of lines]) each of them is inclined towards the other. Hence an angle is made in every part of a circle. No more, therefore, is a circle 'a figure described by a single line' (as in Euclid's definition)[42] than is a four-angled or any other figure that Euclid defines [only] in terms of the number of lines comprising it.

(b) Given two rectilinear angles, each made up of two straight lines DEG and FEG, one inclined to the other, and that the [two] angles total two right angles.[43] According to Euclid's definition, the inclination of the straight lines DE and EF is not an angle, because they lie straight and end-to-end. Hence the two angles added together are equal to a non-angle, which is absurd, for equality can exist only between quantities—and these of the same type, as between line and line, surface and surface, solid and solid, angle and angle.

13. In Euclid, the definition of 'angle' is not the only faulty definition. There are others, which include that of a straight line and that of a plane surface. To 'lie equally between its points, or between its lines'[44] is obscure; it can be understood only if explained thus: 'A line lies between its points' means that every point on it is in the same straight line as the end points are.[45] Hence the defined [here, the word 'line'] is [still] in the definition, which is incorrect procedure.[46] Euclid himself did not use his own definitions in any demonstration;[47]

---

[41] Figure 47.
[42] *Elements* I, Defns. 15 and 16.
[43] Figure 48. For 'DEF and FEG' the manuscript has 'DEF (*erased*), DEG, FEH'. Once or twice in the present work the lettering in the text does not match that of the diagrams. Perhaps Hobbes was transcribing both narrative and figures and forgot to make the necessary adjustments.
[44] Euclid says (*Elements* Bk. I, Defns. 4 and 7–see Heath, I.165–9): 'A straight line is a line which lies evenly with the points on itself', and 'A plane surface is a surface which lies evenly with the straight lines on itself'.
[45] Cf. Ch. VI, Article 3, above.
[46] I.e. the 'definition' just cited does not say what a line is, but gives one of its properties.
[47] Given here literally, this seems a curious statement, unless Hobbes means: 'In the actual demonstrations Euclid did not re-state things he had defined earlier'.

therefore he could not show that a straight line is shorter than any other line having the same limits. Though such a proposition is very obvious, it cannot be proved by the absence of[48] the definition of a straight line.

14. So that I may explain clearly my own feelings on this question I shall first speak of matters relevant to the nature of contact; then of the nature of an angle and how it is produced. The following, we say, touch one another mutually, or are contiguous: two lines, or [two] surfaces, or solids on which there are two points–one point in the one, the other point in the other solid–and no space can be conceived as lying between the points.[49] Things that are contiguous at every point of any line or surface are said to be congruent along that line or surface. Solids, however, if understood as being contiguous at every point, are necessarily understood as coinciding, or as being only one solid twice considered; but it is impossible for two entities possessed of quantity, i.e. two bodies, to coincide. Hence, in fact, if two points, lines or surfaces are said to coincide, i.e. to be in the same place, we must realise that such points, lines and surfaces are not two things but are two concepts of, or two ways of looking at, the same thing. Therefore to coincide is different from to be contiguous, for things are considered to be contiguous only with several things, but 'as a whole' with one thing only.

FOL. 255v

15. Two mathematical points cannot touch one another, each to each. Proclus demonstrated this; I shall show the same, but by a different method. If they do touch and yet do not coincide, i.e. if they are not one and the same point, then they will be two points whose contact is in a third point. Now the point of contact is neither of the points that touch, nor is any imaginable space contained between it and them; therefore the point of contact touches, and this at a point, both those points that are supposed to touch each other. Thus three points of contact are created. Likewise if we were to grant that two mathematical points could touch one another, then the points of contact would be multiplied infinitely. Hence there cannot be mathematical points that are contiguous.

FOL. 256

---

[48] *Defectus*: Possibly Hobbes means 'flaw in'.
[49] This last clause may refer to the solids only.

16. It follows from this that two mathematical lines cannot touch one another laterally. For example, if the straight line AB is moved up to CD so that the points B and D are supposed to touch, and the points A and C to touch,[50] so that, as a result, the whole line AB touches the whole line CD, then, I say, on that supposition it is understood that the lines AB and CD are not interpreted mathematically, for if AB were considered mathematically, the points A, C, B, and D would be mathematical points. Hence, from what has been said, each cannot touch any one of the others–which is against the hypothesis.[51] For the same reason, if we suppose a straight line or a curved one to be moved into contact with the curve GH at a point I, we also grant that those lines are not mathematically considered [as being] without breadth, for if another line, BK, is brought up to the line AB there is no reason why the line should not be said to touch the line KB at the point B.

17. From the above proposition it is clear [a] that two lines touching one another in the manner in which the angle of contact is supposed to be created cannot be conceived of mathematically [as being] without breadth, and hence [b] that they have no point in common, such as I, but that, being concurrent, they end not in a point but in a little line, as they meet ON at MN.[52]

18. If, of two straight lines touching one another at every point of their length, either line be broken at any spot, there will be created at the point of fracture an inclination of the fractured line from the other line. Therefore [a], one can measure the [amount of] inclination–choosing[53] any [radial] part of it –in terms of the arc of the circle whose centre is the point of fracture; [b], [the amount of inclination] is divisible; and hence [c], such inclination is quantity.

In the preceding proposition [Article 16] it was shown that two lines touching one another are considered as having some width. So let there be two straight lines, AB–CD and CD–EF, that touch one another at every point of the length CD.[54]

---

[50] The text says: '...so that B, C and A, D touch...' Figure 49.
[51] I.e. against the supposition made at the start: that AB be moved up to touch CD.
[52] Figure 49.
[53] 'Therefore': text, 'so that'. 'Choosing': text, 'at'.   [54] Figure 50.

Let the breadth of the straight line CDEF be the straight line FD perpendicular to CD. Now imagine that the line CDEF is snapped or torn open by rupture at the point D. Point F will necessarily turn towards G and consequently, if DH be drawn perpendicular to DG, DC will lie at DH. Again, let point G advance [further] to K, by means of the same rupture;[55] if DL be drawn perpendicular to DK, DC will [now] lie at DL. If, next, the arc of a circle be described with D as centre, it cuts DC, DH and DL at the points C, H and L wherever they fall. Hence the [angular] inclination of DH to DC at point H will be in the same ratio to the inclination DL from the same DC at the point [L] as is the arc CH to the arc CL. Since this holds wherever we place the points C, H and L, the ratio of any one declination [x] to any other declination [y] will always be as the arc [x] to the arc [y].[56] Hence that inclination is measurable; and as that measure, namely the arc of a circle, is divisible, that inclination is also divisible and is therefore quantity.

FOL. 257v

19. Furthermore, it is agreed that the straight line FD which touches CD at D does not, when rotated about the centre D, always touch CD in one and the same point [as] mathematically accepted; for point D is both itself rotated and is therefore moved away in a small circle of its own from its contact with CD, since [D] is continually being replaced by another point. Hence the inclination of the two said lines is related to the point of contact itself.

20. Say that two straight lines touch one another along their entire length, but that one is torn away from the other (though one of the two ends remains fixed) such that the line which determines the breadth [of the moved line] is not shifted out of position. The inclination brought about by such dragging apart will not be determinable by one measure, nor will there be any inclination at the actual point of contact. Hence there will be nothing divisible at that point, or any quantity of inclination.

FOL. 258

Given two lines, AB and CD, that touch one another along their entire length, and let CD have some breadth DE.[57] First,

---

[55] I.e. considered subsequently: the same operation is performed.
[56] For '...L, the ratio...the arc [y]' the text has: '...L, there will be a constant and the same ratio of declinations everywhere'. See fol. 242, note 5.
[57] Figure 51. 'Touch'–Mersenne's alteration from 'cut'. DE–*sic* MS.

271

FOL. 258v

let the portion CD be snapped, i.e. broken at F, such that the part CF lies at GF. The latter, because of a similar tearing-away at every point between F and G, is understood as being bent into a circular curve. It is clear that the remainder of FD is up to now a straight line and has no inclination. Let us proceed in the same way by tearing apart HF and forming a curve HI so that all HC[58] has the form of the circumference of a circle and lies at HIK. It is obvious that HD is straight and has no inclination. And whatever part of the same DC is dragged away, the remainder will be straight; it will be adjacent and parallel to AB itself and will be in no way inclined to AB. If the rest [of DC] is all torn apart, i.e. the [yet] unbroken part, at point D, then the line DE must be rotated in the direction of M. Therefore the angle ABE will not be a right angle;[59] therefore AB does not touch the circle DLE, as was supposed. Hence when a straight line is torn apart from another straight line adjacent along its whole length such that [the first named line] is curved into a circle, [then] the point of contact, and hence the whole *width* DE, remains unmoved. The inclination, therefore, is no greater than at the beginning, when they touched one another along their whole length. So there is neither divisibility of inclination, nor is there quantity, at the point of contact. It is clear that there is not the same inclination DLE at each of its parts, for the motion at E away from the point C, from which [E] begins to recede, is greater than the motion at L away from the point H, from which L has begun to move. Therefore no sure measure is equally valid for all the forms of inclination, taken as a group.

FOL. 259

21. As the meeting of two lines is customarily termed an angle, and as they may sometimes meet in such a way that the angle is a quantity, and sometimes such that it is not, then clearly that term 'angle' is ambiguous: only in the first sense has it any connection with geometry, the task of which is to consider the proportions of quantities. Therefore the definition of the word 'angle' must treat an angle quantitatively only[60]–a thing

---

[58] Text, 'HA'.
[59] Because the angle CDE is no longer so, DE being rotated?
[60] Text: 'Therefore it was necessary to have defined (*definiisse*) 'angle' in such a way that it [the definition] contained "angle" as quantity only'. Probably the reason for the use of the past tense is not that the context, here or earlier, demands it (not, 'Clavius and Peletier should have done this'), but rather that Hobbes is already thinking of his next sentence ('Euclid...').

which neither Euclid nor anyone else has done.[61] Now, an angle is, in that sense, 'the amount of the separation, or of the inclination, of two lines, that arises from the rotation of either of them about a common point'. So the measure of an angle is the arc of the circle described about such a centre, the arc being marked off by the two lines. Hence the angle of contact, being created by curvation, is commonly called 'angle', but only equivocally, as is any narrowing of space, for such an angle is neither measurable nor divisible, and is not quantity at all. Hence the angle of a semicircle, which is shown to be greater than any acute angle composed of straight lines,[62] is also a right angle and is equal to a rectilinear right angle; the remainder [of the angle of a semicircle], which is commonly called the angle of contact and is shown to be smaller than every acute rectilinear angle, is also less than any quantity, i.e. properly speaking, it is non-quantitative; and the circumferences that pass through a point of contact and are drawn between a straight line that touches the periphery of a circle and the circle [itself] do not divide an angle; they divide a surface, for the nature of an angle consists in rotation.

FOL. 259v

As the periphery of a circle is circular length, two things must be considered with regard to the circle: length and circularity. Of these, the former is simply length, or a line, but the latter is called an angle.

22. From this it is evident that, even in a straight line which is considered as consisting of two lines placed rectilinearly end to end, such straightness or the inclination through a semicircle has the nature of an angle and is really an angle which is twice a right angle, even in the present figure.[63] Here the [reflex] angle on the other side of the angle BAC is genuinely an angle, namely one equal to the [sum of the] two angles BAD and CAD. Also the line is correctly said, after having rotated and returned to the place it had quitted, to have described an angle, i.e. one equal to four right angles. All this is so clear that there is no need to prove it from what precedes. Now that I have made these observations on the nature of the plane angle and of the

FOL. 260

---

[61] A bold claim, even if true?
[62] The angle of (not 'in') a semicircle is the angle between a tangent and the radius at the point of contact.
[63] Figure 52.

angle of contact, it remains for me to set forth our author's opinion on the same topic.

After seeing the quarrel among geometers about the angle of contact, he first deplores their ignorance of logic and then puts forward a definition of 'angle' he himself has drawn up. Though assuring us that no embarrassment in geometry will result, he yet pleads that the angle of contact is not less than every acute rectilinear angle.

23. 'Let us only examine', he says, 'in what the real and the true quantity of an angle consists. We shall see that it has no quantity other than [that of] a surface included within sides. From this it is obvious that that angle is greater which encloses the greater surface within lines that have been duly accepted as being of equal length. If we propose this, it will be clear that the corollary induced (Euclid, [Book 3], Proposition 16) is false even though the proposition is very sound.[64] From this definition of ours it is manifest without extraneous evidence that no injury to geometry will result.'

FOL. 260v

Such are his words. So that we may dispose of them, let us draw ABC, the semicircumference of a circle, and let the straight line DA touch it.[65] Now, as he says that the true quantity of the angle of contact is that surface included within sides, I ask (*a*) What *are* the sides of the angle of contact made by the circular [line] AFBC and the straight line GA (for AG can be produced to infinity)? If the side is infinite, the included surface will also be infinite; and because the quantity of the angle is alleged to consist in that surface, that angle of contact itself will be infinite, which is not credible. Let it be finite, then; and let a point be taken on AD beyond D, say at G, and any line drawn from G to the semicircle, say at C, so that the surface included is enclosed and rendered finite. The surface enclosed between the lines AG and AFBC is therefore the quantity of the angle of contact; but for the same reason, if DB or EF be drawn, then the surfaces ABD or AEF will be the quantity of the same angle. Hence, by this definition [of White's], the angle of contact will contain quantities infinite in numner, which is absurd. But we should accept the sides as being equal, he says. Very well; let us

FOL. 261

---

[64] Hobbes inserts the reference to Euclid and the Proposition number. After the word 'sound' he omits White's words: 'and that the corollary inferred is false, in that the definition has not been agreed on beforehand'.
[65] Figure 53.

suppose AG equal to AFBC, AD to AB, and AF to AE: isn't the same problem still there ? But equal sides, he says, should be 'duly accepted'. This, unless he explains what 'duly' means, is incomprehensible.

(b) Further, that the angle of contact GAF can be greater than any rectilinear angle he thinks he has proved from the fact that the surface included between the lines GA and AFBC can be greater than that included between GA and AH. With what justice this follows will be seen if we consider that the straight line HA necessarily cuts the circle below point A, say at I. Hence if we take AK as equal to AI, the surface enclosed between the straight lines AI and AK will be greater than that enclosed between the straight line AK and the curve AI. Therefore the angle KAI,[66] which, according to him, is also rectilinear, will be larger than the mixed angle[67] KAI; but this same is the angle of contact KAI. Hence the same angle of contact is both greater, [as White thinks he shows,] and less, [as I show,] than the same rectilinear angle, which is absurd.

FOL. 261v

---

[66] A phrase following in the text: 'and hence the angle K' is apparently marked for deletion.

[67] Either the intersection of a curve and a straight line (only now does Hobbes describe such an angle as 'mixed' if this is the meaning); or the angle KAI consisting either (or both) of the straight line KA and the curve AI or (and) the straight line AI.

FOL. 262

CHAPTER TWENTY-FOUR

❧❧

On Problem 12 of the Second Dialogue:
That the moon is not directly
[*immediatè*] moved by the sun

1. The difficulty of investigating the physical causes of the motions of the moon. 2. A scheme[1] of the motions of the earth and moon and their appearances [*phenomena*], as supposed by our author. 3, 4. That he has not understood [these] phenomena correctly. 5. That in spite of his arguments the moon, if she were attracted downward, would fall upon the earth. 6. That the reason why the moon does not complete her orbit within twenty-four hours is not gravity. 7. That the vapours rising from the earth are not the cause of the moon's orbit's deviating from the ecliptic. 8. That the reasons by which he endeavours to show that the moon moves faster and is lower when outside the ecliptic than when within it are unsound. 9. That the motion by which the sun illuminates, and the motions by which it spins about its axis, both, working jointly, have the power of rotating the air at all distances [from the sun] at an equal speed. 10. That the sun cannot, by the motion of its spin alone, cause the air around it to rotate, but that an 'illuminating-motion' is also required. That, when the

FOL. 262v

air around the sun is moved, the earth must be moved through the ecliptic with it; and that the [earth's] axis is always parallel to itself. 11. On the reason why the [orbit of the] earth is eccentric. 12. That the motion which brings about the eccentricity of the earth ['s orbit] resembles the

---

[1] A plan to show, each to each, the positions of the heavenly bodies. Cf. fol. 62v.

CHAPTER XXIV: 1

vital motion in animals. 13. That the cause of the earth's daily motion also resembles [such motion]. 14. That the reason for the earth's movement through the zodiac, and for that of the moon, is the sun; but for the moon to possess a monthly motion, we must ascribe to the earth a motion of similar influence to that we attribute to the sun, which causes the sun to distend. 15. If the moon's monthly motion were due only to the earth, she would not complete her orbit in less than 56 days, or 2 months, and would not always present the same side to the earth. 16. That the reason for the moon's monthly motion lies partly in the sun, partly in the earth. 17. That the reason why the moon presents the same face to the earth lies in the sun and is the same as the cause of [the moon's] phases. 18. That the reason why the moon is swifter in the conjunctions than at the quadratures is the same as the reason for phases. 19. That the reason why the moon's orbit cuts the ecliptic is a combination of the annual motion in the ecliptic and the daily in the lines parallel to the equator. 20. On the cause of apogees and perigees of the moon. 21. Why all influences do not constitute light.

FOL. 263

1. Such is the irregularity of the lunar movements observed by astronomers that it seems (at any rate, to me) we must despair of ever discovering their natural causes: first, because of the host of bodies acting on the moon from different places, at different distances, and with different degrees of force; and second, because of the diversity of her nature. This makes her differ from other planets and perhaps also makes one part of her differ from another.[2] Consequently she is not affected by any single one of her movers in the same way as some other moved body, diverse in its nature, would be affected by the same mover.[3]

Now, the essence or especial constitution of any body, i.e. that property which makes it appear to our senses to be different from all other bodies, consists in a certain motion of its inner parts. It is self-evident that there are many bodies acting on the moon: both the sun and the earth in particular,

FOL. 263v

---

[2] Text: '...because of the diversity of that very nature of the moon' (or, 'of the nature of the moon herself') 'by which she differs from the nature (sic) of other planets...'

[3] The same mover as (being one of several) moves the moon? Or a mover that moved bodies additional to the body specified?

and the other stars, both fixed and wandering, to some degree. The sun, moreover, [affects the moon,] first in respect of that motion with which it causes both the earth and the moon alike to rotate together every year, and second, too, as regards that motion with which it illuminates everything by means of a certain rectilinear ejaculation in all directions. The earth can exert some power over the moon not only by means of its daily action but also by a kind of influence—as if a magnetic one; and, again, by means of the vapours which are perhaps raised as far as there [in the region of the moon]. In fact, my hope [of explaining lunar phenomena] is no whit lessened by my seeing that the problem has been tackled a great deal and without success by the greatest astronomers and men of science.

Our author asserts that he has found, for the moon's movement, causes of the above kind. As he says on his page 253, 'The following perfectly explain in every way—and this easily and naturally[4]—the phenomena of the moon: in what way the reasons are clear why she always presents the same face to us; why she is whirled round the earth; and whence she derives an annual and a monthly motion, breadth, anomaly, depression and elevation [in her orbit]'. What he has proved we must see in the words that follow; but first of all we must draw up, according to the Copernican method which he himself adopts, the scheme or configuration of the moon with the sun and the fixed stars, and enumerate the phenomena he has tried to account for in the present Problem.[5]

FOL. 264

2. Let the sun be at A, and round its centre let there be drawn the ecliptic of the eighth sphere: Aries, Taurus, Gemini, etc. Now, let the earth be at B, and let the dotted circle CFD around it represent the moon's monthly motion; but let another circle, CED, the unbroken one, be understood as lying in the plane of the ecliptic of the eighth [sphere], this [circle] to represent the ecliptic of the moon's heaven whose radius is BC or BD.[6] Hence the dotted circle CFD, in which the moon's monthly motion takes place, is understood as cutting the circle CED. So let the common line of intersection be CD. Lastly, let the circle described by the radius AB represent the earth's

FOL. 264v

---

[4] The words here given in parentheses are not in White.
[5] With what follows compare *EW* I.432*ff*. For 'breadth', 'latitude'?
[6] Figure 54.

annual motion. Having established this, let us approach the phenomena our author puts forward for consideration:

FOL. 265

[a] That in any one month the moon goes about the earth once, and that she accompanies it in its annual circuit;

[b] that in the [monthly] passage she describes the circle CFD, which does not lie within the ecliptic but cuts it;

[c] that the moon is swifter at the quadratures than in the ecliptic; and

[d] that at the quadratures she lies in a lower position [than at other times].

3. Before I come to the causes adduced by our author to explain these phenomena I should like to know from what source he has discovered that topic [c], i.e. 'the moon is swifter at the quadratures than in the ecliptic', is true, and where he found[7] in the astronomers that the recessions of the moon's orbit from the ecliptic towards the sides on either hand are called quadratures. Let the points of intersection of the moon's orbit, CFD, with the ecliptic, CED, be C and D. These points the astronomers call nodes. One is the Dragon's head; the other the Dragon's tail.[8] The points F and E, in which these two circles are at the greatest angle to one another, are the Dragon's limits and protuberances, but are not quadratures, for the latter are related to the syzygies. Now, these last are the two points in the moon's orbit in which this is cut by the plane passing through the centres of the sun and earth in such a way that whenever the moon happens to be in this plane she must always be either at the full or new. But if she is distant from this position by a quarter of a circle, she is said to be at the quadratures, so that the latter are always the two points a quarter-circle distant from the syzygies.

FOL. 265v

According to Kepler, Tycho Brahé noticed that the eccentric movement of the moon in her circle, i.e. CFD, is always, because of the moon's irregularity [of motion] (which is due to the eccentricity), swifter at the syzygies, i.e. at new- and full-moon,

---

[7] The rendering 'where' takes the manuscript word as *ubinam*. But, if taken ironically, *utinam* of the printed text may be correct: '...true. Would he had found...!' Either way the sense is little affected.

[8] These terms are explained by Drake, pp. 66, 475. In medieval times they were assigned to the two nodes, i.e., as Hobbes says, the points at which the lunar orbit cuts the ecliptic. See also O.E.D., s.v. 'Dragon's head' and 'limit', 2c.

FOL. 266 than at the quadratures. Kepler gives a physical explanation of this matter, which may be read in Book IV of his *Epitome of astronomy*, at page 562 [of the 1618 edition].[9] Now, our author might indeed have read: 'The moon is faster at the syzygies than at the quadratures' and, taking the latter to mean 'limits, or the greatest declination of the moon from the ecliptic', might, through his ignorance of [these] limits,[10] have fallen into the error of thinking the moon to be swifter within the ecliptic than outside. Yet because he says the opposite, namely that she is swifter at the quadratures, i.e., as he believes, at the greatest declination, than at the intersection with the ecliptic, I do not know how he fell into that error, unless perhaps through a lapse of memory. Consequently, after using on his page 248 the words 'Astronomers note and teach that the moon grows faint in the ecliptic but is elevated at the quadratures' he ought to have quoted those astronomers, or any one of them, instead of hastening to congratulate himself and saying: 'I didn't expect of you, Andabata, such skill in astronomy'. Such words are indeed superfluous even [when applied] to

FOL. 266v someone who disserts upon a subject most learnedly; to one who does so only moderately well they are excessive; to one who errs they are shameful.

4. Likewise, I ask, regarding the fourth phenomenon, [d], from where he derived [the fact that] 'the moon is lower at the quadratures, i.e. at perigee'. 'The astronomers say so', he asserts. Which astronomers? If he understands 'quadratures' as defined above it is wrong; for if it were true, an eclipse of the sun or of the moon would never occur when the latter was in perigee, because an eclipse happens only when both luminaries lie in the ecliptic, i.e. when the moon is not in apogee, since if a perigee of the moon occurs at one quadrature, the apogee will occur at the other. An eclipse, then, occurs only when [the moon] is in the ecliptic–unless he takes 'quadratures' as the astronomers do, namely as 'the interval between new- and full-moon'. It would [then] follow that no eclipse could occur when the moon was in perigee or in apogee, for no eclipse hap-

FOL. 267 pens at the quadratures, all eclipses being caused by the conjunction or opposition of luminaries. Again, every calendar

---

[9] Drake, p. 391. See also Appendix II below.
[10] Or, 'of the terms (*termini*) [applicable]'?

proves that apogee and perigee, or the line called apsis,[11] are perpetually altered: they occur according to the sequence of the signs [of the zodiac], sometimes at the syzygies, sometimes at the octants,[12] and sometimes at the quadratures. He ought, therefore, to have named those astronomers who he says support him, lest he should seem to be trumping up his phenomena himself, appearing to make them conform, at his own will, to causes that he himself had held in readiness beforehand.

5. I come now to the causes or hypotheses themselves [of the moon's irregular progress]. They are two: the earth's daily rotation, and the moon's gravity or *conatus* towards the centre of the earth. Now, in order to agree on a reason for the moon's circling the earth, we should recall that in Problem 8 of the present Dialogue our author stipulated that the earth's daily motion carries round–and this within twenty-four hours–not only the air trapped between mountains and in hollows in the earth, but also all the air as far as, and beyond, the moon herself. He now thinks one can infer from this that the moon, floating in the air thus carried which surrounds her, must be carried around with it, but in the space of a month. I omit here the fact that the air cannot thus be carried circularly throughout its daily movement, for I think that this has been sufficiently shown above in Chapter XXI, but I ask: Why is it not carried round within a day, [rather than monthly]? He replies: 'The [motion of] the moon is slower than [that of] the air which surrounds her, because of gravity'; for he supposes the moon to possess a propensity and *conatus*–which we usually call 'gravity'–towards the centre of the earth. To this hypothesis of his I shall note two objections only: First,[13] if the moon were drawn downward, she would fall on to the earth, as every heavy body would. He thinks that on his pages 251 and 252[14] he has sufficiently answered this objection, as follows: 'You see', he says, 'that, because they contain air enclosed, as it were, in their hulls [*quasi ventre*], yachts and ships do not sink. Fish are considered as having been endowed with a similar property by nature, so that they take into the ventricle–which nature has

FOL. 267v

---

[11] Here, the line joining these two positions. See also fol. 276.
[12] The four points where the moon is 45° from conjunction with, or from opposition to, the sun.
[13] The second objection commences Article 6.
[14] See also *De mundo*, p. 100.

FOL. 268

adapted for them for such a purpose–the amount of air necessary [if they are to descend or rise].[15] As for birds in the air, they have outside their bodies that which fish contain inside [theirs]. When their wings are folded, birds sink: when spread, they are carried upward.[16] There is a single explanation for all these facts.[17] Imagine only the moon's effluvia as having been, as it were, firmly attached to her very body (a thing I have no doubt you accept)–so firmly, I say, that the vapours of the earth are not, try as they may, strong enough to disperse them. You see at once that the moon herself must either remain stationary or must move with them.[18] You see, then, that [the analogy of] wings has been applied to the moon. If, so far, you are very uneasy [about this], turn your thoughts to lofty mountains and deep valleys, and consider also the hollows with their inner void that the nature of mountains renders dangerous to us. Not infrequently mountains have the habit of boiling up with subterranean fires caused by over-heating at the base.'

Then he proposes on his page 254 another means by which [the moon] seems to be held [in space], i.e. the fast speed of her motion; and, by means of analogy with swimmers in the water and by comparable instances he tells us that heavy bodies, which otherwise would sink, are prevented by their movements from doing so.

Lastly, he avers that wine floats on water if poured on to it gently; likewise with both oil and water [poured in turn] on water; so also with air on air. How childish and silly this reply is will be apparent if one scans each several example:

---

[15] Material, of doubtful meaning, from the *DM* (pp. 252–253) that Hobbes omits is supplied in the next three notes. For the insertion 'if...or rise' (MS.: *etc*) White has: 'so that they must either sink towards [descend from?], or be borne up to, that place whither their appetite draws them'.
[16] Hobbes omits the words 'motion being the midwife'.
[17] Hobbes omits: 'If the medium that must be cleft, through gravity, during descent be equal to the lighter appendages fixed to a heavier body, you will stay fixed wherever you will; if you go beyond [it], you will be borne upward into a rarer region'. For these words the manuscript has *etc*.
[18] 'I say...disperse them'. Perhaps: 'These effluvia are not strong enough to disperse the earth's vapours when the latter resist them'. But from what precedes and follows we would expect the moon to be able to control her course. After 'move with them' White has: 'Consider now: the bulk of the lunar vapours ought to be as large as that of the terrestrial. Certainly, from the nature of bodies, we are of opinion that the more she is subjected to the sun, as opposed to the earth, the more diffuse the moon is'. For these last words the MS. has *etc*.

(a) Is it not obvious that he is incorrect in declaring that the same reason explains why ships, yachts and fish are supported in the water, and birds in the air? The cause of the first's buoyancy is that if a ship, together with the air contained within it, be considered as a single body, it is lighter than water of the same bulk [*moles*]. The same must be said of fish and of everything that rests motionless in water; but a bird, together with its wings (either folded or spread) will immediately fall by its own weight [*gravitas*], unless it supports itself by flying, because it is heavier than a mass [*moles*] of air equal to itself.

FOL. 268v

(b) The reason why a bird is supported in its flight is not the explanation of the moon's being upheld in the air by her own motion round the earth. The [upward] motion of flying creatures is such that, by means of their wings, they propel towards the centre of the earth the air on each side of them. Hence the air that rises between their wings raises the middle of their body by a similar amount. But this cannot apply to the moon's motion about the centre of the earth.

(c) When he tries to convince us that the moon's effluvia have been firmly attached to the very body of the moon this is tantamount to telling us to consider that effluvia are not effluvia; for effluvia are what they are by virtue of being fluid, and fluids are fluids because they are not firmly packed. Again, if the moon is attracted downward [*gravitat*], she will surely gravitate more than her effluvia do, just as the earth is heavier than its own effluvia, namely the air.[19] So the moon will gravitate within her own effluvia and will sink in such a way that she can no more utilise them than she could wings.[20] Her body, however, is in contact with the *terrestrial* vapours–indeed, she is plunged into them–so that she will drive out of place [those of] the earth's effluvia that stretch [*sunt*] beyond the moon, since they are heavier than the lunar ones. Similarly, water deeper than the rim of any cup pushed down into it will sink into the interior of the cup when the air has been driven out.[21]

FOL. 269

---

[19] The moon will gravitate towards earth because she is heavier than her vapours? But if, as Hobbes says, the earth is heavier than its effluvia, why will not the earth itself gravitate, particularly in a heliocentric system?

[20] Or, 'utilise them as wings (in White's hypothesis)'.

[21] I.e. as soon as a vacuum is opened to a fluid or solid, the latter will enter it (cf. fols 23*ff*); hence if the earth's vapours were to disappear, the moon would crash upon the earth.

Therefore, however you look at it, if the moon gravitates she will fall upon the earth.

Lastly, it is true that wine floats on water, air on air, vapours on vapours and, universally, lighter bodies on heavier ones. But it does not follow that bodies from which extremely buoyant vapours are given off cannot themselves sink within vapours heavier [than these]. For otherwise, if anyone was sweating and jumped into water, he would not sink–because sweat (the effluvia of a body that is perspiring) is lighter than water; his sweat would cling to him just as much as the moon's effluvia to her. Therefore if the effluvia of the moon that is gravitating prevent her from sinking in the air, then far more will the effluvia of a sweating man prevent him from sinking in water, [and clearly they do not].

6. Another objection[22] to the same hypothesis of his is this: If in its daily motion the earth carries the moon round, the latter should complete her orbit in 24 hours. But it is agreed that in order to complete her circle she needs more than 27 times that period. He replies that such retardation of the moon is caused by gravity, just as dense bodies in rivers move more slowly than the stream itself. Why this should be so he hardly says at all; nor, if we consider the matter carefully, will the explanation [he gives] be the true one. It is true that, if a feather and a ship are carried along together with the current, the feather will move ahead of the ship, and yet both will travel at the same rate as does the river carrying them. Yet the parts of the water near the surface–the feather floats on these alone–flow in a course that has a greater declivity and hence is swifter than the underdraw,[23] and consequently is swifter than the rate of flow arrived at by averaging the rates both of the lower and of the surface waters. A large ship would lie in both, part of it extending [even] into deep water; so the feather must be moved faster than the ship. The difference [between the flow at the surface and that deeper down] is not, however, great. That the moon, being supposed heavy, is for that reason moved within the earth's vapours twenty-seven times more slowly

---

[22] The first objection begins at fol. 267v.
[23] If Hobbes means that the surface-waters are at an angle, relative to the horizon, greater than that of the lower waters, he seems to suppose that this is due to the existence of some partial obstacle downstream; cf. fol. 179v.

CHAPTER XXIV: 7

than these effluvia by which she is carried is something quite extraordinary.

7. From the same hypothesis of the moon's gravity he infers the following reason why she is not borne beneath the ecliptic: The vapours exhaled by the earth, he says, are more numerous and [rise] higher [when the earth is] beneath the ecliptic than [they do] elsewhere, because the sun is [then] always directly above them. When they have reached their highest point they are thrust outward towards the sides of the ecliptic by other vapours constantly rising beneath them. (Likewise, rivers drive from mid-stream towards their banks objects floating on them if the objects move only a little more slowly [than the rivers].) Hence the moon, on reaching the ecliptic, does not stay there long but is pushed towards the sides [of the ecliptic], where the earth's vapours are lower-lying.

Since he had said that the moon's movement is brought about by the earth's daily motion, he should–in my opinion, at any rate–have asked: 'Why is the moon not carried parallel [to the celestial equator] beneath the equinoctial?' rather than 'Why is she not carried beneath the ecliptic?' The earth's daily motion takes place in the equinoctial, but during the daily motion every part of the earth moves through either the equinoctial itself or some circle parallel to it.[24] Therefore he ought to have given a reason why the moon, when in motion, is not moved in the same path as her mover, i.e. the earth, is. He should also have shown what her mover was when the moon was in any great declination from the equinoctial, for her daily motion will not carry her there.

FOL. 270v

By way of illustration, let the earth['s centre] be A, and let BC represent the equinoctial.[25] Then let the moon be carried in the circle DEFG, [the amount of] whose inclination [with reference to the ecliptic] is the arc FC (or let DB equal the greatest inclination from the ecliptic, i.e. $23\frac{1}{2}°$).[26] Next, let us

---

[24] Hobbes may mean: 'The earth's daily motion takes place along [the arc of an elliptical orbit parallel to] the equinoctial [circle]'. Here 'equinoctial' seems to mean 'the great circle through the equinoctial points whose axis is that of the earth's diurnal rotation', which can only mean the celestial equator itself. Moreover, 'the earth's daily motion' can mean either that along the ecliptic or its revolution about its own axis. Or is 'equinoctial' a slip for 'ecliptic'?

[25] Here, the line joining the two equinoxes. Figure 55.

[26] The present nomenclature seems inconsistent with that used relative to the figure, since now C lies *outside* the plane containing DF, and E lies *in*

imagine the moon to be at D or F. It is clear that the whole motive force [*vis motiva*] that the earth possesses from its daily motion is directed along the parallels drawn from B to C. How, then, does that force reach the moon at D or F? Obviously this is impossible, except if [*i*] the earth's body were of such magnitude that, if the earth lay between C and F or between B and D in the moon's orbit, it would fill that whole space, which is ludicrous, or at least [*ii*] any rectilinear motion from the earth were propagated spherically in all directions. The latter he has not supposed to be the case; [he supposes] only that the sun raises vapours which, he says, ascend vertically in the torrid zone and, driven upwards from there, coalesce into spherical form. Yet it is undeniable that the earth's daily motion,[27] but not this only, *may* cause the moon's rotation. How this can be so I shall explain later.

Let us now test the validity of the cause (according to our author) of the moon's not being carried beneath the ecliptic. Granted that the rising of the vapours is greatest where the sun's rays strike the earth perpendicularly, i.e. in any part of the ecliptic, what if the moon were in a part of the ecliptic opposite the sun as when, the latter being at F, the moon would be at D? I do not see why she must be driven towards the sides [of the ecliptic] into a lower position; for if the vapours are lower in those places furthest removed from the sun, then the moon cannot find a position lower than the one she occupies, because the sun is furthest from that point of the earth opposite it. Again, if the reason for the moon's quitting the ecliptic were her being driven (by the rising vapours) into a lower position, it would follow that when in the ecliptic or at the nodes the moon was always in apogee, but when in the limits, in perigee. This he assumes to be so, as if it were a true description of the event; but in Article 4 above we have shown that according to the evidence of calendars, i.e. from the experience of every epoch, it is false.

Next, if the moon is forcibly expelled from the ecliptic,

---

that plane; in the figure the converse seems to hold. The angle DCB can be the *greatest* inclination of the ecliptic only if the moon's orbit is considered to be entirely within the ecliptic itself (which would be against what precedes). The greatest inclination would be $23\frac{1}{2}° + \sim 5°$, being the sum of the obliquity of the ecliptic with reference to the celestial equator and the inclination of the moon's orbit with reference to the ecliptic. The only other possibility is that $23\frac{1}{2}°$ should be $5°$.

[27] Again, 'the earth's daily motion' is imprecise.

by what force, it may be asked, does she return to it? For when she approaches, the vapours under the ecliptic should keep her at a distance by the same property through which they drive her forth when she is under the ecliptic. Coming to this question, he says that she returns to the ecliptic because, on the surface of a sphere, the shortest path is by a great circle. But, [I say,] passage along a great circle is impossible except by cutting the ecliptic, because the latter *is* a great circle. So the moon would have to move from any given point to another point by the shortest path. If she had to do so, either [a] she could not find in the fluid ether, through the chord of an arc [*per subtensum*], a path shorter than one through the arc of a [great] circle, or [b] if she had to do both these things, such necessity would be independent of [any] physical cause that our author should have explained here as part of his case. He should have said, therefore, how and why the earth (by its daily motion) not only drives the moon out of the ecliptic but also forces her to enter another great circle, such that the same strength of vapours is required to expel her as is needed to draw her back again.

FOL. 272

8. The reasons he cites in order to account for the last two phenomena, namely 'the moon is swifter at the quadratures' (i.e., as he understands the term, at her greatest declination from the ecliptic) 'than in the ecliptic itself', and 'the moon is lower[28] in these same quadratures', are confuted by the very fact that such happenings do not take place. Let us see, however, whether his reasons do convince us that things are as he says.

He states that a swift motion causes a thickening [of the vapours]. But why and how this happens he illustrates neither with a reason nor by example, but merely assumes it. Because it is when the vapours are beneath the ecliptic that they are the more violently stirred up by the earth,[29] he concludes that such vapours are thicker under the ecliptic than outside it. Yet to most men of science, and not without plausible reason, matters seem otherwise. They believe that the vapours are

FOL. 272v

---

[28] Cf. 'forced downward...towards the centre of the circle' (fol. 272v); 'depressed' (title of article at the head of the chapter): i.e. 'inward,' towards the centre of her orbit, i.e. towards the earth?

[29] Hobbes's word-order makes it uncertain whether the earth is, or the vapours are, beneath the ecliptic.

scattered and thinned by the very same heat that produces them, and hence that within the tropics the air is more tenuous than outside. So even if we grant that (as our author thinks), when the sun is near and overhead, the air within the tropics is also thicker, it will not be more dense when the sun is over the opposite hemisphere. Hence, even though she gravitates, the moon will not sink down closer to the earth at the quadratures (when she is in the same hemisphere [of the celestial sphere] as the sun is) than in the ecliptic (when she is in a different one). Hence it will not always be true that the moon is lower at her greatest declination from the ecliptic than at her least.

FOL. 273

From [the question of] the moon's being forced downward when outside the ecliptic he shifts the discussion to her speed. 'Of this', he says, 'you have two causes: the shortening [of the lunar path] when forced down towards the centre of the circle,[30] and the vehemence of [the moon's] motion beneath the ecliptic. Since [she] moves quite fast, the ecliptic [itself] reduces the difference, and the moon seems to be moved more slowly.'

I really cannot understand these words at all; they seem to contain contradictions. 'The difference is reduced' and 'the moon seems to move more slowly' sound contradictory to me. Furthermore, it is ridiculous that a motion he should suppose to be a daily one (i.e. a motion beneath the lines parallel to the [celestial] equator) is the more vehement under the ecliptic, for every separate point on the latter has a daily velocity that is greater or less according as the parallel circles which cut the ecliptic are larger or smaller.

FOL. 273v

Now, as regards our author's views on the theory of the moon so far, I shall publicly set forth as briefly as I can my own opinions on the matter, not because I think I can account for every detail of the moon's behaviour, but because what I shall say about the motion and functioning of the heavenly bodies is more coherent and less inconsistent with the daily motions than what others have said so far, and also, perhaps, is in closer agreement with the phenomena of the skies. I am ignorant, however, as to whether man can contrive and apply any move-

[30] I.e. the circle on centre A in the figure to fol. 270v. White's first argument, not taken up by Hobbes, seems to be that a body orbiting something will, in a given time, make more circuits the nearer it is drawn towards the centre of its orbit.

ments to moon, earth and sun that will explain the appearances, of which [motions] it would be no disgrace for him to know, one day, the true explanation.

9. First, then, because we see that the sun is everywhere about us simultaneously, we take it that the sun acts everywhere at the same time. And because every action is motion, it is agreed that the sun possesses a certain motion which is continually propagated in straight lines through a fluid medium in every direction into [the form of] a sphere. Such motion is, in fact, termed illumination. Its cause is the continuous thrust by which an outermost part of the sun, however little it is forced outward, causes the part of the medium closest to it to yield. This part, in turn, causes another part, that next this one, to give; and in this way is brought about the propagation of the motion or action by which the sun shines forth in every direction. Again, it can easily be seen that, apart from this motion, the sun can contain within itself a motion called the turning-motion, by which all its parts rotate around its own axis. Being able to grasp this, we may also follow the thought-processes of Kepler, Galileo and other famous astronomers and scientists. From these two motions, one of which, the illuminating-motion, is clearly within [the sun], and the other, the rotary motion, is supposed to be within it, we show that the air, or that liquid medium that extends in every direction from the sun, is carried circularly around it, and this at a velocity which is the same at all distances. Any part of the air, therefore, that is further from the sun than another part is takes more time to complete its orbit, in the ratio of their respective distances from the sun. So the illuminating-motion, which of itself is rectilinear,[31] is initially circular owing to the sun's spin, and in that circular movement it has a velocity (determined[32] by the sun's spin) that is neither increased nor reduced by any other agent. Neither of the following arguments vitiates this [hypothesis of a] circular movement of the air: [a] the imparting [of motion], as a wind is stilled when, after being stirred up in some compacted part of the air, it is transmitted to a looser part; for that motion imparted by the sun is propagated as a whole, in every direction, and in such a way that it cannot

FOL. 274

FOL. 274v

---

[31] *Per se rectus.* Perhaps, 'right motion'.
[32] *Determinata.* Perhaps, 'limited'. With this account compare Chapter IX above.

be transmitted into air that is more loosely packed; and [b] gravity, as the motion of projectiles is arrested [*extinguitur*] as they fall towards the centre.[33] The descent is a hindrance to the motion of projectiles, directly or indirectly. So the circular motion of the parts of the medium, a motion generated by the sun's rotation combined with the sun's illumination, extends at a constant rate to all distances. Hence the circles described by the parts of a medium thus moved are transposable into the times proportional to the distances from the sun.

10. If the sun's rotation be considered on its own and without that radiant motion, it could not move the air circularly; for in every motion, that which moves [something] directly and that which is moved should always be in the same line. The moved, therefore, quits its place because another body tries to assume that same place. Now, if the sun moves some part of the air, this part must always lie in the same circular line with the part of the sun it is in contact with, which is impossible. If the outmost parts of the sun are to drive the air round, therefore, some grip will be necessary by which [the sun] can take hold of the air. In like manner, even if the sun's surface were rough, enabling it to move the air in contact with it, that air could not in turn move a second [area of] air adjacent to itself because of the lack of grip [between air and air]. Therefore the air is moved circularly, not by the sun's spin alone, but through the agency of that hurling-motion [*motus proiectitius*] by which, as I have said, the sun is perceived and by which it shines.

Now, say we accept that the air moves round the sun. The earth, because it rests upon the air, must also be carried round with the air and must describe its orbit in a fixed period. That orbit is called the ecliptic, and the period of motion, the [terrestrial] year. In this way we have the annual motion of the earth within the ecliptic. Again, it has been shown in Article 3 of Chapter XIX, on Problems 6 and 7 of the present Dialogue, that, from such a movement of the air round the sun, the earth floating in it must always keep its axis parallel [to the axis's self].

11. Why the [annual] circular motion of the earth does not have the sun exactly as its centre but is eccentric I do not

---

[33] Owing to the increasing resistance of the medium (air)?

know. Possibly this is due, as I said above at the end of Chapter XXI, to the fixed stars.³⁴ Perhaps it is also due to a property of the earth's nature, thus: Any solid body, i.e. a hard one, whose parts cannot be torn apart or separated one from another save by [the action of] a great motion from without, will owe this property to the internal motion of its parts, namely to that motion of material which determines the body's form. So the earth also possesses such internal motion of its material as causes it to be earth. Again, if any part of a hard body [a] is set in motion by another body [b] that collides with the first body, body [a] has the property of self-adjustment according as its hardness or the force of the colliding body [b] is greater or less.³⁵ And, if it cannot do otherwise, in freeing itself it recoils, so as to retain its own specific characteristic, not as does a sense-endowed agent that knows what it can preserve, but as a top which is made to spin quickly. This, knocking into a wall, at once bounces back, not with the purpose of retaining its circular motion but from the force of that self-same motion which has moulded the top's material into a definite shape that could not hold together [*consistere*] on [the top's] colliding with the wall.

FOL. 275v

The earth, also, has that property: If any agent acts on it so that the earth's internal motion (which makes it earth) is hindered by such an action, it at once recovers itself. Now, the earth may possibly be heterogeneous and its parts be of different natures, so the heat of the sun against it is more direct, or fiercer, from one direction than from another.³⁶ Hence one of its parts being nearer to, and another further from, the sun, the earth best indulges its own nature. If this is the case, we have the reason for the earth's³⁷ orbit's eccentricity. If that is not the cause, perhaps the fixed stars are, except that this movement seems to be contradicted by that of the apsis; for today the sun's apogee and perigee are under different fixed stars than previously. It may also be that of the moon and the earth each is the cause of the other's eccentricity. The first reason, i.e. the diversity of the constituents in parts of the earth, seems to me to be the most plausible, because similar effects

FOL. 276

---

³⁴ Hobbes has made no such statement, there or elsewhere in this work, but cf. fol. 201v.
³⁵ I.e. one or more of four conditions is or are involved.
³⁶ Or, '...in one part (of the earth) than in another part'.
³⁷ The text, which reads 'sun's', passes uncorrected by Mersenne.

derive from similar causes in other phenomena. Now–to omit magnetic properties–what is better known than that in winter men turn towards the hearth, even if doing something else? That is, when the parts that are turned away from it have their vital motions clogged by the cold, they move these up to the fire, but when, by being kept there too long, the men have grown hotter than the same vital motion can endure, they move them away again. Further, if any part of a body is so exerted that it can the less endure heat, then when that part has to be exposed to the fire they stand further back than they would if the part were healthy. The same seems to apply to the earth: the sun's heat is less in one region of the earth, as befits the nature of that region, but greater in another; so when that area for which a lesser heat is suitable has to be turned toward the sun, [the earth] draws further away; when another, it draws rather closer [to the sun]. In this approach and recession, therefore, lies [the explanation of] the earth's [orbit's] eccentricity.

12. Perhaps someone will reason as follows: Instances deduced from animals are relevant only if earth were also an animal.[38] I declare, therefore, that even stones, logs and any terrestrial body will rotate in face of the sun unless prevented by gravity, i.e. unless another, stronger motion, such as would impede such rotation, prevent it. The whole earth (comprising stones, logs, metals, soil and other things of different kinds) is rotated in daily motion facing the sun; so one cannot doubt that if any part were carried off from the earth into orbit in any direction the remainder would go on rotating with the same motion; and it would rotate in the same way if still more were taken away until there was left but a single stone or a globule of metal or of wood at the [earth's] centre. Hence any heavy body would be rotated in face of the sun provided that it were free from gravity, as is the earth as a whole. Indeed, the first principle of that motion by which animals seek what is fitting for them is not the will or appetite; for such appetite is the inner motion of animals by which they recruit themselves when oppressed by a motion contrary to their nature and bodily make-up, and by which they draw near to those agents that strengthen and invigorate this same motion of theirs. Although the terms 'pleasure' and 'pain' are applied only to living things, inanimate creatures

---

[38] Drake, p. 271.

CHAPTER XXIV: 15

also possess such qualities, i.e. the strengthening or the obstructing of the specific motion.

13. Such specific motion of the earth's inner parts may be understood as producing [the earth's] daily motion, by which it moves in turn some of its parts towards, and some away from, the warmth of the sun. Let, then, these be our comments on the reason for the movements attributed to the earth.

14. As regards the moon, the sun seems to act on her in the same way as it acts on the earth. Therefore just as, in its annual motion, the earth is borne round by the sun, so the moon is also. Herein lies the reason why [the moon] always accompanies the earth. But just as people suppose that the sun is rotated about its centre in order to draw the earth and the moon jointly around itself, so some people think that the moon is carried round the earth by [the latter's] daily motion. Let us see how this can be done. I have demonstrated above in Article 10 that the sun cannot, by its spinning only, have driven the air circularly without the addition of a hurling-motion by which it moves the air all round it as a whole and propagates a certain rectilinear motion into every part of [its] sphere. It can be shown that, for the same reason, the earth cannot move the air [stretching] as far as the moon unless it also has within itself some similar motion. This latter motion is what some might term 'influence', for 'influence' either is motion or is non-existent. Given, therefore, such motion in the earth (how it differs from illumination I shall say later [at folio 281]), let us see what the consequences of this are as regards the moon.

FOL. 277v

15. First, it is clear that the moon in her monthly motion will circle the earth in the same way that the earth circles the sun, and at the same speed as that of the [earth's] daily motion. Hence, unless there is some other reason for the moon's monthly motion, she will complete her circle in a period whose ratio to one [mean solar] day is as the [ratio of the] distance between the earth's centre and the moon to the earth's radius. Again, the earth's axis is always carried round the sun parallel [to itself]; likewise the axis of the moon, if the sole cause of her motion were the motion she has from the earth, would always be borne parallel [to itself], so she would always turn towards

FOL. 278

the earth not the same face, but ever a different one.³⁹ Two points must therefore be queried here: (1) As, according to the astronomers, the distance of the moon from the earth–if you take the mean between the apogee distance and the distance at perigee–is about 56 radii of the earth, then the moon completes in each day only one fifty-sixth of her orbit, i.e. 6° 24′, but, the astronomers say, she completes daily 13° 10′ and more. If, by virtue of the earth's daily motion, [the moon] passes through only 6° 24′, one queries by what property she completes the other 6° 46′.⁴⁰ (2) Because of the earth's daily motion, the moon ought to present to the earth successive parts of her surface. By what property does she always present the same one part, i.e. the one with flaws?

FOL. 278v

16. Let us refer again to the diagram in Article 2, [i.e. that on folio 264v]. Given the sun at A, the earth['s centre] at B and the moon['s centre] at C, let us suppose that owing to the earth's daily motion [in the plane of the ecliptic] the moon is carried circularly, following the order of the signs, from Aries to Taurus through any given space in any time. Clearly in the meanwhile the sun acts on the moon in the same way in which it acts on the earth, i.e. warming her by means of its irradiation. But such heating was the cause of the earth's daily motion, which enables the earth to move its parts one after another up to the sun. It is therefore reasonable to believe that, through the same irradiation, the sun is the cause of the moon's approaching the sun's focus, until she enjoys its heat as much as suits her and then moves away.

Now let us suppose that the moon, either because she seems to possess a nature resembling that of the earth, or for any other reason, lies at a distance from the sun that suits her

---

[39] It will be recalled that in this paragraph Hobbes is describing what he considers will happen under White's reasoning, not what is actually observed. For the second 'to itself' we should perhaps understand 'to the earth's axis'–but this might suggest that the moon presents ever the same appearance to us on earth.

[40] The system of reference is insufficiently defined here. In fact, with reference to the background of stars, the moon moves 13° 10′ per day as observed from earth. With reference to the line joining the earth and the sun, her mean motion is about 1° per day less; with reference to her own axis, she rotates through one revolution per month, or rather less than 12° per day. But it is unclear what Hobbes's figures 6° 24′ and 6° 46′ precisely signify.

nature best. When she is lying at the same distance from the sun as the earth is, i.e. during the quadratures, it will follow that the same moon is nearer the sun at new-moon and further away at full-moon than suits her nature,[41] i.e. when the earth, through its daily motion, forces her into an unsuitable position. Hence, driven by a continuous motion from the position where she enjoys the correct warmth, she has to return to that place, also by a continuous motion. So when she is in C, rather further from the sun than is fitting, the moon will be moved by the sun's warming-property from Aries to Taurus. When she is at D she will be moved towards Scorpio.[42] But she was also being borne along the same path because of the earth's daily motion. So there will be two reasons for the moon's movement: the earth's daily motion and the warming-power of the sun. But it was supposed that the moon was advanced[43] by 6° 24' only, by the force of the earth's daily motion; therefore the remaining 6° 46' will be due to the sun's rays.

FOL. 279

17. From this we may also suggest the reason why the moon always presents the same side to earth. Let us suppose that the moon is moved round the earth by the latter alone, in the same manner that the earth is moved round the sun, i.e. such that she will always keep her axis parallel [to itself]. Just as the earth happens, as a direct consequence of this, to display different parts [of its orb] to the sun, so the same is true of the moon with respect to the earth. Let us now add the property of the sun's rays. Since this virtue has caused the daily rotation of the earth about the sun,[44] the same sun will also impart a similar rotation to the moon. The part of the moon, therefore, that was turned away from the sun when the moon was in C will, as the moon moves forward through CED, become different step by step until she is placed in D; the part that was turned away at C will now be turned away at D. Hence the same face will be turned towards the earth when the moon is at D as was turned when she was at C. By this means the cause that produces phases [of the moon] is the same as that causing her always to present the same face towards earth; and the property

FOL. 279v

---

[41] Cf. fol. 201.
[42] Mersenne: 'See the figure in Article 2'.
[43] Text (accepted by Mersenne): *supponebatur luna promotam fuisse*. Probably the copyist missed the upstroke on Hobbes's *a* in *luna* (denoting *-am*).
[44] For 'about the sun' (i.e. the annual motion) perhaps, 'to the sun', in accordance with the analogy on fol. 276.

by which the sun brings about the lunar month is the same as that by which the sun determines the terrestrial day.

18. The same cause is responsible for the moon's being moved rather faster at the syzygies, i.e. at new- and full-moon, than at the quadratures. At the quadratures she is equally distant from the sun as the earth is, i.e. at a distance that is proper and suited to her nature; but at the syzygies she is at a greater or a less distance (for her distance is greatest at full- and least at new-moon). There is no doubt, therefore, that from each of these positions, according to her need, she draws closer to the sun or withdraws further from it.

19. It seems easy to explain why the moon's orbit does not correspond with the ecliptic, along which she is guided by the sun's movement in its annual path, or with the [celestial] equator, along which she ought to be moved by the earth's daily motion; for when two movers combine they bring about a mixed motion that differs from them both. Let us again consider the figure to Article 2.

Given that CED is the ecliptic and CGD the equator, and that the moon is at C, where these meet. The sun in the ecliptic tries to drag the moon downward and the earth tries to raise her above it, in the signs from Aries to Libra in order. Hence the moon will glide between the two circles, say in the dotted circle, so that she will appear in the signs bearing a northerly latitude. Likewise and for the same reason when in the signs running from Libra to Aries she will appear in a southerly latitude.

20. As for the moon's apogees and perigees, these have been observed by astronomers as being twofold, for the moon can approach the earth and recede from it, [her] orbit being eccentric. Another reason why she can do so—let me quote the ancients—is the epicycle in which the moon is carried sometimes closer to the earth, sometimes further away, according as she is on this or that part of the epicycle. But leaving out of consideration the solid spheres of the old astronomy, we may explain the cause of this variation as follows: (1) We said that the earth's orbit could be eccentric because one part of [the earth] would, by its natural constitution, enjoy the solar rays that are closer, another in those further away. Similarly the

moon['s orbit] could also be eccentric because one part of the earth (for the moon always displays the same side to the earth) affects her more strongly than another. Moreover, when that part of the earth which acts most strongly on the moon is turned away [from her], this is at [*fit*] apogee; when the one that acts most weakly, at perigee. (2) Since it is supposed [by me] that the earth craves the sun's rays, which are closer on the one side but further off on the other,[45] the same may be presumed of the moon [relative to the sun]. Therefore when the illumination of a half-hemisphere[46] [of the moon] invariably coincides with that of a [terrestrial] hemisphere which seeks the sun when this last is more remote [from the moon], she will be at apogee; when the converse, at perigee. Hence there will be two apogees and two perigees, the former due to the diverse active powers in different parts of the earth, and the latter due to the diverse passive powers in different parts of the moon. Phenomenon (1) has as its efficient cause the earth; phenomenon (2), the sun. Phenomenon (2) follows[47] the diversity of the [lunar] phases, but (1) is connected with [the moon's] periodic course only; but when both are combined the moon is at greater apogee or greater perigee. These terms have shown that our author would seem to have based an inconsequential problem on this question in order to disparage the astronomers' belief in observations.

FOL. 281.

21. We said in Chapter XI, Article 2, that the nature of light, both the solar and that in any other luminary, consists in the fact that the sun and every luminary dilate on every side spherically and in every direction at the same time. Again, in the present chapter we also ascribed a similar action to the earth. So the sun, by means of its light, together with its spin about its own axis, not only draws the earth circularly in the latter's annual motion but also causes parts of the earth to be turned sun-ward one after another in the daily [axial] motion. Likewise the earth draws the moon circularly by means of its influence in conjunction with [its] daily motion.

Now, someone may ask: 'Why is the earth not, as the sun is,

---

[45] Presumably on the sides of the earth severally turned towards, or away from, the sun at a given time. There is no reference implied to an elliptical orbit of the earth.

[46] 'Half-hemisphere', Mersenne; text: 'different hemisphere'.

[47] 'Is in accordance with'; but the meaning may be 'succeed (in time)'. For 'phenomenon (1)'...'phenomenon (2)', perhaps 'apogee...perigee'.

FOL. 281v

lucid of itself? *If* light (which is the influence of the sun) is created only by an expansive-motion, and if the earth indeed exerts an influence on the moon through a similar expansion, that expansive-virtue of the earth will also be illumination'. We must say, therefore: The influence of those bodies whose nature is (as I may put it) mathematically *homogeneous* through all [their] smallest particles is light, and the light is strong or weak according to the degree of velocity at which [the bodies] swell. But consider truly *heterogeneous* bodies, e.g. the earth and any other body whose parts possess specific motions proper to a body itself.[48] When these motions conflict with those of adjacent particles—though any part, and hence the whole, has that motion we called expansive or dilatant—such motion, [one must say,] is not light but is an influence different from light.[49] But bodies differ essentially from one another because of the internal motion of their parts. There are countless specific differences between bodies, so the differences between their influences will be infinite too. Hence the influence of any one part of the earth will differ not only from the influence of homogeneous bodies such as light, but also from that of the other parts [of the earth]. Therefore the influence of metals is different from that of stones and plants; and the influence of a magnet will be different from that of other stones.

Finally, transparency consists in there being no conflict between the motions of the parts that coalesce into one body; light, likewise, is[50] the influence of a body in whose parts there is no conflict. Hence, if every translucent body possessed a motion that caused it to dilate, it would shine; but if opaque bodies possessed such a motion they would not; they would act at a distance and yet would retain their several influences upon themselves, though these would differ according to the bodies' forms, i.e. their different internal motions.

---

[48] Translator's italics.
[49] [One must say,]: The original, in *oratio obliqua*, lacks a main verb. 'Such motion' perhaps means 'the resulting or compounded motion'.
[50] Text: 'consists in the fact that it is'.

CHAPTER TWENTY-FIVE

FOL. 282

❧

On Problems 13, 14 and 15 of the Second Dialogue:
That astronomy is liable to error. That
[White's] supposition concerning circular motion
is wrong. That the motions of the remaining
planets [other than earth] can be calculated
from data to hand

1. Why bodies appear larger in thick air than in thin.
2. That from two bodies moving circularly and uniformly neither a circular nor a uniform motion can be created.
3. The reason why the sun appears circular is not its spin about its own axis.

1. That astronomers are prone to error, both for the reasons listed by our author, namely infrequent observations, the absence of apparatus, or the shortcomings in it, and also their ignorance of dioptrics, we shall be the last to deny. Yet as regards the errors that our author says have been made concerning refractions, i.e. that astronomers think refraction to take place, not around the zenith,[1] but only in a convex surface of vapours, here, I say, he is incorrect. 'Those who consider that refraction takes place only in a convex surface of vapours and not around the zenith', he says, 'are wrong. It takes place in the middle of the air as well.' Anyone, however, who promulgates a new theory that is contrary to those held by every astronomer, scientist, and student of optics must set forth an absolutely

FOL. 282v

---

[1] Perhaps loosely, 'in the upper air'.

clear reason for his opinion. [White] has given the following one: 'I remember that in a very thick mist in London I once thought that a boy approaching me was a man, so long as he stood a little distance away; and when I was staying at Lisbon I saw that some riders who were proceeding at a walking pace in the night air on a mountain that faced me seemed on a narrower view to be giants. I also hear that, when at mid-depth, divers think that all objects are larger [than they really are].' Yet how there follows from this the conclusion: 'Therefore refraction takes place in the middle of vapours' I do not yet see, nor have any astronomers, scientists, or students of optics seen it. Up to now, they have all asserted that refraction occurs at a surface or on entry to a transparent medium of different [optical] density. All of them have noted, however, no less than White has, what happens in a mist, namely that a boy appears as a man; nor does such an observation presuppose outstanding intellect. On the contrary, that conclusion [i.e. concerning the alleged refrangibility of light within one medium] does not follow at all; for I shall show how, even though there be no refraction in the middle of the air, things must appear larger when seen in thick air than when in thin.

Let AB be the eye and CD the object (or the boy) seen in the rarer air by means of the rays CE and DE.[2] Now, since the air is supposed to be rarer than the aqueous humour of the eye at E, the rays CE and DE will be refracted at the surface of the eye at F and G. Hence EF and EG if produced will pass, not through the points C and D, but between them, say at H and I. So CD will subtend the angle HEI.

Further, say the air is thicker than normal—as dense as the eye's aqueous humour, for instance. As the air's thickness is equal to the other, the rays CE and DE will not be refracted. So the object will subtend the angle CED, which is greater than HEI. Again, when the straight lines CE and DE, and HE and IE likewise, are produced as far as the fundus of the eye, a greater part of the latter will be filled by the first lines than by the second. Therefore the object will appear larger in dense air than in rare, i.e. a boy will appear a man, even though no refraction is postulated as taking place in the middle of the air. For the same reason, objects that are in the water will appear the larger

[2] Figure 56.

to underwater swimmers, but objects in the air will appear smaller [to them] than they really are.

2. Now for Problem 14. Though I readily concur with our author that the motions of the stars[3] are neither perfectly circular nor uniform, I disagree when he says that any motion which is sometimes slower, sometimes faster, cannot be brought about by several bodies moved circularly. 'On the other hand', he says, 'given any motion made irregular by a continued changing of velocity, who is so foolish as to believe that such a motion can be represented by several circles, [points on which are] carried at a constant rate?'[4] I for one reckon myself among those he calls mad. I believe–indeed, I know–that, if the earth is moved circularly about its own axis and at the same time is carried circularly around the ecliptic, any one of its parts that is so moved is [borne] sometimes quite fast, sometimes quite slowly [relative to the sun]. Likewise if upon [the circumference of] a sphere rotated uniformly and circularly there is [placed the centre of] an epicycle on which any star is also carried circularly and uniformly, I still declare that this star is moved sometimes more swiftly, sometimes more slowly. These matters are so obvious that whoever has asserted the contrary should be considered as out of his mind.

FOL. 284

3. Lastly, as regards the closing Problem of this Dialogue, i.e. that the ratios of the [motions of] other planets, [each to each,] can be calculated from the motion of the earth around the sun, I agree with our author; the sun drives round the other planets by the same virtue by which it drives the earth. Again, just as I have shown it to be false that the earth is moved by a wind aroused by the sun, so also I believe it to be untrue that the other planets are borne round, each of them, by winds of their own, but true that they are carried round by the sun's spin in conjunction with its warming-property in the same way as the earth is. But I simply would not dare to say, as our author does, that the sun spins because it seems to be of perfectly spherical shape and lacking that haze visible around the fixed stars. Rather would I ascribe the spin to the effect [of the sun];[5]

FOL. 284v

---

[3] The title of the chapter and Article 3 below deal mostly with the *planets*.
[4] Text:...*hunc [motum] pluribus circulis constanter latis imitabilem.*
[5] But which effect? White attributes the sun's spin to the sun's figure and the absence of haze; Hobbes ascribes it to the sun's own virtue or energy.

for only if we suppose [there to be] some rotation of this sort will any necessity for the planets' moving about the sun emerge.

The argument by which he attempts to deduce the sun's motion about its own axis is not only of no value but even goes against what he said earlier. He reasons thus: 'Since the sun's nature is fiery and boiling, then if it were not carried circularly[6] the sun would take on a kind of variegated appearance as the flames rushed forth. These, being deflected backwards because of the circular motion, flicker around the body [of the sun], and it appears to us as an orb of light'. Granted that (a thing I believe to be true) the sun is of a fiery nature, there is no reason why those flames he endows it with should not blaze forth on every side equally, for, in the fire familiar to us, the reason, why the flames dart up like pyramids lies in the material being burnt, which is not equally consumed by the fire on every side. But this does not apply to the sun; so the argument relating ordinary fire to the sun is worthless. Again, the fixed stars seem to consist of [flaming] pyramids of different kinds, not because the fixed stars really are of such a shape but because, owing to the smallness of the angle they subtend [at the eye], their light reaches the fundus of the eye through a number of points in the pupil, and gives the appearance [of there being] more stars [than are in fact viewed].[7] Now, the area common to all these seems to be the body of the star, whereas the points of light which are not common do not look like stars but look like the halo round a single star. Hence the argument: 'The sun itself rotates because its light forms a circle' is worthless.[8]

Furthermore, our author had affirmed that, the earth being rotated on its own axis, the whole air must be carried round with it, and at the same time. For the same reason, then, when the sun is rotated about its own axis it simultaneously drives its flames round. These therefore will appear of circular form, and there is no reason why (as he claims), moved backwards as a result of that motion of the sun's, they should flicker about the

---

[6] I.e. 'rotated'; White does not imply a sun orbiting the earth.
[7] Hobbes tries to explain an optical illusion. The rays from one star penetrate the eye at several points, the effect being that there are several stars. These images are close to one another and are partly superimposed one upon another. The parts of the total image produced which are common to several single images are seen as the body of the star; the parts not common are seen as a *capellitium* or haze.
[8] For the words of the argument, perhaps: 'The sun's light—and this alone—moves circularly because it appears circular', (or with 'the sun' for 'it').

CHAPTER XXV: 3

solar body. So this [concept of a] counter-motion of the flames contradicts what he laid down earlier about the motion of the air around the earth. Such an argument is, therefore, as I have said, not only unsound but also contrary to his former assertions. Let us pass now to the Third and last Dialogue, which deals with the efficient causes of the world.

CHAPTER TWENTY-SIX

❦

On the Introduction to the Third Dialogue

1. Concerning what is disputed by our author in the First, Second, Third and Fourth Problems of this Dialogue: whether the first sentence of the Apostles' Creed is true, namely that God is the omnipotent Creator of Heaven and Earth. That to undertake proofs of dogmas in this way is (2) unphilosophical, (3) untheological, (4) irreligious, (5) contrary to Providence, and (6) inequitable. 7. What the philosopher should do when he encounters a question of faith.

1. In this, the Third Dialogue, the first of the questions argued [a] is: Does God exist? It is phrased thus: 'Has the motion of the universe an external cause (for everyone says that the cause of the world and of its motion is God)?' Question [b] is: Was the world created? Thus worded: 'Was motion always in being?' Question [c], on the other hand, is: Has God always existed? And [d], Is God the cause of the world? Question [e] is: Could He have created a world better than the one He did make?

All these questions may be combined in this one: Does there exist, and has there existed, an omnipotent God, Maker of Heaven and Earth, as the Creed of our faith proposes we must believe? Does He exist, or has He existed or not? In the Introduction to the present Dialogue, the Third, our author promises us that he will argue the question out. On his page 267 he asks: 'From what I have said, do you think one can deduce the existence of God; that He has established and controls the world; and to what end [he does so]?' [One cannot,]

CHAPTER XXVI: 2

he replies, [make such deductions] 'merely from what I have said, perhaps; but I shall rely on it as an aid in this speculation I am to undertake'.

2. But our author, or anyone else who promises demonstrations of this kind, is not proceeding correctly. As a start, [to make such promises] is unphilosophical. The only people you will have seen making them are the ones whose ignorance stares forth in all their other writings; for philosophers know that the truth of any proposition consists in this: within the meaning of the predicate must be included that of the subject. Hence 'Man is an animal' is true, because the word 'animal' embraces and includes whatever the word 'man' really means. Further, they know that [in disputation] the proposition is demonstrated when, as has been said, by means of explanations or definitions of the terms it is made clear that the subject is contained within its predicate. Therefore demonstrable truth lies in logical inferences; and in every demonstration the term that forms the subject of the conclusion demonstrated is taken as the name, not of a thing that exists, but of one supposed to exist. A conclusion, therefore, has a force that is not categorical, but is merely hypothetical. For instance, if someone demonstrated some property of the triangle, it would be necessary, not that the triangle should exist, but only that [the demonstration] should be hypothetically true: if the triangle does exist, then it possesses such a property. For someone to prove that something exists, there is need of the senses, or experience. Even so, the demonstration is not thus established, for someone rigidly demanding the truth from people who say that Socrates lived or existed will tell them to add: 'Unless we have seen [Socrates's] ghost or spirit', or 'Unless we were asleep, we saw Socrates; so Socrates existed, etc.'[1]

FOL. 287

Under these circumstances there is no doubt that those who declare that they will show that God exists or that no body at all has existed at [a time] more than six thousand years ago, or at least at any period in the past, act unphilosophically. Some believe that a consequence of the definition of 'body' is that body was created, or that a consequence of the definition of

FOL. 287v

---

[1] Words within quotation-marks are translated as if in the plural (text, singular); otherwise there will be in the translation confusion between the questioner and the person addressed.

'incorporeal' is that it has existed for ever. (Such must be the necessary consequence if they wish to show either that a body has been created or that the incorporeal is timeless.) If so, they do not adequately understand what 'body' or 'the incorporeal' is.

3. These same people err against theology; for theologians hold that the world was created in such a way that it could have been eternal if God had not willed the contrary. But he who would demonstrate that the world was created should also show that this was necessary, i.e. that it could not have been otherwise; hence that God, although desiring that the world be co-eternal with Himself,[2] was yet unable to perform that which the theologians consider would belie His omnipotence.

4. These people sin against religion. When a demonstration persuades us of the truth of any proposition, that is no longer faith, but is natural knowledge. Just as knowledge is 'being convinced through reasons arising from what is under discussion', so faith is 'being persuaded by the reasons derived from the authority of the person who speaks'. Therefore as soon as any proposition is demonstrated it is no longer an article of faith but is a theorem in philosophy. The inescapable consequence is that, in matters determined by the authority of the Church, as philosophy is acquired, so to the same degree is faith eroded. Articles of faith are, however, the limbs of religion; therefore when the articles of faith desert religion for philosophy, religion cannot but be gradually weakened.

5. Next, they sin against Providence, for in my opinion there is not one of them believes that through knowledge he can merit his own salvation: for the promises of the Divine Mercy have been given not to philosophers but to the faithful, that is, to those who accept the authority of the Church. To what end, then, should one wish to employ natural reason in enquiring into the articles of faith? For if reason convince us upon matters that conform to faith, we lose the grace of faith; if upon matters contrary [to faith], not only will you be forced to say one thing and to think another, but also you will be less

---

[2] Cf. fol. 374.

CHAPTER XXVI: 7

obedient to the authority of the Church; [and to be thus] is against the salvation of the soul.³

6. Lastly, it must not be thought that the articles of faith are [philosophical] problems; they are laws, and it is inequitable for a private individual to interpret them otherwise than as they are formulated. For a private person to call for a re-examination of matters that have once and for all been settled and determined by the authority of the Supreme Power is absurd and directly counter to the reasons for the Church's peace and unity. Nay, rulers must strive with the greatest zeal not to allow men to argue out any article of faith, for the belief of countless other Christians will be endangered by the mind of one man. Up to now, some have declared that they have demonstrated the existence of God, the Creation of the world, and the immortality of the human soul. Their reasonings, however, have only led weak men (such is the nature of the masses) to consider these things false, because the people who wished them to be true could not show that they were.

7. Perhaps someone will ask: 'What then, will the philosopher not be allowed to investigate the cause of motion?' Or, if this is not the case, 'What is it, then, that we shall assign to philosophy as her proper function?' First, I reply that nothing may be fixed as true or false by natural reason, except on supposition, because terms and names are acceptable only inasfar as we understand them: every reasoning-process advances when the meanings of the terms have been [already] settled. If, therefore, either the meaning is changed, or the thing which has been named cannot be grasped by the mind, every power to syllogise falls to the ground at once, and the only conclusion we can reach is that we do not realise how the thing could be other [than it is]. Hence the conclusion 'I do not know in what way this is true or false' is correctly inferred; but the other, 'It is neither true nor false', incorrectly.

FOL. 289

Second, I say that the philosopher is indeed free to enquire into the nature and cause of motion, but that as the investigation proceeds he will stumble upon a proposition that is now held by the Christian faith and that seems to contradict a conclusion he has established earlier. He can infer (if he has

---

³ Hobbes may mean: '...authority of the Church, because this authority militates against the soul's salvation'.

FOL. 289v

previously reasoned correctly): 'I do not understand under what meaning of terms that proposition is true'. So, for instance, he says: 'I do not see, or it is beyond my grasp, how that which is not moved moves something else, or how that which exists is not in a [given] place, or how something incorporeal sees, hears, understands, wills, loves, hates, etc.' This is the attitude both of a balanced mind and, as I have said, of one that reasons correctly. But he cannot conclude that it is false; for how can anyone know whether a proposition is true or false that he does not understand? Whoever, then, has followed this way of proceeding will not impinge upon the Church's authority, which he acknowledges and conforms to, nor will he therefore philosophise the less freely, being one who has been allowed to advance as far as correct reasoning leads him; nor, in sum, will he, in his efforts to buttress his creed, have to contrive unsuitable arguments, and paralogisms. These two last may cause him to impair both his own and others' faith.

The above points, then, on reasoning about the Christian faith from principles that are merely from nature,[4] have been made at the outset. I come now to the question asked in Problem 2: Is the motion of the universe due to an external principle?

---

[4] So the title-page of the DM.

CHAPTER TWENTY-SEVEN

※

On Problem 2 of the Third Dialogue:

That the motion of the universe
is due to an external principle

1. The meanings of the terms *ens*, body, material, *esse*, *accidens*, essence, form, subject, act, and nature,[1] and (2) of agent, patient ['thing acted on'], to act, to be acted on, integral cause, cause *sine qua non*, efficient cause, material cause, effect. That there is no final cause or formal cause. 3. Definitions: potential, full potential, the potential of the agent, passive potential; and the difference between potential and cause. 4. That an integral cause, or full potential, is necessarily followed by an effect. 5. That no body can perform anything upon itself or be moved of itself. 6. Definitions: the universe, external, and external act. That two bodies are not understood as being in the same place, and why. 7. Motion is not correctly defined as 'the act of *ens*', [being,] 'in potential'. 8. A comparison between motion and rest, according to nature, time, and efficacy. 9. If a body moves, it is moved. 10. That every action is motion; and that rest cannot be the cause of any effect. 11. The motion generated by a cause does not need a cause to continue it. 12. The term 'the principle of motion' distinguished, and in what sense a principle is internal or external. 13. The question proposed [for solution] by our author is [that of] the motion of the universe, but he discusses a different one, namely the motion of its parts. 14. From what our author says here, and from what he said above in the

---

[1] 'Act, and nature'. Mersenne's correction, in agreement with the text below, of Hobbes's 'as act and material effect'.

Third Problem of the First Dialogue, faith in the existence of God is destroyed. 15. The argument of our author refuted, and its consequence shown to be that [the occurrence of] the Creation was impossible. 16. Even granted that the argument is a sound one, the conclusion does not conform to what was set forth to be proved. 17. The refutation of an argument by which he wishes to prove that motion needs causes if it is to be continued. Here it is shown that motion is a continuation of itself. 18. For the reason given (according to our author) by the ancients for their inferring a mover to be some incorporeal creature, they could equally have inferred that that mover was the movable. 19. On the action of the soul in producing animal motion. 20. Our author's philosophy on [this] same action of the soul. 21. Surface is not the equivalent of lines. 22. Our author's argument implies denial that the Creation [took place].

FOL. 291 1. In the question advanced for discussion, 'principle' is substituted for 'cause', and on several occasions elsewhere we find that other words connected with 'principle'[2] are [used] sometimes in one sense, sometimes in another. This certainly ought not to be done, for the language of philosophy is accurate and clear-cut; on the other hand, nearly every error arises from too much freedom in the use of tropes. It is therefore my view that, most important of all, I should define and explain the terms I shall use in the argument to follow, so that their meanings shall everywhere be fixed.

We shall begin with those terms most frequently met: *ens* and *esse*. Two types of *ens* are recognised. [a] There are the *entia* of which we retain some kind of picture in the mind. For example, we can conjure up a man, an animal, a tree, a stone, in fact (since the image we conceive of every body is space, a space of a given kind or size corresponding to a body of a given kind or size), any object at all. [b] There are other *entia* of which we have no picture in the mind, so that a man is quite unable either to perceive them or to imagine them. God and the angels, the good as well as the evil, can be neither perceived nor embraced within our imagination. So God is very good, very

FOL. 291v great, and cannot be understood; neither He nor an angel can

---

[2] Or, 'other principal' (i.e. main) 'terms'.

CHAPTER XXVII: 1

have dimensions, or can be circumscribed, either in the whole or in part, by space, not even in the mind.

But since philosophy is in no way allowed to decide or to dispute in matters outside man's capacity, and since we shall not define the *ens* which we cannot conceive and which is usually termed 'incorporeal substance', we shall define only the *ens* that is conceivable. In this sense, therefore, *ens* is everything that occupies space, or which can be measured as to length, breadth and depth. From this definition it appears that *ens* and *body* are the same, for the same definition is universally accepted for 'body'; hence the *ens* under discussion we shall always refer to as 'body', and this is the word we shall use.[3] Next, as 'body' is 'that which has dimensions or which occupies a space in the imagination', then whether it is thin or thick, rare or dense, has nothing to do with its being body. [For it to be body] the only criterion is that it occupies space—that it occupies space, I say, but not that the occupied space itself *is* body.

If, therefore, the spirits are *entia* that can be perceived, as can the air, the ether, animal spirit, or some other thinnish humour, they are bodies. So the images seen in mirrors or in dreams, because they are spaces and merely imaginary, are not spirits, as they are commonly believed to be. 'Body' and 'material' are names of the same thing—this is because 'thing' is interpreted in different ways. When considered *simpliciter*, an object that exists is termed 'body', but when considered as capable of assuming a new form or a new shape the same body is called 'material'.

FOL. 292

Now, although with the grammarians *ens* is indeed called a name but *esse* a verb, yet, in the way in which it is made the subject or predicate of the proposition, *esse* is indeed a name.[4] For example, in this proposition: 'To be a man is to be an animal', [the word] *esse*, both in 'to be a man' and in 'to be an animal', is a name.[5] In this way names are formed from proposi-

---

[3] For '...and...use' the text has: 'namely using the Latin term'.
[4] Or, for 'name' (each time), 'noun'. For a useful account of the term 'names' in philosophy see the Port-Royal *Logique*, 1662 (edn 1684, facs. 1964, Part I, Ch. XIII*ff*).
[5] Here Hobbes may be using the word *esse*, not as a noun in the technical sense he has been describing, but as the infinitive verb: 'In "To be a man is to be an animal", "to be", both in "To be a man" etc.'

tions through the joining of the copula[6] to the predicate. When, for instance, we say: 'Man is an animal', those who query the truth of these words must enquire not only what 'man' and 'animal' are the names of, but also what that 'to be an animal' is the name of, in order to know what is meant when these two things are linked by the word 'is'. Now, people

FOL. 292v

form a proposition when they say (for instance): 'Stone is opaque', 'Iron is opaque', 'Glass is transparent', or 'Socrates is seated or walking'. So they say that those bodies–iron, glass, Socrates–are [respectively] opaque, transparent, seated (or walking), in order to show that in their opinion something is happening to those bodies that causes them severally to be perceived in different ways, and causes the same body to be perceived in different ways at different times. That is to say, they think that different things *happen* both to iron and to glass, such that the latter is seen as transparent, the former as opaque; and to the same Socrates, in that on one occasion he appeared seated, on others walking. The result is that *esse* is nothing but an *accident* of a body by which the means of perceiving it is determined and signalised. So 'to be moved', 'to be at rest', 'to be white' and the like we call the 'accidents' of bodies,[7] and we believe them to be present in bodies, because there are different ways of perceiving bodies. 'That accidents are present and inherent in bodies' must not be understood in the way that we understand 'Body is present in body as a part in the whole', but in the way that there is motion in a moved body. So *esse* is the same as *accident*. (I do not take *accident* as meaning 'something fortuitous' here, or as belonging, as Porphyry

FOL. 293

places it, among terms that can be used in the predicate;[8] I take it as the *accident* that is called predicamental or is distinguished by contrast with 'substance'.) Aristotle himself recommends this. He states [i] that *accident* is the same as *existence*, and *ens* as *that which exists*, and [ii] that to say 'Socrates is one who walks' is the same as saying 'It so happens that Socrates walks' and 'the *accident* of Socrates is to be one who walks'.

---

[6] Normally the copula *is* the joining-word, but sometimes, as seems the case here, it stands for 'the present tense of the verb "to be"'. Cf. folio 316v and the note.

[7] Compare: '...*lumen et colorem non objectorum accidentia, sed phantasmata nostra esse*...' (*LW* II.7).

[8] The many works of Porphyry (d. 304 A.D.) include a commentary on the Aristotelian *predicaments*, which was popular during and after the later Middle Ages in a Latin version.

CHAPTER XXVII: 1

We must also note that the names of accidents do not always include the term *esse*. Sometimes this latter is included in the verb infinite, as when we put 'to live' for 'to be a living creature', and sometimes it is in the pure name [or in a name divorced from time],[9] e.g. when 'to flourish is life' is put for 'to flourish is to live'. Hence 'to be white' and 'whiteness' are synonymous, as are 'to be moved' and 'motion', and many other examples of that kind. Indeed, a great part of philosophy consists in distinguishing, after a name has been announced, whether that name virtually includes the term *esse* or not. For this is the same as distinguishing whether the thing signified by that name is body or accident. For example, a person who finds out whether the word 'illumination' includes the term *esse* is discovering whether 'illumination' is body or accident. The question is disputed by philosophers to this day.

Esse, then, is usually called 'essence' when the term *esse* denominates a body. For instance, when any body is termed 'man' because it is a 'rational animal', that 'to be a rational animal' is called the 'essence' of man. So 'to be a man', or 'humanity', is the essence of 'man'; to be body, or corporeity, is the essence of body; and to be white, or whiteness, is the essence of white. The same *essence*, as far as it is produced or generated, is usually called 'form'. Hence the same *esse*, when considered *simpliciter*, is called 'essence', but when considered as having been produced or introduced into the material, it is called 'form'. Likewise that which is called *simpliciter* 'body' is called 'the subject' when considered as having within it some *esse* or accident.

FOL. 293v

Now, body or matter can be created from nothing, and reduced to nothing, by the Divine Power; but how this can be done man cannot conceive. Yet because we do not understand how it can be done we do not rightly infer that it is impossible [to do]; nor, on the other hand, can a man be shown that it has been done; but we know that accidents, essences and forms (corporeity excluded) are produced and perish every day. Rightly, then, did Plato distinguish two kinds of thing: one, namely *tò òn*, he said existed but did not come into being; the other, *tò einai*, did not exist but did come into being,[10] unless

FOL. 294

---

[9] The interpolation is Mersenne's. 'Divorced from': dissociated from, independent of (*abstractus a tempore*)? On *abstrahere* cf. fols 77v, 357, 393v; see also 295v.

[10] τὸ ὄν, *res existens*, a thing that exists (and has not become); τὸ εἶναι, *existentia*, being, existence.

313

tò òn could also be made by God, and in the Creation was made, by a method we do not understand. Further, for *esse* people very often say, by an accepted convention of speech, 'nature', as when they say 'It is by nature that, of some bodies, some are warm and others cold'; and for the same reason in the schools [of philosophy] they usually say 'act', unless they use the latter term because they want to distinguish *esse* from *posse* or from potential (which is the same thing).[11] So for *ens* and *esse* I shall put on every occasion 'body' and 'act' [respectively].

2. Every act, other than corporeity itself, may be understood as being generated or produced and then as passing away, as when we see something to be white that previously we had seen to be black, and when we feel something to be cold that previously we had felt to be hot. We cannot, however, see how such a coming-to-be and passing-away of acts can be brought about except *by* some body and *in* some body. Therefore everyone is in the habit of calling that body which produces or destroys some act the 'agent'; but that body in which the action is produced we call the 'patient' [the thing acted on]. Hence 'to act' is the same as 'to produce or destroy an act', but 'to be acted on' is the same as to incur or to lose [an act]. Now, a body already possessing any act can neither produce nor incur any act you like—fire does not produce coldness while it possesses that act of being fire; and water does not take upon itself dryness while it is water. So there must be present in any body clear and defined acts that render it fit and suitable for producing, nourishing and destroying [other] acts. Of the first-named acts, one (choose which you will) is required for the production or destruction of any [other] act, which cannot occur without it. These acts, taken as a group, are all called 'an integral cause'; and an action produced inasfar as it *has* been produced is called the effect of that cause. Each act singly, among those acts combining to produce an integral cause, is termed a cause *sine qua non* or 'something hypothetically necessary'. Some causes *sine qua non* or hypothetically necessary lie, as I have said, in the body that acts [the agent], some in the body acted on [the patient]; so all properties present in the

FOL. 294v

---

[11] The text makes it clear that Hobbes means, not that potential is a synonym of *posse* (though the two are connected) but that to distinguish *esse* from *posse* is the same as to distinguish *esse* from potential.

agent, if taken collectively, are termed 'an efficient cause', and those causes in the patient, taken collectively, are called 'a material cause'.

Both an efficient cause, then, and a material cause are but a part of an integral cause, or of a cause that produces an effect. Only that which is necessarily followed by an effect is rightly termed a cause, for cause and effect are related things. Where there is no effect, anything that exists or that will have existed is not called a cause. A cause, therefore, whether *simpliciter* or integral, I define thus: 'A cause is one act or more acts through which another act is produced or destroyed'. An efficient cause is 'the acts of the agent, taken collectively, through which another act is produced or destroyed'. A material cause is 'all the acts of the patient, taken collectively, that produce another act within the patient itself'.

FOL. 295

It is usual to point to two other kinds of cause: the formal and the final.[12] The latter, as far as man can understand it, is exactly the same as an efficient cause: from something pleasant arises the thought of enjoying it; from the thought arises the notion of the path to secure it; and from the notion of the path arises the progression towards the object desired. In this series of products[13] the object, or the goal, is the agent, because the act of the object that is the goal is the efficient cause of our movement towards the goal. The former, i.e. the formal cause, is properly, i.e. correctly speaking, not a cause. An example of a formal cause is something like this: Given any figure, say a square: because it is a square, the consequence is that all its angles are right angles, for 'to be a square' is said to be 'a cause' because [the square's] angles are right angles. But since in the material world [*natura*], as well as regards time, 'being a square' is simultaneous with 'having right angles', though the cause precede the effect, people see that a square does not create that 'right-angledness' of the angles by any action of its own. Hence they call such a cause not an efficient, but a formal one, because the angles' being right angles is a consequent of the square's form, not of its act. Correctly speaking, [however,] that the angles are right angles is not a consequence of the square's form; rather, our understanding of the

FOL. 295v

---

[12] *EW* I.131–132. Aristotle, *Meta.* xii.7 (994b).
[13] Product: that which results from the operation of a cause. With this passage cf. fols 308vff, 341v, 428.

proposition: 'Those angles are right angles' follows from our knowing that proposition: 'That figure is a square'. In other words, property derives from form, but awareness of property arises from a knowledge, by reasoning, of form. The result is therefore that where they ought to say: 'Our awareness of form is the cause of our awareness of property', people say, mistakenly: 'Form itself is the cause of property itself, but such a cause is a formal one'.[14]

3. When an act (or any number of acts taken collectively), if postulated, makes it necessary for another act to come into being, the original act is called 'potential' with respect to the production of the future act. For example, if we notice a hand's being moved towards a ball–a movement that must result in [the arousing of] motion in the ball–the motion of the hand is called the hand's potential in respect of moving the ball. Thus if one wheel moves another, the motion of the first is its potential by which it can move the second. The same motion of the hand is the cause of the motion in the ball, and the motion of the first wheel is the cause of the motion in the second; so 'the potential of the agent' is exactly the same as 'the efficient cause'. This latter, however, is called by different names for different reasons: 'cause' is used only when an effect has followed [it]; and only if [the effect] has followed will [the cause] be called a cause. Where there is no effect there is no cause, but [instead, the cause] is called 'the potential' [to bring about an effect], because one believes that an effect *can* result from it, even though sometimes [the effect] does not follow. So the hand, because it is moved, will be said to be able to move the ball, i.e. to have the potential [to do so], even though it has not moved it; but if it has not shifted the ball it [the hand] will not be termed 'a cause' of its motion, because the hand has induced no motion in the ball. Likewise, the 'capacity' of a body (or its potential to be acted on, [passive potential,] which enables [the body] to produce an action within itself) is the same as a material cause, but the names are different because of the reason I have just given. So all actions, both in the agent and in the recipient, that are necessarily required to produce an effect are called 'an

---

[14] Unclear. Possibly (1): 'Form is the cause of a property of itself', etc. (i.e. the form of things' properties is its own cause); and (2) '...itself, but of formal property'.

integral cause'; likewise the same acts can be called 'full potential'. So 'full potential' and 'integral potential' are the same; 'efficient cause' is the same as 'active potential', and 'material cause' is the same as 'passive potential'.[15]

4. An integral cause always and necessarily produces an effect, but only if that effect be absolutely possible. Being possible, an effect can be brought to pass; if, therefore, it is not, either the agent or the patient lacks some act necessarily required to produce it. But those acts thus needed are called, individually, causes *sine qua non*; [yet if the effect is not brought to pass] some cause *sine qua non*[16] is lacking. But an integral cause is defined as 'all the causes *sine qua non* taken together'. There is not, therefore, as was supposed, [if the effect is not brought to pass,] an integral cause, because if a cause is integral the effect cannot but ensue. For the same reason it can be proved that where the potential is full the effect must follow, for, as explained above, full potential is the same thing as an integral cause. It should be remembered, however, that we are here discussing a cause only as to its being the act of a body, i.e. [the act] of *ens* that is discernible; for of *ens* that is not comprehensible the acts, potentials, and reasons for acting or for bringing to pass are also incomprehensible.

5. No body can produce any act within itself. If it so happens that a body does produce its own act, then the body had within itself all the acts (i.e. an integral cause) necessary for producing that act at the very instant that the latter was produced. Either, then, [the body] had, or it had not, the same acts before the effect was produced. If it had, then the act was produced not at that instant but earlier, because, as was shown in the last article, as soon as a cause is integral, the effect is produced at once. Conversely, if [the body] did not have all those acts previously, then it produced within itself all those acts at the same instant as the effect [was produced]. If this is so, the body then had within itself not only the integral cause of the effect to be produced, but also the integral causes of all those acts of which that integral cause consists; and things proceed in sequence till the body is said to have produced within itself

FOL. 297

---

[15] For the last, cf. fol. 386v.
[16] For 'some...non', perhaps 'a certain' cause [*aliqua causa*] *sine qua non*'.

all its acts, and hence its very corporeity. From this it follows that body has created body, i.e. that it has created itself, which is neither conceivable nor consonant with faith. Hence it remains that, as was to be proved, a body cannot produce its own act.

The corollary of this proposition [of White's] is that either the world did not have a beginning or that it had its beginning in something incorporeal or in imperceptible *ens*. As has now been proved, [the world] cannot have begun of itself or [originated] in another body, because [to postulate] a body [as lying] outside the world is inconceivable, the world being the aggregate of all bodies.[17] From this it follows that whatever is produced is produced from[18] something else; for what is moved or produced, and not by itself, is moved or produced by another.

6. I come now to the other terms in our author's proposition that have to be explained. When he says that the movement of the universe is due to an external principle he seems to have interpreted the words 'universe' and 'external' incorrectly. 'The universe' means 'the aggregate of all things'. If, therefore, he understands 'universe' as 'the aggregate of all *entia*, incomprehensible as well as comprehensible', it is absurd to say that the motion [of the universe] comes from without, for there is nothing external to the aggregate of all *entia*, [if these be considered] unconditionally; it is also absurd to say that the universe is moved, for it is infinite and therefore is immovable. I take 'universe', then, in that sense in which philosophers have been permitted to dispute about it, i.e. as 'the aggregate of all bodies', i.e. as 'the aggregate of all comprehensible *entia*'.

We say that something external to a body is body, whatever the external thing and whatever the body, except that a part is not external to its whole. For example, though some water has been enclosed in a flask, to the flask itself it is an 'external' body even though contained in it; although it is not 'external' with respect to the aggregate of the flask and the water taken jointly. In bodies 'the external' will be defined as 'that which is

---

[17] For '[to postulate]...inconceivable', perhaps, 'except the world, no other body is conceivable'.
[18] Or, 'by'.

in another place'; hence 'the external' means the same as 'something else' or 'something different'. Yet bodies are counted by the mind no differently than are the places they occupy; therefore a man cannot conceive of two bodies in one place, and can conceive of several bodies only in terms of an equivalent number of places. Now, bodies not in the same place are said to be 'external' each to each, any one to any other. An act, therefore, said to be external *to* a [given] body is an act *of* an external body or not *of* the same one, as the whiteness of paper is 'external' to a wall, though the latter be white [too]. It is difficult to guess what our author means when he says 'external'; for he here endows the world with an external mover, yet in Problem 3 of the First Dialogue above, where he wished on those grounds to demonstrate that there cannot be several worlds, he asserted that outside[19] the world nothing existed at all. To him, therefore, 'external' and 'nothing' seem to mean the same.

7. It is also uncertain what he wishes to be understood by the word 'motion', for earlier, in the Introduction to the Second Dialogue, he says at his page 130: 'Correctly speaking, "motion" properly means "change of local relationship" or, to put it more clearly, "change of position"'. As, when a boat sails past a tree situated on the river-bank, then, with an equally truthful and correct form of speaking, one may say that the boat and the tree are moving. Or, to use our author's example, either the earth is moved when the sun is at rest, or the sun is moved and the earth is stationary; yet equally truthfully, either the sun or the earth will be said to move, because, under either postulate, the position [of each] ever varies accordingly.

FOL. 298v

But below, in Problem 18 [of the Third Dialogue], at his page 435, he defines motion as 'the act of *ens* in potential, inasfar as it *is* in potential'. As regards the first definition, I have said in its place above [at fol. 131] what has been observed [about it].[20] But his second definition, Aristotle's, is obscure and imperfect; yet Aristotle conducts an accurate enquiry into the nature of motion in his *Physics*, Book [III], Chapter [1].[21] He considers that motion by which the stones required for building a house

---

[19] Perhaps, 'apart from'; but the reasoning is unaffected.
[20] For 'I have said', perhaps, 'I have given my opinion'.
[21] Lacunae in the manuscript.

FOL. 299

are moved. They are not yet a house, but they can be, i.e. they are a house in potential, with the aid of the motion that the builder has imparted to them when he collects and assembles them, as is required by the well thought out plan he has made to build a house. But from this [instance Aristotle] has derived, not a definition, but only a property, of motion. Why? Is not, [I say,] a body at rest capable of motion; that is, has it not the potential of assuming motion? So will not a body at rest possess some act through which it is in potential to move? Now if every act of that which is in potential, inasfar as it is in potential [to move], be motion, then the act of a body at rest [is motion too]—because the body is in potential to assume motion, *inasfar as it is in potential to assume motion*. In other words, the act of a body that is in fact at rest but yet is capable [*capax*] of motion as much as it can be, is motion.[22] Therefore the act of a body at rest, is motion, which is the same as saying: 'A body at rest is moved to the degree to which it can be'.

All that Aristotle could legitimately infer from the case here argued is this only: 'All active potential is motion'. This is true for bodies, i.e. for perceptible *entia*; for when at rest a motionless body cannot act, as will be stated below. In order that, from this stage on, our discussion of motion and rest be self-consistent, we must remember the true definition of both, given in Chapter [V], Article [1]:[23] 'Motion is the continuous quitting

FOL. 299v

of one place and the acquiring of another'; 'Rest is the occupation of the same place at any time'. It follows from these definitions that what is moved is not, through any period of time, however small, in one and the same place.

8. The essence of body, or corporeity, consists (as has been said earlier [at fol. 291v]) in the body's possessing dimensions or (which is the same thing) in its occupying place. Now, what occupies a place either passes from that place to another one or, as we may accept with equal ease, it stays in the same place. So the property of being moved, i.e. mobility, is as much to do with the essence of a body as the property of being at rest is, i.e. quiescence. Rest is no more a deprivation of motion than is motion a deprivation of rest. Nor must we infer from bodies themselves any reason why rest should precede motion, or

---

[22] Translator's italics.
[23] Hiatus in the manuscript each time.

motion rest, [or ask] whether all bodies were created with motion or with rest (or some with motion, some with rest); whether the whole world, which we believe to be finite (it being considered to be one body), has always been at rest or has always been in motion from the time of the Creation; or whether rest or motion is the more sought-after by bodies. Whether rest is nobler than motion is a ridiculous question [to ask], for nobility is the renown of men, [deriving] from lineage, riches, civil power, virtue, and the like. 'Nobility' may, however, be taken as meaning 'a special power of efficacy' or 'the potential to act', in which case motion is indeed more noble [than rest], for every action of bodies is motion: how bodies at rest are acted on can easily be seen, but how they act may not, as will soon be shown. But perhaps rest seems nobler than motion because we do not ascribe motion to God. I do not hold, however, that rest is customarily attributed to God. The Holy Scriptures attribute both [motion and rest] to Him, yet not in the same sense in which they are attributed to bodies, but in a manner that we cannot understand. No term in nature is applied to Him except [the statement] that He *is*; or, negatively, [in designating Him] 'infinite' and 'incomprehensible'; metaphorically, or in order to honour Him, 'good', 'just', 'wise', 'blessed' and the like. The Epicureans apart, not even the pagans [*ethnici*] termed their gods, to honour them, 'calm' or 'placid' ('at rest'); they called them 'indefatigable' ('tireless').[24]

FOL. 300

9. We can conceive of only one efficient cause that sets in motion any body initially at rest: the motion of an adjacent body. As the commencement of motion is the quitting of place, we can see that the only reason why a body leaves its place is that another body standing adjacent replaces the first by moving forward. If this happens, the first body must recede into another position, i.e. it must be moved. Hence the whole origin of motion derives from the push of an adjacent body. Indeed, we see that sometimes a body is even drawn out of its position by another body that in no way tends towards, but rather moves away from, the place of the body it draws. But we cannot see

FOL. 300v

---

[24] For an instance of the Epicurean point of view compare Horace, *Carm.* 3.3.35–36: '[Romulum] adscribi quietis / ordinibus patiar deorum'. For the general concept cf. fol. 396; on *ethnici* ('heathen'), fol. 27.

how this can happen unless we first understand how bodies cling to, and are connected with, bodies. Cohesion of this kind between a body at rest and one in motion is quite infeasible save in terms of pulsion.[25] Again, why a part that is drawn follows a part that pulls it is inexplicable unless the drawn part and the part that draws possess within their parts some motion common to them [both], and curvilinear. Through such motion the attracted part is by-passed [by the striking part] and struck in the rear. This matter has been sufficiently argued out in Chapter [...], Article [...] above.[26] Given, then, that in the topic under discussion there may be two reasons why a body at rest begins to move: propulsion and traction, both exerted by an[other], contiguous body (for any reason other than these is inconceivable). It is established, therefore, that no body can be thought of as moving another body without first being moved itself. If, of bodies in contact, one is driven by the other, it is driven by the one that moves into its place from an adjacent place, i.e. by the one that is moved. Again, if a body is drawn by a body in contact with it, the body that is drawn takes the place of a body that draws it. Hence–for it is inconceivable for two bodies to be in the same place at the same time–the body that draws quits its own place; but what leaves its place is moved; therefore anything that moves [anything else], either by pushing or by drawing, is itself also moved.

FOL. 301

10. Next, to prove what I have said above–that one cannot understand how anything acts except through motion–we have this, from the first proposition immediately preceding: 'We cannot understand how, without motion, anything moves'.[27] If, therefore, I have shown how, without motion, one body cannot conceivably change another, then the whole of the assertion 'Every perceptible action is motion' will be convincingly established. We must therefore examine [the fact]

---

[25] And even then the cohesion is only momentary? This important argument of Hobbes is given prominence by Gassendi in his *Syntagma* (*Opera*, 1658, facsimile 1964, I.345, 385).

[26] Hobbes seems to mean: 'A body at rest requires, to be moved, the thrust of an adjacent body', as stated at the head of the article. In the present work *he has nowhere discussed attraction in terms of repulsion*, so one leaves unfilled the two blanks in the text.

[27] The clauses here parenthesised may be dependent upon 'we have' that follows.

CHAPTER XXVII: 10

that it is only through the senses that man can perceive change in bodies.[28]

Man thinks that those things which are seen to be of a different colour or shape from before are changed in colour and shape; otherwise he takes it that they are not changed. We understand likewise that things which sound to the ear, tang to the taste, savour to the smell, and are felt by the touch differently from before, have been altered as to the sensible qualities they have; otherwise they have not been changed. Now, say all the senses remained in a similar state as previously, such that any given particle [of a man] was at rest and in the same place as it had been; alternatively, that all its particles were in motion when we noticed [the object] first. Then, if they were similarly in motion when we perceived it a second time, they lay as much in the object (which is the agent) as in the observer, who in both acts of viewing is acted on in the identical way. Hence in each instance action and being acted on will be the same; but where action is similar to action throughout, and passivity to passivity, here the effects are also similar, i.e. the second perception of an object will be similar to the effect, namely to the first perception of the object. So a change in the object will not be perceived when the senses remain in their [same] state, unless the parts of the object are moved should they be at rest or, should they be in motion, unless they are moved in a different way. Hence not any body can be seen to change unless there is a new motion either in the object or in the observer.[29] If, however, there is such new motion in the percipient he, and not the object, will be said to change, because if it is the object [alone] that is changed we perceive this only if we have noticed the motion in its parts. So, as was to be proved, whatever is perceived either to move something else, or to alter it, moves it or changes it by means of its [the mover's] own motion.[30]

FOL. 301v

Yet because we realise that 'to act' is no different from 'to move something else' or 'to change it', it follows that every action we can perceive is motion and hence that rest cannot be the cause of any effect. On the other hand, the efficient [cause] of rest is motion, i.e. the motion in a body that resists [motion].

---

[28] Or, 'why bodies change' [*corporum mutatio*].
[29] So fol. 129.
[30] Cf. fol. 338.

Any body's coming to rest after moving is attributable to the action of some agent, but, as has now been shown, every action is motion.

FOL. 302

11. Again, though it is impossible to understand how motion can originate without [the intervention of] an efficient cause, once [motion] has come into being it can be thought of as continuing without [the aid of] a cause, and of an agent; for the act to continue, it is enough that it shall have been set in train. Further, we cannot conceive of motion as diminishing, or being changed in direction, or disappearing altogether, without [the intervention of] a cause, i.e. an external agent. We shall see, therefore, that what is once moved is always moved at the same time—unless there intervenes an agent which, by its resistance or by a contrary motion, reduces and at length annuls [the body's] own motion. No-one, I think, seeing something at rest, seeks a reason why it persists at rest; rather he expects such immobility to continue for as long as there is no agent to push or pull [the thing] from its place. The same logic applies to motion: we must not imagine that if a movement is continued, many new movements take place which need causes [to bring them about], any more than we should believe that during the prolongation of rest there are many new states of rest. In continued motion, it is true, new spaces [*loca*] are continually being traversed; but it does not follow that new *movements* take place, for [philosophical] reasoning would advance from 'new places' to 'new motions' only if 'place' were the same as 'motion'. Again, that the prolongation of an effect does not need the continuation of a cause has, I think, even been confirmed by experiment both in other examples and also in the motion of projectiles which, being shot from the hand or from a catapult (from which they have derived the principle of motion), do

FOL. 302v

not cease being moved by the force initially impressed, and are brought to rest only with difficulty by external agents that meet them.

12. There remains for explanation what, in the subject under dispute, 'the principle of motion' is, for the term is equivocal. In every context 'principle' means 'the first part', i.e. the one from which we start reckoning, hence 'the principle of motion' may be taken as 'the first part of the motion of the body in which [the principle] is'. So the principle of motion

CHAPTER XXVII: 14

in a hurled stone is that part of the motion by which it crosses the first interval traversed on leaving the hand or the catapult. In this sense the principle of motion lies in the very subject of the motion; so we shall call it 'the internal principle of motion'.[31] 'Force' may be defined as 'the main part of causation' or, as it were, the first link in a series of several causes, in which one cause depends on another. In the hurled stone the principle of motion lies in the catapult's cord, if we commence our reckoning with a proximate cause, or in what moves the cord, or, again, in what moves the mover of the cord–according to whether our reckoning from the causes to the effect begins with a more remote, or whether with a more immediate cause. In this sense the principle of motion is always external; for it lies in the agent, and this latter, as we showed above [at folio 298], must be external.

13. Having settled these matters to do with the agent, the patient, cause, effect, potential, act, motion, rest, and other words necessary to explain the present question, let us now examine our author's disputation.

FOL. 303

Let it first be noted that the question is not phrased in words that make clear his meaning. Here he is not arguing from what principle the whole created world is moved, i.e. how [a] it advances from its own place to another. He holds that it cannot do so, because in Problem 3 of the First Dialogue [at fol. 17] he supposes there to exist beyond the world no place in which another world could be established or into which the world, when formed, could move. [b] [No; he argues thus:] 'Of the parts of the earth that are moved, does one owe its motion to another, and this unceasingly? Or do we finally arrive at a part that owes its motion not to another part of the world but to some external mover? That is, have the world's parts commenced moving, or have they been moved for all time?' That is what he wishes to propose. The question advanced, however, concerns the motion not of any part [of the world] but of the universe.

14. Let us suppose it true that the world is finite, which is credible, and that outside the world there is nothing. He has made this claim in the Third Problem of the First Dialogue at

---

[31] Or, 'the principle of motion will be said to be an internal one'.

his page 28, where he says that there cannot be more worlds than one because there is nothing outside the world. 'If this be true', he declares, 'what can be more obvious than that an object located in nothingness has no place? Hence those two worlds, if set apart [one from the other], have neither of them a place and consequently cannot possibly be separated as regards place.'

In sum, if it *is* true (I believe it to be very true) that that mover which brought about the first motion in the world was God, it follows that, according to our author's precepts, there is no God.[32] For, since it is true that the principle of the universe's motion is outside the universe itself, and that the principle of the universe's motion is God, it is easily inferred that God (being infinite) is outside the universe, which is finite. So if we add to these remarks what we quoted above from the First Dialogue, namely that beyond the world (i.e. the finite world) there is nothing, don't they imply that God is nothingness?

But the arguments offered against the possibility of [there being] several worlds (and from these White has had to affirm that beyond the world there is nothing) I believe I have sufficiently refuted where they occurred. Meanwhile it is worth observing into how serious errors must fall those philosophers who are ashamed to admit that there is any *ens* or any act that they do not understand or the properties of which they themselves cannot demonstrate. *I* believe, however, that those who worship no god but the one they do understand are not Christians; and that those who think they can demonstrate any attribute of something[33] they do not comprehend are not philosophers.

15. This is the demonstration he employs on his page 272 to prove that the motion of the universe has an external principle: 'Say the number of bodies in the universe is three: A, B and C. Given a time, DF, divided at E, so that C moves during the time DE and is at rest at E.[34] Unless A, B or C be extrinsically altered, it follows that C will always be at rest.' But this 'C will be at rest, unless C is changed' is absurd, [I say,] because, believing as he does that 'motion' is 'change',

---

[32] Or, '...he was not God'.
[33] Or, 'Him' (God).
[34] Figure 57.

CHAPTER XXVII: 15

he could equally well have inferred: 'C will be at rest *unless it is moved*'.[35]

That A, B and C being unchanged, C will be at rest he proves thus: 'The same causes being postulated', he writes, 'the effect must remain the same; but let it be laid down that there are no corporeal causes other than A, B and C. This granted, we have also granted that C is at rest. Clearly, unless any of these causes be changed, [C] will always be at rest'.

First, I think, we must deny that dictum universally published: 'The same causes being postulated, the effect must remain the same'. Although this is true for corporeal causes within our comprehension, I do not think it is true for the First Cause of all effects; for the First Cause of the whole variety of effects in the world is God. God must not, however, on that account be spoken of as movable or mutable, but He remains ever the same and unchanged. Moreover, if we concede that [dictum] 'all causes remain the same', it does not follow that when A, B and C are unchanged, C will be at rest–unless A, B and C are taken, as all causes, *simpliciter*. But they were not taken as 'all causes, *simpliciter*', but as 'all corporeal causes'.[36] So [White] commits a paralogism by using in the one proposition 'all causes' and in the other 'all corporeal causes'. Hence if we amend the reasoning as follows: 'If all causes postulated are the same, the effect will remain the same; A, B and C are all causes; *ergo* when A, B and C remain the same, the effect at C will remain the same', then the reasoning will be legitimate. Because, however, the major premise is false, the conclusion will not be a necessary one, for among the causes (all of which are A, B and C) there is in fact one which remains the same and yet can produce motion at C.

FOL. 304v

Perhaps he will say: 'It is incomprehensible that causes which are the same in all circumstances do not produce the same effect'. Indeed, what is unintelligible is impossible, but this [statement] should not be made [as though] of universal application: a human and finite understanding does not grasp that which is unintelligible; but we must not say that it is unintelligible to the Divine Intellect. Further, say it were true that, given that all the causes laid down were the same, the result would be the same. It would follow that C is moved only if A

FOL. 305

---

[35] 'Be at rest' in these three sentences should perhaps read: 'come to rest'.
[36] Figure 57.

or B is, B only if A or C, and A only if B or C; hence that there has been motion from all time; and this is to deny [that] the Creation [took place]. So we see that every time a philosopher disguises what is true, i.e. [the fact] that he does not understand what is being disputed, and [in so doing] makes use of what is false, namely that a subject [for disputation][37] is itself an impossible one, he falls into some dogma contrary to the Christian Faith.

16. Say [White] had argued thus: 'Given that all corporeal causes are the same, the effect dependent on those causes, and on no others, will remain the same; but all the causes that can move are A and B; therefore, unless these[38] or one of them have or has been changed, C will always be at rest'. He would have thus pleaded correctly and proved that a body cannot be moved by [another] body unless the latter be moved. Yet he would not have proved what he had proposed to prove, namely 'the motion of the universe lies in a principle external [to it]'; for although each of the separate bodies of the universe is external to every other body, none of them is 'external' with respect to the universe.

FOL. 305v

17. Next he shows on his page 274 that motion cannot be continued without a new efficient cause, thus: 'Given all bodies to be A, B or C. Say C is to be changed. It must be changed either of itself or by one of the remainder. [It will not be changed] by itself (for it is clear that, if A and B remain as they are, C continues doing what it does). So C brings itself to the point E, but [it cannot be changed by] B, for it is obvious that if A and C stay as they are, [B][39] makes C lie at E, and the same applies to A and to all of them as a group. You see, therefore, that, unless motion is initiated from a non-corporeal principle, and, as it were, as the motion of an individual particle, there can be no motion at all'.

To grasp the force of this argument we must imagine a body

---

[37] The meaning is unclear. For 'a subject [for disputation]', perhaps 'the subject [being disputed]', or merely 'it' (i.e. the Creation).
[38] (1) 'And on no others': The text has merely 'only'. The periphrasis is necessary if ambiguity in the English is to be avoided. (2) For 'all the causes...are A and B' the text has 'A and B are [represent?] all the causes that can move C'. (3) 'These': the reference seems to be to 'A and B'.
[39] Not in the MS. Supplied from *DM*.

to be, when in motion, elsewhere than when at rest.[40] When at rest it is in the same position for some given [length of] time; but when in motion it is in the same place for no time-interval at all and hence may be said, not to be at this or that time in this or that place, but to have passed through a place without any pause at all.

Now, given that C is at the point E. Either C is at rest at that point, or C is being moved through it. Say C is at rest, and let the cause of its being at rest be A or B, or both of them. It will be true that if A and B are unchanged C will stay fixed at E, because C's principle of motion depends on the motion of the agents A and B. If, however, C is at the point E and is now in motion through E, then, from the very fact that C *is* in motion, [C] must move into another place. Therefore A and B, or either of them, need not be changed because of this.

On the other hand, as A and B are postulated as being the reasons why the existing C is moved towards E,[41] then A and B are themselves moved (for they are given as being bodies). Further, in order that C's motion be continued, A and B (which were in motion) must be changed; hence they had to come to rest or move along a different path, and neither alternative can be the cause of C's moving forward. Because White supposes that C must be changed, his argument proves that C is changed only if A or B is. But it has no bearing on the prolongation of motion, for the continuation of motion is no more a 'change' than [is] the continuation of rest. As, then, the presence of body C at the point E constituted not rest but the same motion as was present before and after [C's movement], that supposed change has not existed, and C does not need the assistance of A and B if its motion is to continue; and all existing motion is the continuation of itself. But if, from time to time, some body begins moving and is continually urged forward by new causes, its velocity will be continually augmented.

Now if, by this argument, we have correctly showed that C's motion cannot be continued after A and B have come to rest (because A and B are understood as being bodies), he ought to have concluded that a motion can be prolonged by a corporeal

FOL. 306

---

[40] Perhaps, 'When in motion a body must be thought of as being "in a place" [in a sense] different from that [applicable] when the body is at rest'.
[41] The clause can be read: 'why C, standing at E, is moved', but the context seems to demand the interpretation chosen.

FOL. 306v property alone. So what he appends is nonsensical: 'You see, therefore, that unless motion is instigated by a non-corporeal principle and is, as it were, as the motion of an individual particle, there will be no motion at all'. And that expression used by those speaking confusedly–'as it were'–why is it interposed? For if the continuation of motion indeed requires a new cause, he could have said explicitly: 'All the possible motions of a particle need new causes, or rather, they need a continual change of causes'. What follows next–a kind of corollary: 'Everything will come to rest when the first thing movable ceases'–falls to the ground together with the very demonstration it depends on.

18. In the next place our author shows us how Aristotle was led to acknowledge the same truth, [i.e. that continued motion is impossible without the renewal of a cause,] namely by means of two axioms arrived at by induction. The first was: Whatever is moved is moved by something else; and the second: That which moves[42] is moved. After seeing that the bodies constituting the finite world are finite in number, Aristotle derived from any one of them that was moved the motion of the second, and from the motion of the second that of the third, until he reached the last. Now, in accordance with Axiom (2), this last body was agreed to have been moved because the next-to-last has moved, and, in accordance with Axiom (1), to be moved by another body. Hence he saw that he must postulate an incorporeal mover.

FOL. 307 Aristotle ought to have added, however, that from Axiom (2) this incorporeal mover is also moved; and, from Axiom (1), that it is also moved by something else. Either he should have shown how incorporeal things move, or else he should have changed Axiom (2) to this: 'Every *body* that moves [another] is moved'. Because of its length and its absurdity alike I dismiss [the discussion of] that induction which, on his page 275, White says the philosophers have employed in order to draw together the two Axioms; and I come to the example he himself uses to show how incorporeal things, even though not moved, can yet move, namely the instance of the soul in animals, on the action of which a few remarks must be made here by way of preface.

[42] Here and at the close of the article it is not clear whether this is transitive or whether intransitive.

CHAPTER XXVII: 19

19. Take cognizance, therefore, that every perception [*sensio*] is brought about by the action of objects, in the following way:[43] vision when the motion from a shining or illuminated body is continually propagated through a medium from part to part, through the eye as far as the brain, and even to the heart itself; hearing likewise through a motion which is set up by bodies crashing together or being violently torn asunder and which is continually propagated from part to part of the medium and to the brain, thence by the ears, to the heart; scent by the odour's continued motion carried to the skin inside the nostrils, which comprises countless little nerves by which that motion is continued to the brain, thence to the heart; taste through a motion originating in the buds of the tongue; and touch through a motion, commenced or continued to any part of the skin of the whole body, thence continued by the nerves, as above.

FOL. 307v

Reproduced [*procreati*] and sent back, by the reaction and resistance of the heart, to the outer limits of the animal [concerned], such motions constitute impressions [*phantasmata*] which seem to be external [to the body] and which, if they are to do with vision, for instance, we call light or colour; if with hearing, sound; if with touch, the sense of heat, cold, roughness, smoothness, etc.; if with smell, scent; if with taste, flavour. In sum, every perception is a motion in the parts of an animal's body; these, though they are called 'animal spirits' and 'vital spirits', are nevertheless [themselves] bodies; and the motion is aroused by objects, which also are bodies. So up to now we need have no recourse to an incorporeal mover. When the object that makes itself noticed is removed, however, the motion it excites in the brain and heart does not immediately cease. (It was shown above [at fol. 212v] that contrary to our author's opinion, the cessation of the cause is not the reason for the disappearance of the motion it arouses.) Let the word 'perception' [*sensio*], then, be applied only for the duration of an object's action, for the motion remains the same, but is now called 'a mind-picture'. The latter becomes gradually weaker and at length, after a long time, it disappears unless renewed by a motion similar [to that described].

FOL. 308

---

[43] The following passage and that commencing fol. 338 below are included, closely paraphrased, in the preface to Mersenne's *Ballistica* in his *Cogitata physico-mathematica* (1644), and are reprinted in *LW* V.309–318.

Again, mind-pictures are so fleeting, even immediately after perception, that as soon as we turn aside we almost forget a shape once seen. This is due, however, not to the weakness of the motion [emanated], but to the present action of other perceptible things that always affect people lying awake. In people asleep, however, when no external action hinders the 'mind-pictures' called dreams, the 'mind-pictures' are as vivid as the perceptions themselves. As, therefore, a 'mind-picture' is nothing but a prolonged perception, a 'mind-picture' is (just as a perception is) a motion in parts of the body, and it originates in corporeal things; so up to this point we need not summon forward an incorporeal mover. No-one can doubt that memory is a 'mind-picture'; hence that from the recollection of one thing we are led to recall another. That is, the sequence of image [*imago*] on image which is called 'the mind's discourse' must also be a motion of the same [bodily] parts, and it has the same corporeal mover [as the memory has], namely a perceptible object; for when several different perceptible objects operate in succession, the unbroken motion continued by them must arouse several different images.

FOL. 308v

Now if, from a 'mind-picture' of procreation, a bird indeed arrives at the idea of laying eggs, and then of building a nest, of spreading out chaff and feathers, and of going to the places where these are, all such 'mind-pictures' are motions in parts of its body, namely the heart, the brain, and the body's spirits (called the animal and the vital). These motions, [when] continued into the nerves, move the arms, the feet and the other limbs. Hence the progressive animal motion is wholly a motion set in train by a perception, and continued, by the mind-picture, into the sinews and legs; it is absolutely unnecessary, whenever a bird makes a nest or a spider a web, to appeal to an unmoved mover.

All this can be seen to result from the action of bodily objects: a fact which our author, on his page 277, not only admits, but even declares that the ancients knew, for he writes as follows: 'They also, as regards animals, knew well that an animal's every motion derives from the senses, i.e. from the action of objects'. In what immediately follows, however, he shows that he has not used those words in their correct meaning, but has chanced to state a truth he has not understood. [Let me re-phrase it:] 'As for man, endowed with a rational and an immortal soul, he too is moved in that way in which the

FOL. 309

other animals are; for the appetite that is the internal principle of animal motion, and the ranging of the imagination from the end to the means, and from the means to the end, are common to man and to the beasts'.

20. Let us now hear, from the instance of the soul (White, page 278), how unmoved things yet move [others]. 'Our mind, the only spirit we have control over, plans a building—how the foundation, the walls and the roof may be adapted to the whole that is being contrived—but does not enter the path [to implementing its design] unless it has established that [the path] leads to a definite goal.' He could have added: 'The mind of a bird plans a nest—how the base, the sides and the shape may be adapted to the whole—but does not enter upon the path to collecting feathers and chaff before deciding to build a definite nest of definite form and shape, in a definite place, in order to rear young'. 'These and similar things', [he resumes,] 'the soul would not be strong enough to do were it not itself, finally and utterly, a long series of actions, or did it not possess such a series within itself.' That is to say, the soul that plans a house—the foundations, the walls and the roof, and the path by which the material is to be gathered—must itself, I [i.e. White] declare, *be* the building of the house: the placing of the stones in the foundations, walls and roof; the taking steps to collect the material, and, finally, the fetching of the material. Indeed, [he believes,] the soul must *be* the determination of the method [of building], the collecting of the material, the placing of the foundations, the erection of the walls, etc., all these acts being considered as a whole and as once and for all. The soul, then, is [he says] a series of actions; not a series [whose constituents follow each other in time] but an instantaneous one.

FOL. 309v

The consequence, he says, is that, 'the soul's nature being unmoved, namely in accordance with one *esse* and act, it spreads itself abroad into material things'. That is, [I say,] the soul's nature is such that the soul is unmoved. 'To spread itself abroad' is 'to be moved of itself towards many material things', and this by means, not of several motions, but of one and the same motion. This was like saying: 'An unmoved soul is moved by one motion towards several things'.

Second, he infers this: 'You see, therefore, that, as we consider the soul to be spirit, this ought to be conceded[44] as regards

[44] I.e. that the unmoved soul is yet moved? Line 31, text: 'mind's nature'.

333

FOL. 310

spiritual nature', i.e. we may freely pronounce contradictions about spiritual nature. As regards corporeal nature this is not permitted. Why? Because, he says, 'the first law of bodies is this: we make sure, by testing, that several bodies cannot occupy the same space at the same time'. That is, he hints that two *souls* can be in the same place at the same time [but two bodies cannot]. But he says: 'Of two bodies within our imagination one must disappear in order to yield its place to the other'. It seems to me that he here contradicts what he said earlier, namely that the *soul* is a series of actions, all considered together as a whole; and by these last words he seemed to me to have wanted to say not only that a whole series of 'mind-pictures' is simultaneously *in* the soul, but that they *are* the soul. How Lucian would have laughed at such metaphysics if in his day there had lived metaphysicians, as there did scientists and moral philosophers! Not only because of this, but because of the whole mob of the philosophers, O Lucian, would you were alive now![45]

21. From what he has argued earlier, and having laid it down that the motion of the world will last for ever, our author thinks that, in consequence, the mover of the world must also endure everlastingly. He denies, however, that in consequence this same mover must be infinite. But he states an argument for the infinity of the mover, which, he says, is not well-based but yet is irrefutable (his page 281): 'Because it understands and controls many [of their] parts simultaneously, spirit is the dominating principle of motion for bodies. That spirit, which works to make the motion last for ever, i.e. infinitely, should therefore embrace the whole of such motion. It will not, however, as spirit would,[46] base its action on the intellect and on the inclination of the intellect. On the other hand, who can doubt that [spirit], which knows the infinite entirely and at a glance, is infinite?'[47]

FOL. 310v

This reasoning from that term ['infinite'], however, our author rejects, because that which at one and the same time knows the infinite is not, he believes, necessarily infinite.

---

[45] In this sentence Mersenne changes the word 'philosophers' to 'metaphysicians'.
[46] Here White seems to mean 'spirit' as understood physiologically in his day.
[47] Perhaps, 'Who can doubt Him to be infinite who understands wholly, and at a glance, the infinite?'

'Don't you see', he asks, 'that one wholly finite surface is equivalent to an infinite number of lines, and the smallest body to just as many surfaces?' That is, he assumes here not only as true but also as obvious what is very false; for no number of lines laid side by side is ever equivalent to a surface. Any one line is individually without breadth; therefore any number [of lines] you care to take as a group will lack breadth, because, starting with zero, multiply as many times as you wish, the sum effect will be zero. Hence, because a surface has some breadth, but lines none, how will a surface be said to be equatable with lines? The same would have to be said of surfaces compared with a solid. Now, a line without some breadth cannot be drawn, so [a plane surface] can be equal to lines drawn on paper, but not to lines infinite [in number], however thin these be, for, properly speaking, they are not lines but plane figures. 'Line' means merely length, even though this never exists without breadth. However, it may be assumed to be without breadth, as their discipline requires geometers to assume.

FOL. 311

That the movement of the world will continue for ever and that, in consequence, the mover is eternal, he indeed admits but does not prove, because (I think) he imagined he had already shown motion to need an efficient cause not only to begin it but also to continue it: we have confuted this above [at folios 302ff]. What he adds, 'Were the world eternal, its movers [sic] would be infinite', can be proved by no argument known to me, nor do I know what his purpose was in pronouncing it; for I do not yet see why that one mover that has brought the world to its present stage cannot make the world everlasting if it so wishes.

22. The last question in this Problem is: 'Does the everlastingness of the heavens proceed from [their] immortality?' Or, (to put it more directly), 'Supposing that the heavens and sun will endure for ever, does it follow that they have existed from all time?' In his view it does follow, and his argument (page 284) goes thus: If the heavens did not exist from all time, there must have been some beginning, in time and place, at which their motion commenced; this fact must have derived from the nature of the firmament; but the nature of the skies and sun supplies no reason why their motion should take its genesis from one place and time rather than from another; therefore their movement could have had no beginning.

FOL. 311v

In this reasoning he in no way derives a [valid] argument from the supposed future duration of the heavens and the sun; one may equally infer that they have existed for all time or that they will endure for ever or that they will perish. It is well that the argument is flawed, for otherwise it would completely remove all belief in the Creation. The flaw consists in that consequent [of his]: 'If the sun and heavens have not existed from all time, then there must have been some beginning of time and place when their motion, deriving from the nature of the sun and heavens themselves, was set in train'; for they could have derived the commencement of their motion from a time and place disposed by the Divine Will.

CHAPTER TWENTY-EIGHT

※

On Problem 3 of the Third Dialogue:
That the world has not existed from all time

1. Time and eternity: what each is. 2. The definition of *time* confirmed. 3. A refutation of the argument by which our author proves that motion has not existed from all time. 4. On *esse* and the names of accidents. 5. What 'to exist' is, how it differs from 'to be', and why these have been taken [by him] in different ways. 6. There is no succession of necessities for existing.[1] 7. 'Necessity' in [philosophical] propositions is not 'the lack of potential in things themselves which prevents their being something different [from what they are]'; it is our own lack of potential, which makes us unable to conceive that certain consequences could be false. 8. Our author has proved beyond doubt that an infinite succession is feasible. 9. His argument against the eternity of the world tells equally against everything called 'eternal'.

1. As regards this problem, two points are put forward for proving concerning the eternity of the world: (*a*) that the world itself began [to exist] at some [particular] moment, no body having existed from all time; and (*b*) that [the world] began to move, i.e. that its motion has not always existed.[2] He says that point (*b*) is demonstrated by the same arguments as those he used in the First Dialogue, Problem 2, to prove the world finite. And certainly, if those arguments against the infiniteness of space or of magnitude were valid, they would also count

---

[1] As will be seen from the relevant article below, Hobbes is going to refute the claim that a prerequisite to existence is a series of causes.
[2] For 'no body having existed', perhaps, 'and its body did not exist'.

against the infiniteness of time, i.e. against eternity. So just as, in order to examine those [arguments], it was necessary to explain what was understood by the word 'space', so we must explain here what is understood by the word 'time'.

To start with, everyone knows (I think) that [a] we sometimes see in our mind the images of those bodies that once existed before our eyes as objects but now no longer do so (this constitutes imagination, or the memory of things past);[3] or [b] such images are kinds of space with shapes and colours when at rest or when in motion, just as those bodies when in existence possessed shape, colour, motion, or rest. Now, inasfar as they represent bodies, these images are called 'imaginary' which are magnified or diminished as we choose, through our adding to, or taking from, them; for it is we, and not the things themselves, that determine the size of a space we wish to visualise. Hence a space is said to be 'potentially infinite' because we can add as many other spaces to it as we choose, and these of any kind. Inasfar as they represent finite bodies with definite bounds, the images are spaces possessing shape or are imaginary figures; and inasfar as [the images] represent coloured bodies, the colours are imaginary. In sum, if they represent bodies in motion, or bodies traversing any space one after another, these same images are imaginary motions, or the images of motion, or the images of a passage from first to next, though a body itself we see actually in motion is [in fact] no longer moving; but there is nothing to stop us from letting our imagination rove as far as we choose. Hence the 'image' of motion, or imaginary motion, is determined not by the motion of the bodies themselves but by our will. Therefore, just as the astronomers invented a place beyond the visible world in which to site the spheres they had contrived, so any poet can fix on an order [of events] in accordance with which he can adapt whatever narrative he has himself devised [relevant to a time] before the beginning of the history we now have.[4] These points seem so obvious to me that, in my opinion, anyone not accepting them has never observed his own thoughts and mental creations. Now, 'the image of motion', or 'imaginary motion', or 'an

---

[3] For the last phrase the text could be read *rerum praetentarum* ('things laid out before us'); the translation reads *praeteritarum*.

[4] Hobbes means that in writing a fiction about prehistory, or set in mythological times, a poet can utilise a time-scale, or points of reference, which is, or are, not known to have existed in fact.

imaginary succession', or 'the image of succession' (all these mean the same thing) is what we all call time. Just as a body is said to be infinitely large whose magnitude we cannot match, however many of the imaginary spaces we add together, so a motion whose limit we cannot reach, however much we extend the train of thought in the imagination, we call infinite in time, or perennial. Likewise that state with whose limit no mind-picture of motion can be equated we call perpetual rest.

FOL. 314

2. The definition of time: 'Time is the mental image of motion', can, I think, be built up of many constituents. First,[5] if the nature of time consists in [time's] being some real succession, time would not exist. A real succession that has passed exists no more, because it has passed; and a future one does not yet exist, because it is a future one. There is no 'present' save the instant, in which there cannot possibly be a succession. It remains, then, that time is not a motion in things themselves, outside the mind, but is mere imagination.[6]

Let no-one interrupt here that 'mere imagination' is nothing;[7] for imaginary motions, inasmuch as they are imaginary, do not have that existence they seem to have. In someone imagining, however, they *are* real, for 'the imagination of motion' is the same as 'motion in him who imagines'.[8] So also a face appearing in a mirror is imaginary and real at the same time: an imaginary face,[9] truly, but a real 'imagination'; for the imagination is the real motion of the organ with which we imagine.[10] Just as what is apparent differs from every real thing outside the mind, so it presupposes an inner reality, since if there were nothing, there would be no appearance.

FOL. 314v

I say, therefore, that time is the image of motion and has the same relation to real motion as has the reflection in a mirror to the true face and a mirrored shape to a real shape. This definition in no way differs from that of Aristotle, who

---

[5] Since this is the only 'constituent' that Hobbes especially singles out, it is not possible to say which the other ones severally are.
[6] Hobbes is not being pejorative: his mood is neutral. 'Time is to do with the imagination only, i.e. it is a matter of how we visualise things, and is independent of external criteria.'
[7] Presumably these words, given literally here, mean either 'Imagination is of no value' or 'imaginary motions *per se* do not exist'.
[8] Or, '...the same as "motion" in him who imagines'.
[9] For 'face', perhaps: 'image' (*facies*).
[10] Or, 'Imagination is the motion of the real organ...'

defines time as 'the number of motion according to first and next'. The reckoning of motion is understood only in terms of number, because calculation is [performed] not in the things counted but in the person who counts them, and it is performed by imagining the addition of something being counted to something else that has been counted. Hence to Aristotle time is nothing but motion in the imagination. But if he has understood 'number' as 'something in the things counted', this definition [of mine] differs from his considerably.

3. The reason why our author denies that any motion could have existed for all time is that there cannot, [he believes,] be infinite time. 'From the point we are at', he says, 'some hour must lie distant by gradations [*media*] infinite [in number]. If this hour is similar to our own hours, it must have been limited [*determinatus*] as to its existence and individuality, and it must lie between two hours equally delimited,[11] one designated "before" and the other "after". Now, between the hour under discussion and the hour that slips past jealously while we are conversing, may there not be enclosed what infinity they wish? My faculties are becoming very dulled.'[12] The force of this consequence will be made clearer in fewer words, thus: 'If time can be infinite, then any delimited part of it is infinite'. What could be more obvious than the falsity of this inference? Yet those who think that infinity ought to be represented in the mind by means of an image as though it were something positive can speak only in such words.

Indeed, [he says,] we use the word 'infinite' not to reveal to our mind the infiniteness of a thing but to [permit us to] concede [that] limits [exist] in the mind's faculty itself. This, [I reply,] is like saying that a real motion has lasted within a time that is longer, or that some real space is larger, than that which

---

[11] Perhaps 'specified'. 'Equally': 'likewise'?
[12] The words of White are not difficult; the meaning is less clear. It seems to be, at least in the first sentence, that if we fix on an hour infinitely distant from the present instant, it must be, like the hours we know, preceded by an hour and followed by another hour. Both these must lie, however, at finite distances from a fixed point of reference, the present. Therefore no hour can be infinitely distant from a given hour. The sentence 'Now, between the hour...they wish' seems to be an objection that could be raised by those advocating the infinity of time. 'Infinite gradations' (a number of infinite gradations, or an infinite number of gradations?) may mean that the space between the two given hours is divided into smaller units or space-intervals.

can be equalled by an imaginary movement or by imaginary space augmented as much as we wish. Now, time has always been, whatever one's wishes. If it be true, however, that motion has not always existed, because (as he argues) there is no such thing as eternal time, then for the same reason we shall conclude that nothing has been at rest, or even in existence, for ever. Further, if anything were indeed to have existed for all time, whether in motion or at rest, or (if this third possibility be granted) neither in motion nor at rest, then time must have flowed by infinitely, which he says cannot happen.

FOL. 315v

A universal rule in [philosophical] demonstrations argued legitimately is that the theorem for proof be inferred from the definition of the subject, or from some property of the subject. Now, that 'motion is not everlasting' he proves neither from a definition [of it] nor from any of its properties, but from the fact that time is not infinite.[13] He does not prove that 'time itself is finite because its nature consists in motion', and he therefore denies everlastingness not only to motion but to every act in general; to existence; even to God Himself. See how, almost inevitably, those who subject to their own metaphysical speculations Divine matters beyond our understanding come at every step into conflict with the Christian faith!

4. Before turning to the argument by which he strives to prove that 'the world has not existed for all time' and in which he says much that is hard to grasp about essence and existence, I thought I must here explain those two terms to the best of my ability. We showed above in Article 2 of the last chapter that accidents, or the acts of bodies, derive their names from the coupling [*copulae*] of propositions. In these last, when both the subject and the predicate are names of bodies, the link [*copula*] with the predicate becomes the name by which the *esse* or act of that body is expressed, e.g. when the proposition 'Man is an animal' is true, then 'to be an animal' is said to be the *esse* or act of man. Certainly it was necessary to posit names in this manner; for unless men turned [away] from [using] a substantive word (and from other words in all of which the force of a substantive word is hidden), the joint-signification of time and the virtue of connecting being re-

FOL. 316

---

[13] 'But...not infinite' is perhaps to go with 'does not prove' in the next sentence; the punctuation of the original does not make this clear.

341

moved, to [using] names, it would be wholly impossible to explain, in terms of distinguishing actions from those that perform them, the variety of concepts which arises from the action of bodies on men's senses.[14]

FOL. 316v

Hence arises the fact that, whereas we say that something 'is' in this or that particular [philosophical] mode, the word *esse* (which is sometimes called 'essence') has been admitted into the number of names, and has also fitted the matter under examination, i.e. the means by which a body makes itself perceived. For example, 'to be white' is the name of some mode or property in the body viewed that shows the latter to be of one appearance rather than of another. Now, for stylistic effect, most of such names are changed into others, as when we replace 'to be white' by 'whiteness' and, in the [philosophical] schools, 'to be everywhere' by 'ubiquity' (and Cicero has said of Lentulus, 'lentulitude'). All the names of accidents or of acts, however, have a meaning equal to that of the infinitive of an [English] verb, in which the words 'to be' always occur. But when, in a [philosophical] proposition, two names of acts are joined by means of a copula,[15] it is not necessary this time to name what were previously the acts and essences of bodies 'new acts and essences of acts and essences'. So that, just as from 'Man is righteous' we form the name 'to be righteous' or 'righteousness', so from 'Righteousness is a virtue' [righteousness] is received as a name 'to be a virtue' or 'a virtue'. [To do] this, however, is seen to be false from [the practice of] those who attribute to an act both act and essence.[16] When they see a copula attached to the names of accidents in [linked] proposi-

---

[14] This sentence, a difficult one, is given as literally as possible. It would seem to be a plea for the abandonment of substantive words (nouns? names of substances?) in favour of using Hobbes's doctrine of names. '...When both the subject...becomes the name...': this proposes the deletion of a comma in the text. Perhaps: '...When both the subject... names of bodies, being a link (of subject) with predicate, we form the name...' On 'the joint-significance (*consignificatio*) of time...being removed' cf. fol. 293, note 9.

[15] Cf. fol. 292 and the note.

[16] For 'Man is righteous', perhaps 'A man is righteous'. The clause beginning 'so that', which the manuscript separates by only a comma from what precedes, seems dependent on 'two names of acts...a copula' and not on 'to name...of acts and essences', to which the words '[To do] this' refer. In 'we form' *factum*, as at 318v, is read for the manuscript's *falsum*; but *falsum* is retained in the sentence following. If *factum* is there read also, we have: 'This, however, has been done by those', etc.; but the sense is little affected.

tions—instead of [recognising] another 'is' by which names are coupled—as when it is said that 'Virtue is a state of mind', 'Whiteness is a colour', 'Accident is separable', they are positive that 'virtue' and 'whiteness' and, in fact, any accident, is *ens*. Hence, in their opinion, *ens* being that which possesses *esse*, *esse* will always possess *esse* and the latter *esse* [will possess] further *esse*, and so infinitely. From this error of those who confuse *ens* with *esse*, or (while we are on the subject of perceptible *ens*) 'body' with 'accident', surely derive those sayings of the metaphysicians: 'An act acts', 'The intellect comprehends', 'The will wills', 'Accident is *ens*', etc. These mean [respectively] the same as 'An action is the agent'; 'The intellect is he who comprehends'; 'The will is he who wills', 'Accident is that of which it [accident] *is* an accident'. And in stumbling at the very threshold [of philosophy by using such modes of speech they have rendered so absurd and nonsensical every term relevant to the higher reaches of philosophy that most of the judicious loathe this discipline's very name.

FOL. 317

To return to the [present] proposition, however: *esse* or essence ([by the latter] I understand 'perceivable essence' throughout) is, I say, none other than the means (I might have said, more clearly, 'the action') by which a body, acting in different ways on those [severally] perceiving it, appears sometimes in one manner, sometimes in another, or it makes itself noticed as being sometimes white, sometimes black, sometimes square, sometimes round, etc., as was shown above in Articles 1 and 2 of the last chapter.

FOL. 317v

5. But 'to exist' means the same as 'to be' or 'to be *ens*', *simpliciter*. For when, in a proposition, the predicate is explicit, e.g. in 'Man is an animal', and when it is included, e.g. in 'Man is', then, in the former [*prior*] proposition man is not said to exist; but if he does, so also does an animal, for that is what is meant by the proposition 'Man is an animal'. This proposition, being necessary, is also universal. In the second [proposition he] is said to 'exist'; for this is what 'Man is' means, and is equivalent to saying: 'Of the bodies constituting the universe, at least one is man'. The reason why our metaphysicians have seen essence and existence as different things seems to be that they have not distinguished the consequences of essences from essences themselves. So because, in the present proposition, 'to be an animal' follows, and will always follow, from 'to

FOL. 318

be a man' (for if we say something is man we also say it is animal), it is true for all time that man is an animal; this they accept as if essences were everlasting. Now, as they believe the essence of man or of another body to be eternal but existence not so, they must inevitably take 'to be' and 'to exist', as does our author, as different things, and must not believe that the one depends on the other.

6. I come now to the argument by which (on his page 287) he denies eternity to the world. His first proposition is: 'Whatsoever is, cannot, while it is, not be' (i.e. it cannot be and not be at the same time). The proposition is true, but that statement 'cannot, while it is, not be' is ambiguous. Even during his lifetime a man is in potential to die, since he may perish, and the same can be said of all corruptible things. Therefore [to speak of] those things as being and existing, now, and at the same time as being corruptible, now, is not only ambiguous but also absurd, unless the possibility of dying is also understood. Potential always refers to the future, and we are said to be able to perform, or to be acted on by, not that which has been done[17] but that which will come to pass. But let us leave this kind of speaking to the metaphysicians.

FOL. 318v

The second proposition is: 'The same thing, before it is in that place, is able not to be[18] in the same place'. The wording is obscure and, indeed, unsuitable, for he means: '[At a time] before it was, whatever now is was able not to be now'. This proposition he lays down, not unconditionally, but with this qualification: 'at least to the degree that it is compounded of pure essence or of a combination of *ens* and existence'. This is unclear, but is tantamount to saying: 'The existence of *ens* does not preclude *ens* now existing from existing later, howsoever the things now existing may encounter (from external causes) the necessity of existing afterwards as well'. As thus limited, this proposition is also true.

From two propositions, therefore: 'What is cannot, at the same time, be and not be', and 'Before it has been, what now is has been able not to be' he infers a third: 'The impossibility of not-being comes progressively to a body, in proportion to the amount of connection it has between its *ens* and its exist-

FOL. 319

---

[17] Cf. fol. 316v, note 16.
[18] This is not, of course, the same as 'is unable to be'.

ence'. That is to say: 'The existence of *ens* does not prevent the impossibility of not-being' (that is, the impossibility of not existing, or the necessity of existing) 'from coming to the body progressively'. This conclusion contributes nothing to proving that the world has existed for ever. However, I should like, if possible, to examine it in such a way that one can understand, and hence know, whether it is correctly deduced from the premises or not.

Let us therefore put for the first proposition this: 'While he is man, man must be man'; for the second, 'Before he was man, i.e. when he was clay or red earth, man could not be man'. (For within the existence of that clay there lay no necessity for him to become man, but [this necessity] came from an external cause, namely from God.) Therefore there was no necessity that the clay, when it was clay, should become man; but after man was created, the same clay possessed the necessity to become man because the necessity progressively entered the clay, i.e. the clay's first need was to be clay, its second to be man. From what has been said, I see indeed that it has been sufficiently proved that there takes place a kind of progression from necessity to necessity: the clay necessarily was clay, and after a man was formed from it, than man necessarily was man; after man has become a corpse, that corpse will necessarily be a corpse; and so on as often as the process of change continues.

FOL. 319v

There are two things in the conclusion that are not found in the premises: [a] This necessity is a necessity of existing. For if the clay is at first necessarily clay it will not necessarily be man; hence 'necessarily to be clay' comes first, and 'necessarily to be man' follows. Therefore 'necessarily to be clay' must be said to precede 'necessarily to be man' or 'to exist', because at the same instant that something is clay it *is*, *simpliciter*, and it exists.[19] So this argument cannot be applied to the world. For that second proposition: 'What is, was able not to have been before it was', if applied to the world, will stand: 'Before it was, the world was able not to be'. But how do we attribute to the world the potential either to exist or not to exist? When nothing existed,[20] from what could [the world] have been made?

FOL. 320

---

[19] Re-worked, perhaps not to advantage, by Mersenne: 'For if in the beginning the clay is necessarily clay, the same clay will not necessarily then be man, so "necessarily to be clay" precedes "necessarily to be man". We must therefore say: "To be clay"' (or, 'that there was clay') 'necessarily preceded "necessarily to be" or "to exist"'.

[20] Perhaps: 'when it was nothing' or 'as it *was* nothing'.

Or how can we apply that limitation–'to the degree that it is compounded of a combination of *ens* and existence'–when there existed no *ens* that could later constitute the world?

[b] The second thing is that the succession of these necessities is in a body itself, for he says in [his] conclusion: 'The impossibility of not-being happens progressively to a body', i.e. to the world itself. Yet this cannot be deduced from the premises and is untrue. [That which is] a necessity is that which cannot not be. Now, according to our author, there is no necessity [arising] from the actual existence of things; therefore anything cannot not be; therefore every necessity is from all time. Though there is a necessity for man's being man or the world's being the world, [I say,] this does not stem from man's or the world's existence or from without; for just as something external cannot make a man not-man as long as he is man, so also it cannot cause man to be man. It has, indeed, caused clay to be man–which was not necessary–but it did not cause man, once made, to be man, which is necessary. Hence there is in no way a chain of such necessities within *entia* themselves; and the necessities which follow one another, each after each, are not necessities for existing. He had undertaken to prove [that they were].

7. But someone will say: 'Why, if that necessity is not found among things in man and [among those] in the world, is it necessary for man to be man or the world the world?' I say that propositions of the following kind, 'The world is the world', 'The world is', 'Man is man', 'Man is', and 'Man is an animal', where the predicate necessarily follows the subject, are called 'necessary' and that the necessity of the consequent is the one incorrectly attributed to the things themselves. Because if the world exists, we cannot conceive of its not existing, or if man exists, we cannot conceive of his not being an animal, or that there is any possible alternative to our calling the world 'the world', we call such propositions 'necessary', for we term 'necessary' that which we cannot understand as being other than it is. So that phrase: '[The world] cannot be understood otherwise' represents not the inability of things but our own inability; hence 'the chain of necessity' is nothing but a succession of concepts or conclusions in our own mind. Say, for instance, it first has this thought: 'It is impossible for clay not to be clay'. Next, 'When man was created from clay' is followed by another concept: 'It is impossible for what

was made man not now to be man', and so on. The succession White is speaking about, then, is a succession not of things themselves but of our thoughts.

8. Yet from the premises stated by our author two points legitimately follow:

(*a*) There is a progression from necessity to necessity which has no end. This is true, but it counts against him; for that progression does not lie in external things but is created within the imagination. Consequently there is an imaginary progression which may be continued as long as we wish by the addition of one progression to another, i.e. without any termination fixed by the things [themselves]. This is called [an] 'infinite' [series]. Hence time may be understood as being perpetually prolonged more and more to infinity.

(*b*) Time is understood [as existing] before the world was established, and hence as anteceding every real movement; for if, before it was made, the world could (as he says) indeed not have been, or even not have been thought of, then [time] can be extended [backward] even to pre-Creation.

FOL. 321v

9. Having stated the conclusion he thinks he has proved, namely: 'If the world has existed eternally, there has also existed eternally some sequence [of causes]', he adds the conclusion he laid down earlier: '[There is] no sequence [that] has existed eternally'. From this follows the conclusion he wished to prove: 'The world has not always existed'. Were his argument correct, however, it would remove not only the eternity of the world but also the possibility of eternity [at large]; for he draws his argument not from the nature of the world or of body but from the impossibility of time's being infinite. This tells equally against eternity as a whole, which is also against faith,[21] not only the Christian but that of every race, as though metaphysics, unheard of by the ancients[22] and, under the name of knowledge, sent forth into the world by I know not what evil spirit, were nothing but a freedom to utter rash words about God.

[21] Perhaps: 'Moreover, the argument which is also against faith tells equally against eternity as a whole, not only...'
[22] Hobbes presumably means 'metaphysics as taught since classical times'.

FOL. 322

CHAPTER TWENTY-NINE

❦

On Problem 4 of the Third Dialogue:
That there is something which has
always been in existence

1. A list of the questions comprising the Problem, together with a gloss of certain terms in which they are worded. Examinations of the arguments or reasons by which [our author] infers (in (2) and (3)), or proves (in (4) and (5)), that there is (2) something that has existed from all time; (3) that the first thing that did change was not made; (4) that nothing can be made of itself; and (5) that, of an infinite number of existing things, some one thing is independent [of all the others]. 6. How 'Given [that something is] "infinite in number"' is to be interpreted. A scrutiny of (in (7)) the reason, and (in (8)) the argument, by which he proves (7) that 'something which is eternal is the cause of its own existence', and (8) that 'no body has existed from all time'. 9. That 'existence' and 'the essence of *ens*' are the same thing.

1. In this disputation he devotes the first part[1] to the proof of the proposition which heads the Problem, namely: 'That there is something which has always been in existence'. Then he proves the following: (2) that which has existed from all time, if it be a single thing only, was not made by something else; FOL. 322v (3) if several things existed from and for all time but [were] finite in number, not all of them could have been made, but only one or some; (4) if the things that were eternal were infinite

---

[1] Perhaps, here, 'he gives pride of place'. Titles: The Latin repetitively spells out each article.

in number (i.e. surpassing every [possible] total) these also were not all made; (5) 'something infinite in number' cannot exist; (6) what exists of itself has some intrinsic basis for existing from which it derives its existence from all time (about this, he says, the ancients were doubtful); (7) no body has existed from eternity; and (8) to *ens a se* existence is a form of essence.[2]

In [White's] demonstrating the above topics, the following expressions frequently occur: 'to be of itself [*a se*]', 'to be from another', 'dependent on itself', 'dependent upon another', 'to have been made' and others of similar force. We shall therefore explain them before proceeding further, so as to avoid using them ambiguously. 'To be of itself' means the same as 'not having had a cause because of which it exists', and it may be understood in two ways, just as we put the word 'is', or [the words] 'to be', sometimes *simpliciter* and absolutely, sometimes in a restricted sense to mean 'to be such and such' or 'is such and such'. For example, when we say 'Man is', the 'is' is put *simpliciter* and means the same as 'Man is *ens* or is existing'. When we say 'Man is of himself' [*a se*] this is the same as saying: 'There has never been any efficient cause that has made man out of *non-ens* or non-existence into *ens* or existence'. So when something is said 'to be of itself' it is said to have existed from all time, because how *non-ens* could begin to be *ens* without a pre-existing cause is incomprehensible, for the following reason. Say we state a proposition thus: 'Man is man' or 'Man is an animal' (or any other proposition where the term 'is' serves only to combine the terms of the proposition). Next we say: 'A man is a man *a se*', or 'Man is an animal of itself'. Here the statement that man himself is the efficient cause that has made him, when not yet either, a man or an animal, means either [*a*] that at least there is no other efficient [cause] of this, i.e. that man has existed for all time, or [*b*], that at least the material from which he is made has formed that man of human shape.

FOL. 323

FOL. 323v

But to be brief: doubtlessly nothing can make itself; so 'to be of itself' is the same as 'to be eternal'. 'To be from another' [*ab alio*] is the same as [*i*] 'to owe to something else, as though to an efficient cause, the fact that [something] is what it is'; or [*ii*] 'to have been brought to pass'; or [*iii*] 'to have been

---

[2] Perhaps by *essentialis* ('of essence') Hobbes means 'essential, absolutely necessary'. Cf. fols 334 and 359v.

made'. But if [a] on the one hand we are talking about 'to be, *simpliciter*' or 'to be *ens*' when we say that 'Man or a thing'[3] is 'from something else', [then] we mean that he has been made, *simpliciter*, and not out of material already in existence; i.e. that he has been created. If, however, [b] we are speaking about determinate *esse*, saying 'Some man is "from another"', we mean that he has been begotten or changed and that he previously did exist, although not yet in the form of a man.

That which, when its cause ceases, no longer exists is said 'to depend on something else'. Nothing, then, depends on anything other than on the cause of its continuing-to-be, in which sense we say that a stone hangs from[4] a wire in the air because in the wire lies the cause why the stone remains suspended and at rest, despite its ['heavy'] nature. But 'to depend on itself' is applied to a thing outside which there is no cause for its continuing-to-be; and in this sense the motion of any body, once commenced, 'depends on itself' and, if unhindered by anything else, will continue for ever. So—again to be brief— 'to depend on itself' means 'to persist in being that which is now without a cause, and which does not need one'. 'To depend on something else' means 'to persist in being that which it is, according as a cause has persisted or ceased'. 'Something made'[5] means 'that which has begun to be, because of a cause'. Say the thing has in fact commenced to be, *simpliciter*, i.e. that from *non-ens* it has started to become *ens*. Then that 'made' is the same as 'created'. Suppose, on the contrary, that something has commenced to be, but not *simpliciter*. If (for instance), from non-man, man began to be man, that thing is now said to be, not 'created', but 'changed' or 'begotten'. From the words I have used, then, [the expressions] 'to be of itself' and 'to depend on itself' are employed quite unnecessarily and have introduced the greatest obscurity into the language of metaphysicians; they are not [correct] Latin; and [correct] Latin terms are readily to hand to replace them, since we can say [in Latin] 'to be eternal' for 'to be of itself', and 'not to depend [on something else]' for 'to be self-dependent'.

2. The demonstration of each of those two propositions must be taken next. 'There has been', says our author, 'a

---

[3] The text has 'or another thing'.
[4] Or, 'depends on'.   [5] Or, 'done'.

certain change before which there was no other change.' (This is correct, though not demonstrated by him.) But 'Therefore from this we have inferred that something has existed for ever' does not follow. [True,] something that itself has been changed must have existed before the change [took place], because as often as there is change something must be changed; we shall not say, however, that *what was first changed has existed from all time.* That which was first changed is the world, or part of the world; consequently, [if we believe true the statement italicised,] the world or any part of it has existed for ever; and this goes against faith. Certainly, it is no less impossible to conceive of motion or of change as having been commenced without material than [to conceive that this happened] without an efficient cause. So those who infer from the fact that, once the commencement of motion or of change was set in train, its[6] efficient cause was eternal could by the same argument have concluded that [it was] the material [that] was everlasting. For if anything is to be moved or changed, why is it necessary to suppose an efficient cause by which, more than material in which, the effect must be produced? The fact that we consider God, but not the world, to be truly eternal the metaphysicians themselves wish to attribute, not to faith and to the authority of Holy Writ, but to their own knowledge. But through their paralogisms they make [both] God and the world eternal.

FOL. 325

3. It being now demonstrated that 'something may be postulated as eternal', [the question] is proposed: Can that be the only one such thing, or are there more [such things]? If more, are the things finite or infinite in number? He says first: 'If that which is eternal be a single thing, it could not have been made'. In the form in which it is proposed, this point is absurd, unless perhaps he means: 'The only thing that could not have been made was the one that brought about the first change'.[7] He seems to have intended this meaning, since that is what he tries to prove. On his page 289 he reasons thus: 'If that which now exists was the only one thing that had been made by something else, then surely there was [once] another thing now extinct. Therefore that extinction anteceded that change we have supposed to be the first'.

FOL. 325v

---

[6] Probably, 'motion's'.
[7] Or, 'Alone that which brought about the first change could not have been made'.

Now, the extinction of one thing can be the generation of another, such that one does not precede the other; hence the extinction of warmth and the generation of cold are one and the same change, created at the same instant. So I ask: Why does he suppose that the first thing that changes[8] is now extinct? Or rather, why will he suppose that only one thing exists? But he supposed that there exists not one thing *simpliciter*, but a single first thing that changes–which does not prevent there also existing simultaneously [a] one thing or [b] more first things, that has or have been changed. If so, the first thing that changes is not extinct, and the change which was held to be the first will be the second.[9] He could have demonstrated his conclusion (for it is correct) if he had spoken thus: 'If the first thing that changed was made by something else, then the changing of the first thing that changed preceded the changing of the first thing that was changed'. Hence the change he has supposed to be the first was [in fact] the second.

FOL. 326

4. Next he proves that 'If more than one thing (but these things [to be] finite in number) were eternal, then one or more of them, but not all, has or have been created'. But as stated above, 'eternal' and 'created' are contradictions, and 'something in existence that has not been created' is the same as 'something eternal'. In the way he puts it, then, the proposition is this. 'If more than one thing (but these things finite in number) are [in fact] eternal, one or more of them will be eternal but not all'. To me these words seem ridiculous; to others, perhaps, metaphysical.[10] What he claims to prove is, indeed, no other than that nothing can be made of itself, indirectly or directly.[11] This is indeed true, but is not what was to be proved, which was: 'There cannot be more than one thing that is eternal'. Moreover, the proof itself is not a legitimate one. 'That is a cause', he says, 'by which something else has been established through the existence or from the force of the cause.'[12] If, then, something could change itself from non-being

---

[8] Perhaps Hobbes intends the transitive: 'that changes something', and so in the sentences following.
[9] 'Second': text, 'first'.
[10] Or, 'to others, perhaps metaphysical'.
[11] For 'can be made' Hobbes may intend: 'can become'. 'Indirectly', *mediatè*.
[12] This retains Hobbes's text, but the *De mundo* has: 'That is a cause by

into being, non-being would be the cause of *ens*. Therefore non-being will be that which, being posited, is different from that which itself is non-being, presumably from the fact that not-being is *ens*: which is absurd. True, if the definition of 'cause' is correctly given, reasoning proceeds correctly; but [here] the definition of 'cause' is the actual conclusion inferred. For when 'cause' is defined as 'that from which something else [proceeds]', and not [explained] indefinitely: 'from which something [proceeds]', then something bringing itself into being is excluded by the definition itself [of 'cause']. So a *petitio principii* is committed. If our author wished to prove legitimately that nothing is made of itself he should have demonstrated it, and so this could not be the definition of 'cause'. That is a cause by which, the cause having been postulated, something is because the cause itself is.

FOL. 326v

But perhaps he will say: 'No one believes that this consequent: "If anything is, then it itself is" is the consequent of the effect in respect of the cause, or of *ens* as regards *ens* (for *ens* does not follow itself), but [it is the consequent] of perception with regard to perception [*intellectio*], because we understand the same thing twice: under the name "something" and under the name "itself". Everyone calls a "cause" that which is followed by something, and this last will therefore be something other [than a cause] *because* it follows; for nothing can precede or follow itself'.[13] Therefore so implanted in the very definition of 'cause' was the conclusion: 'Nothing is made of itself' that to construct a syllogism there was no need to assume any other proposition. So reasoning of this kind is a mere *petitio principii*, however true the conclusion 'Nothing can be made of itself' be. There was no need, either, to add 'indirectly or directly' and to strive to show that (as he says, [but] metaphysically) 'otherness' is greater in indirect than in direct causes, although 'otherness'[14] is required in every bringing-to-pass.

FOL. 327

5. He next puts forward for proof, recording it in the margin,

---

which, it being established, something else exists because the cause exists, or by its [the cause's] force'.

[13] Since Hobbes is hypothesising about what White *might* have said, there is no means of telling where the 'quotation' ends. Perhaps the argument is intended to end at 'itself', even though, as here given, the remainder is in reported speech and is presumably dependent upon 'understand'.

[14] See fol. 330v, note.

that everlasting things, even though infinite in number, have not all been created. This is the same as saying: 'Even though everlasting things are infinite in number, some of them are everlasting'. For what he proves (if indeed he does so) is this: 'If *entia* exist that are infinite in number, one of them must be considered as being something extraneous, [i.e. that] which depends on none of them, and on which all the remainder depend'.

To demonstrate this he assumes, first, that [the concept] 'infinite in number' is impossible, which is to exclude the very same thing he has assumed; hence no conclusion at all is possible. Even though this proposition is true, and if *entia* infinite in number do exist, then there is one extraneous *ens* on which the other *entia* depend, but not the contrary.[15] But the conclusion remains uncertain.

Second, when he says that one [of the *entia*] must be assumed extraneous, we cannot see how [this] *one* can be added to an infinite number, or that anything is extraneous to something infinite. The demonstration, however, goes thus (his page 292):

FOL. 327v
'Because existence and bringing-to-pass have been accepted as being from all time and as preceding every change, it is clear that everything they do is eternal'. But this is false. Something can be understood to exist before every change, but one cannot *perform something* prior to every change. Doing, [or] acting, or bringing-to-pass [*operatio*] is each either motion, or change, or causing-to-be, or, in sum, creation, i.e. a progressing from being to not-being,[16] and [White] includes them all in the term 'change'. So action cannot have preceded change. But he adds a reason: '...[Even] if there was lacking any bringing-to-pass of anything, some change would now have been brought about'. If, perhaps, some bringing-to-pass had existed and had then ceased, some change would have followed; but this is not necessarily so. But if there had existed something at rest and no bringing-to-pass took place, no change could have resulted, for every change is brought about by something that brings-to-pass. But from his statement 'The acts of eternal things are timeless' he infers a different conclusion, namely: 'Therefore all things have begun and are continued, at the same time [as

---

[15] For '*ens...entia*', perhaps, 'thing...things'.
[16] Surely this must be an error, yet the text is specific: the correct expression appears at fol. 354.

CHAPTER XXIX: 5

one another]'. This contains all too evident a contradiction, i.e. that the actions that bring-to-pass, and which are independent of time, had a beginning. Who is unaware that an act which is called 'eternal' has never begun?   FOL. 328

Next he adds that any bringing-to-pass of separate things depends on higher (or earlier) acts-of-causation [*operationes*], and that all the lower (or later) instances depend on that [bringing-to-pass]. Again a contradiction, this; for a bringing-to-pass that has been from all time cannot depend on anything else. Nonetheless, from it [this conclusion of his] he infers that every [act within a given] number of acts-of-causation depends on some act-of-causation outside that number, which in the end is true but is not inferred from the premises. Having established the above position, he has woven a most subtle argument, as follows: 'See', he says, 'what I have thought out, [no matter] whether number be either finite or infinite'. His reasoning (page 293 of his own book) is expounded very obscurely and is also more prolix than it need be. I shall try, however, to explain it here so as to make it readily intelligible, not adding anything I can confute later, but [writing] in a way such that he himself will indeed accept his own meaning [as being fairly given] by my words.

In a series of acts-of-causation, therefore (these, he has said, are without beginning or end), let there be taken a number of them as large as you please. Of these acts, let us call 'No. 1' the one which immediately precedes the change posited above   FOL. 328v
[at folio 327v], this being the first occasion when all the acts suffered a change. In this number [of acts,] [I say,] the number of those acts that are dependent on others ought to be equal to the number of acts depended on, because acts-of-causation severally depend on [other] single ones. Now, he had previously concluded that in the whole number of acts there was not one act independent of another (for he had postulated that the series of them was without beginning or end). So, [he claims,] the number of dependent acts becomes larger than the number of those depended upon; for all, from first to last, are dependent, but there is one act, the last, on which no other depends. Unless, [he continues,] by increasing the number [of acts-of-causation] some act is at length reached that is not dependent upon another, the number of dependent acts will exceed by one the number of operations it depends on. So he concludes that, for this [to occur], i.e. for the dependent acts to be equal [in

355

number] to the ones they depend on, it is necessary to assume [the existence of] one extraneous thing on which dependence is placed, though the extraneous thing itself is independent [of others].

FOL. 329

I do not know whether it is worth while either refuting or transcribing reasons such as this; but in order to avoid puffing up a solitary mental aberration into a sample of metaphysics as a whole, I cannot help doing so. I consider therefore, the following incongruous: (1) Eternal acts-of-causation existed before every change [did]. (2) He declares that, however large a number of acts-of-causation you take, any one of them depends on some earlier act. Yet he concludes that to this number you must add one [act] that does not depend on an earlier one. (3) After stipulating that the first change postulated above [in (1)] depends upon the act most nearly preceding it, he then says that no other act depends on this one. (4) He considers it beneficial to things, or that the Creator of things has disposed, that dependent acts-of-causation be equal to the acts they depend on. (5) Wishing to prove something to be independent, he has cited [as his example] something from which it follows that nothing is independent: i.e. that the chain of things dependent upon one another is unending and that, whatever the number of such causes that is taken, any one of them depends on some earlier one. From this it is clear that, whatever cause is taken as the first, it is not independent.

FOL. 329v

6. Upholding the possibility of [there existing] infinite number in no way helps him; for who is unaware that every [single] number is finite? For in order that [a single number] be accepted as independent it is necessary, not only that every collection [*multitudo*, of numbers] be finite, but also that there should not be anything infinite in number. He has in fact correctly proved that there cannot be a collection of numbers that is infinite; but what it would have been useful to prove, according to his purpose and to what in itself he had undertaken to prove, namely that 'there cannot be anything infinite in number', he has not proved, because 'an infinite number' and 'infinite in number' differ considerably. To understand this better we should be aware that things are bounded [*finiri*] in two ways: by means of a mind-picture [*imaginatio*, of them] and by number (or measure).

CHAPTER XXIX: 6

Things are delimited[17] [a] by means of a mind-picture when, for example, we set the boundaries of a field by the mind-picture of a river or of some visible object that borders it, or when we define a day by remembering the sun's rising and setting; and [b] by measure or number, when we measure a field by the number of roods or acres, and time by the number of divisions the shadow in a sundial crosses. So we say 'to delimit, or to end something' is the limiting, not of the thing itself, but of a person imagining, measuring and counting it. So because a person who imagines, measures and counts is bounding and limiting everything by imagining, measuring and counting, likewise whatever is actually conceived of or measured or counted must in fact be delimited. And whatever can be conceived of or measured or counted as a whole must be potentially finite. On the other hand, what no imagination can grasp and no number of units can encompass we call, *simpliciter*, infinite, i.e. unbounded or 'impossible to compass'. It is therefore clear that the number and measure by which we delimit actual things are themselves finite. But say we have something infinite in number, i.e. that which no number can encompass and bound, whether a number of yards, or miles, or parasangs,[18] or the earth's or the world's radius. That which is called infinite in magnitude and which cannot be encompassed by any number of hours, days, years, centuries or, in sum, Platonic years–that which is called infinite in time, or eternal–in no way discounts any of the pleas here advanced, or elsewhere adduceable, that infinity of number is impossible. Whence comes it, I wonder, that men have so tormented themselves in explaining this term 'infinite', when other terms meaning the same thing cause them no trouble at all? When someone says something is 'uncountable' we believe we understand that term. We do not, on hearing it, go mad or play the metaphysician, as we do on hearing the word 'infinite', yet both mean the same thing: 'infinite in size' is exactly the same as 'cannot be counted using number that is finite in size'; and 'eternal' or 'infinite in duration' is the same as 'cannot be reckoned in terms of a number of finite times'.[19]

FOL. 330

FOL. 330v

---

[17] Or, 'defined'.
[18] Parasang: thirty stades. The length of the stade was not universally standard in ancient times.
[19] Last word: 'units of time' or 'occasions'?

7. After establishing that there is but one thing eternal, he asks: Does it possess any intrinsic cause for existing from which it draws its existence now and always? By these words he is asking, in fact, whether [something] eternal possesses within itself an efficient cause of its existence, from which cause it begins to exist. ('To draw existence' is said only if [a thing] derives its existence from some cause or source.) But perhaps he has forgotten what he postulated earlier [at folio 328v]: 'Nothing is its own cause', and in every effect 'otherness'[20] is sought, he being unaware that every cause precedes its effect and hence that things which are eternal can have no cause or origin. The demonstration itself is based upon three propositions: (1) That which was, at the very instant that it was, must have existed; (2) That which existed at every instant must have existed perennially. (From these two propositions he infers that [something] eternal has necessarily existed for ever, i.e. that [something] eternal is of necessity eternal.) (3) This necessity (which, changing the term, he calls 'a repugnance to not-being') is not from something extrinsic, for nothing can be extrinsic to the only everlasting thing that exists: thus [the latter exists] from something inside itself.[22] Hence within *ens* that is eternal there is an intrinsic necessity of existing, i.e. a source from which it draws its existence.[23] Here, on the other hand, from 'necessity', [which is] the consequent of the predicate to the subject in the proposition 'What has been has been', he brings the consequent of the effect to the cause.[24] To do so is as if, because 'Man is man' is a proposition true for all time, one inferred that man himself has always existed, and not only this, but also that man has always begun to exist. 'A thing that is eternal has a reason to exist or has achieved existence' is the same as 'A thing that is eternal has begun'.

FOL. 333[21]

---

[20] *Alietas*, like *hocceitas* and *aliqualitas* (of which latter pair the former was used by Duns Scotus to signify the 'principle of individuity'—the topic of Chapter XII above—and the second came into use as an alternative to *quidditas*), is one of the accretions of mediaeval logic ridiculed by Gassendi (*Syntagma*, I.i.13). Cf. fols 16v, 316v, 377v.
[21] Here the leaves are again bound in the wrong order: fol. 332v is blank.
[22] For the second 'extrinsic' the text has 'intrinsic'.
[23] It is not clear whether this sentence is part of White's alleged argument or is Hobbes's comment on it.
[24] Hobbes may mean: Necessarily, the consequent of the predicate to the subject in the proposition 'What has been has been' is related to the consequence of effect from cause.

CHAPTER XXIX: 8

8. Now, since the *ens* that is eternal has an origin within itself, by which it necessarily exists, and assuming (from what he has said before)[25] that it is natural for everything corporeal to be able to perish, he concludes that no body has existed from all time. The reason why it is natural for it to perish (which our author has stated previously at his page 287) depends on this: no succession can be eternal; yet a body has the property of being able to exist in successive states [or at successive times]. From this ability to exist he deduces that a body cannot be eternal. He said that an everlasting sequence cannot exist, because no instant is far from the present by an infinite number of hours. We have refuted these reasons where they occurred [at folios 314v*ff*].

FOL. 333v

Next [he says that] a body which is everlasting cannot endure, because we do not grant [the existence of] a succession that [itself] is everlasting; for the same reason nothing else will be perennial. This is untrue and sacrilegious. Even if it be given as true that no body will last for ever, it does not follow that no body has existed from all time. He wishes the necessity of eternal existence to proceed from that necessity which declares: 'Whatever has been, has been by necessity'.[26] But there is no necessary proposition which says: 'Whatever has been, has been by necessity'. Nor, [if there were one,] would it lead him to infer that a body *did* not exist for ever because it *will* not endure for ever. Indeed, I wonder why he first persuaded himself, from [the instance of] motions, that all body will perish at some time or other. This implies the annihilation of faith, for neither can we conceive how this [perishing] occurs, nor has [our author] ever witnessed or heard tell of anything brought to nothingness; nor do, I think, the Holy Scriptures or the Church propose that we must believe in the destruction or annihilation of the world. The bodies of the saints in heaven, are they not immortal? Is heaven to be destroyed? If so, where will those bodies be [found]? If they are bodies, [they will be found] in a place–but according to our author, who does not think body can have a place in nothingness, this cannot be so.

FOL. 334

9. What in the last instance he turns to prove: 'Existence is

---

[25] Or, 'from its antecedents'.
[26] Or, 'Whatever has been of necessity, has been'; or, 'Whatever has been, of necessity has been'. In the preceding sentence the punctuation of the original is corrected.

FOL. 334v

essential to *ens a se*', I freely concede, and also what is more: that 'the existence of all *ens*, created and uncreated,' is the same as its 'essence'. 'To be' and 'essence', 'to exist' and 'existence' are synonyms, all meaning precisely the same. When we say, *simpliciter*, 'Something is', the predicate being understood in the copula 'is', we mean the same as 'Something is *ens*', or 'Something is that which exists', for, used *simpliciter*, *ens* means the same as 'existing'. So *ens* and *that which exists* have the same essence; but the essence of something that exists is existence, as is the essence of *ens*.[27] 'Existence', therefore, and 'the essence of *ens*' are the same, be that *ens* 'of itself' [*per se*] or of something else.[28] However, the statement by which he proves that existence is essential to *ens a se* is highly metaphysical; but there is no time or need to refute it logically, since I admit the conclusion.

---

[27] This last phrase (*sicut entis essentia*) is elliptical, and seems to mean: '...just as it (the essence of an existing thing) is also the essence of existence (itself)'.
[28] Hobbes seems to have in mind the relationship, which he discusses elsewhere (fols 293, 377v and 384v), between existence (*tò einai; esse*) and an existing thing (*tò òn; ens*).

CHAPTER THIRTY

❧

On Problem 5 of the Third Dialogue:
That *ens a se* is alone of its kind and
is the cause of everything else

1. The heads of the subjects disputed in this problem. 2. An examination of the argument by which he proves that *ens a se* is alone of its kind and is the cause of everything else. 3. On perception. 4. On the imagination. 5. On the memory and on time. 6. Why the imagination is weaker than perception is. 7. On fictions. 8. On the discourse[2] of the imagination. 9. On reasonable [*ordinate*] and unreasonable discourse [of the imagination]. 10. On remembering, synthesis and analysis. 11. On experience and on the future. 12. On expectation and prudence. 13. On signs. 14. On comparison and notes. 15. That the nature of man differs from that of the beasts in that it finds and uses notes and names. 16. On 'name' and the universal. 17. On the [philosophical] proposition, [its] truth and falsity. 18. On the syllogism. 19. How the discourse [*discursus*] of the imagination is transferred into discourse of speech or of language. 20. How it can happen that the discourse of speech is often incorrect when that of the imagination is not. 21. Of comprehension. 22. Of reason, reasoning, and right reason. 23. Of pleasure and disgust, appetite and revulsion, love and hate. 24. On good and bad, the beautiful and the shameful. 25. Of the connection between good and bad things, and of the seeming

---

[1] This leaf is misbound; fol. 335rv is blank.
[2] *Discursus*. It would appear preferable to render this as 'ranging', 'roving', but within the context of Article 8*ff*. Hobbes relates the term to verbal discourse. The Latin word is inserted in brackets where this is useful.

good. 26. Of appetite and disgust [arising] in turn;³ or, of deliberation. 27. Of the will. 28. That the will and deliberation are proper not only to man but also to other animals. 29. Of voluntary [action]. 30. Of liberty. 31. Our author's opinion concerning the intellect. 32. A scrutiny of his opinion. 33. That the consequences of his opinion are that man understands in the same way as God does; that the Creation was not a passage from not-being to being; and other absurdities. 34. Our author's reason for denying that the time of the Creation was settled by the free-will of God. 35. An inspection of his explanation why God created the world at that time especially, i.e. that, before the Creation, time did not exist. 36. Whence, for metaphysicians, arises the difficulty of making themselves clear.

FOL. 336 1. The principal topic in the present Problem is not the one stated, but is this: That God understands and wills. As is fair in every demonstration, [White] tries to show this through an explanation of [terms: here] the terms 'intellect' and 'will'. He therefore describes for us what the human intellect is, and likewise what the human will. He is, however, unaware that, if he has reasoned correctly, we must necessarily infer: 'God has a human will and human intellect'. This, I think, runs counter to what he has said [before]. I shall scrutinise his proofs; but first I shall also put forward something on the nature of the soul's faculties myself. This I shall do very briefly; but to start with we must examine the metaphysical argument by which he proves that *ens a se* is something of itself and the cause of everything else.

FOL. 336v 2. In propositions, the predicates are sometimes accidental; for instance, when we say 'The man is white' the predication is accidental,⁴ it not being *necessary* that the man be white. From this, [our author] assumes—how, I do not know—that those accidents unconnected with the essence of a thing are merely arrangements [*dispositiones*], or steps in a certain order that relate to the existence of *ens*, and that, of those steps, the more relevant any one of them is, the more essential [the accident] is.

---

³ Reading *alternis* ('de appetitu & fuga') for *alterius* ('Why, of two things, one attracts and the other repels').

⁴ 'Accidental', concerning accidents, or things that happen to bodies. 'Predication', an assertion concerning the subject of a proposition.

CHAPTER XXX: 2

Next he says: 'The very reason for existence chiefly applies to *entia* whose entity is the same as their existence'. He infers that existence is the most essential [*sic*] part, difference, or notion of such *ens* (i.e. of *ens* whose entity is the same as its existence).

But these ideas are what he had established in the previous problem as an apparatus to an apparent demonstration, and from them he [here] deduces that nothing essential can befall existence. He argues as follows: If there are more than one *ens a se*, *ens a se* will be a class [*genus*], and will be said of several things[5] essentially different; but the existence of *ens a se* possesses no essential differences; therefore, *ens a se* is not a class and hence will not be said of several essentially different things. Therefore there is no more than one *ens a se*; there is one only and, in consequence, all other *entia* depend on it.

What should have been shown, then, has been shown, though very metaphysically. He says: (1) Accidents are previous dispositions to existence, i.e. accidents are present in something not yet existing; (2) Existence is very much connected with the *ens* whose essence is the same as its existence, i.e. the same thing is related to itself; (3) Existence is a part of *ens*, i.e. because every part of *ens* is *ens*, existence is *ens*–just as if he said 'to be' or 'to exist' is *ens* or 'existing', or that 'to run' is 'running'; (4) Anything essential cannot, he believes, be added to existence–as though there cannot be attached to an existing seed the essence of a tree or to an existing tree the essence of a bench; (5) 'Essence'[6] is the same as 'existence' in *ens a se* (i.e. in *ens* that is everlasting), but not in *ens ab alio* (for it is clear that whatever *is* or is *ens* also exists or is existing, and that, conversely, whatever exists is *ens*); (6) Because *ens a se* possesses no essential differences,[7] one cannot, he infers, apply that term to several things that differ in number; because it had to be proved that *entia a se* are not plural in number, for this is the meaning of 'to be the only one of its kind'. (7) (I omit the metaphysical crudity [of his argument]): all his narrative consists of words which, being used only in the schools, should first have been explained and defined. [I say, however, that] they have never been defined either by him or by anyone else; which [shortcoming] is against the nature of [philosophical] demonstration.

FOL. 337

FOL. 337v

---

[5] Or, 'of several *entia*'.
[6] 'Essence'–Mersenne; 'Existence'–MS.
[7] *Differentiae* here seems to signify 'means by which it can be differentiated from *ens* proper'. Cf. the preceding folio.

Moreover, there lurk within his mode of arguing other flaws, and these are not infrequent. It is enough, however, to have noted the few I have dealt with.

I think that, in proving truths in this fashion, we must proceed by the following, a preferable and more Christian method: 'There is no reason why the Church should have wished to deceive me. Now, she has received without controversy these dogmas from the Apostles that handed them down, and there is no reason why the Apostles should have wished to deceive the Church of days later than their own. They held the same dogmas that Christ Himself taught; and whatever He taught, He confirmed with miracles that supported [it] to the full. Since, therefore, the dogmas of Christ were proved and transmitted from His times to our own without deception, they must be true'. This reasoning [of mine], one concedes, is not demonstrably very accurate, being a reasoning partly dependent upon faith; yet it approaches far more closely to the laws of [philosophical] demonstrations than does any argument we have so far heard from the metaphysicians.

FOL. 338

3. About to enter on the disputation concerning the Divine Intellect, [our author] questions correctly, saying on his page 307: 'Tell me what "to understand" is'. But he continues, incorrectly: 'We must enquire among matters proper to man [only]', for he himself does not think that the Divine Intellect resembles the human. Since, however, he tells us to enquire concerning the latter, we must now do so.

We must begin with *sense*. If it indeed be sufficiently accepted and familiar that there is nothing in the human intellect that was not previously in the sense (for sensation takes place through the action of objects even, as he says, upon the sensoria or the organs of perception), then every action and every being-acted-upon is motion, as was shown above. Hardly anyone denies this, except someone accustomed to philosophise, not from the contemplation of things themselves, but under the duress[8] of [using] propositions already rashly conceded.

Perception is therefore a motion in the inner parts of the person perceiving, and is brought about by the motion of the object acting on the sensorium. Hence vision is produced by the shining body's motion propagated through a diaphanous

---

[8] This last word, a contraction, is doubtful.

medium and continued through the tegument of the eye (itself also diaphanous) to the retina or base of the eye, and next through the optic nerve into the spirits or the material—whatever it be—in the brain, and, by way of the matter that reaches as far as the spirits, to whatever the matter is in the heart. Likewise the advance or continuation of motion is produced from a medium that is compressed between two bodies which collide, or that infiltrates between two bodies suddenly dragged apart; for that medium which is moved pushes or drags the medium it is in contact with, from which the continual driving or seizing is propagated to the ear, thence, by the nerves, to the brain, and finally to the heart. In scent, the [extra-bodily] medium is moved as far as the lining of the nostrils which is composed, all of it, of minute nerves; the motion is continued thence, as before, to the heart. In taste and touch the same thing happens: the nerves are pressed and agitated, and then the motion is propagated by the nerves between [the skin and the brain] to the brain, thence as far as the heart. Every motion in the heart assists or hinders the vital motion—indeed, any motion coming upon any [other] motion either strengthens or weakens it. What kind of effect results [from this] I shall state later; meanwhile we should remember that whatever is acted on or moved reacts or resists at the same instant. It must follow from this that the action or motion in every act of perceiving is propagated and continued in reverse towards things outside [the percipient]: from the heart to the brain and thence into the nerves as far as the outer surface of a body, from which reaction arise two principal effects. One is fancy, which is none other than a motion in the brain, but yet appears as something external, making us see things in places where they are not, such as stars beneath the water, and [hear] a voice where there is an echo. The other effect is that the observer's body and its thicker parts vibrate to and fro in sympathy with the fancies, the nerves being the connecting link. This motion is termed animal motion. In sum, let the gist of the present article be this: perception, and every mind-picture, is a motion of the very kind we have spoken of.

FOL. 338v

FOL. 339

4. The motions thus excited by objects remain, even when the object is removed or in any way ceases to act; for just as, for motion to be produced, an agent is required, i.e. an external mover, so also an agent, i.e. an external mover, is needed if

FOL. 339v

something in motion is to be brought to rest. Rest generates nothing; it does not change anything within something else, because nothing can annul a motion except an equal and opposite motion. What can arrest the motion of liquid bodies (such as the spirits within animals; yes, the blood itself), except gravity in the way in which the motion of the air and of water is stilled, it is hard to imagine. But neither the spirit within an animal's body–at least, within a body that is active–nor, perhaps, the blood, gravitates at all; but it flows round in a kind of circular course to serve the several limbs. There is, therefore, no reason why the motion in which consist the fancies should be halted, except after a long time.[9] So, when a pebble is thrown into water, the water set in motion does not stop moving as soon as the pebble rests on the bottom. We must hold similar views concerning the motions of the spirits–motions aroused within the brain by objects; that is, these do not cease as soon as the object is removed or otherwise ceases to act.

So, from the many motions generated as I have described, the one dominant in the heart [at the present instant] constitutes the impression within us [*praesens phantasma*]. This last, so long as the object itself is acting, is called by different names, because of the difference in the organs through which the action was created. If induced through the eye, it is called light or colour; if through the ears, sound; if through the nostrils, scent; if through the tongue, taste; if through the surface of the body only, it is called hot or cold, rough or smooth, etc. The being-acted-on is itself termed perception. But if the object that is the agent is removed, and the same motion or impression [*phantasma*] remains, it is customarily called 'a mind-picture'. So 'a mind-picture' and a perception are the same thing; and as long as it is concerned with the object, 'a mind-picture' is called sense. If, however, [the object] moves away, [a mind-picture is called] by a name taken from the [mental] images, i.e. a notional mind-picture [*in visione, imaginatio*].

FOL. 340

5. But a 'mind-picture' is motion, and every motion consists in a succession; so the mind-picture must also consist in a succession and must always contain something that precedes the present time. For this reason, when we are *not* considering the past, we call the motion a mind-picture, but every time we *do*

---

[9] Mersenne changes to: 'except for a long time afterwards'.

CHAPTER XXX: 7

wish to think about the past, we call the same motion 'the memory'. Likewise a mind-picture of the past (leaving out of consideration its 'impression'), namely an imaginary succession, is called 'time'.[10] Hence [in my scheme] one and the same motion of the mind has now received four [different] names for four different points of view.[11]

6. Here we must not pass over the reasons why all the images and impressions are vivid during perception, but afterwards, when they are called 'mind-pictures', i.e. when the object that aroused them moves away, they are very dim. This occurs because fresh objects take the place of [the first] and act not only on the organs of one sense but on those of all the senses, and cause the motions to be no stronger at that time than were the motions previously aroused. Hence [the earlier motions] do not vanish through time, but by comparison [with the present ones] they are hidden. The evidence for this is that, when all the external sensoria are shut out and every approach to the brain has been closed except to motions proceeding from the heart (which can happen only during sleep), all the mind-pictures are equally as sharp (namely, when we dream) as they were in perception itself. We must also note that the impressions which people have when they sleep have originated in perception and are [in fact] these same motions.[12] Therefore 'a dream' is a second name for 'fancies'.

FOL. 340v

7. As in every liquid whirled round by different motions, so also in the spirits and the brain it necessarily happens that two or more motions are compounded into one, i.e. of the 'impressions' (these being the motions under discussion) two or more form a single one. This happens when, after seeing a mountain and some gold, we dream of golden mountains, or after seeing a man and a horse we visualise a centaur. The noble acts we have seen or heard of in heroes or others we can, with empty glory, assume as our very own, and in this respect we

---

[10] Hobbes means that we abstract the sense-impressions (or images) and retain only the succession.
[11] I.e. in accordance with four different functions performed by the same motion within the mind. These names are *phantasma, imago, imaginatio* and *memoria*.
[12] 'These same motions' are those produced by an external object, which have been transmitted to the brain and eventually to the heart.

can assign to mind-pictures a second name: 'fictions, or figments of the mind'.

FOL. 341

8. As in every liquid, so also in the motions of the mind it is reasonable to suppose that a part that is moved shall draw an adjacent part. Just as, wherever you draw with your finger one portion of [a film of] water freely spread over a flat surface, the other parts also follow it; so one 'impression' arises from another, neighbouring one. Now, those 'impressions' are adjacent that in an act of perceiving follow close upon one another, each to the rest. Hence arises that continuous chain of mind-pictures usually called 'the discourse of the mind', in which two 'impressions' always cohere that will at some time or other be joined, each to each, in the sense.

9. So this 'discourse' (or collection of mind-pictures) is sometimes orderly, sometimes inordinate. When the latter, it seems as though random, as someone's thoughts might wander from deaf-and-dumb language to [the word] 'faba' ['a bean'], from 'faba' to 'fabula', and from 'fabula' to 'Aesop'; such is the 'discourse' of those dreaming or in delirium. Orderly discourse is that governed by some purpose or goal. Whoever wishes to reach it is drawn either [a] from any point of departure or [b] from a point suggested by his very mind-picture of the goal. For an example of an ordinate discourse that advances from any starting-point, or chance happening, take people who want to find something that lies hidden because it is very small. They start from wherever they please and then progressively scan a whole area. Another instance is in hounds

FOL. 341v

hunting partridges; starting at any given point, they range over all the level land. Likewise also those composing [Latin] verses employ 'discourse' when they look for suitable words to fit the metres being used. Whether or not this kind of discourse of the imagination is distinguished from the others by any special name I do not know.

10. The principle of discourse is derived from the end to which tends the agent discoursing, when the mind-picture of the goal is succeeded by the mind-picture of the path to the goal–the path by which (take the starting-point where you will) the series of mind-pictures is continued through a chain of causes and effects, and this either from cause to effect or from

effect to cause. If the progression indeed proceeds from the mind-picture of a cause to the mind-picture of an effect, and so thus to the goal (which is always the final effect), the mind's discourse is called a collecting-together, a 'synthesis'. If, on the other hand, it proceeds from effect to cause and hence to preliminaries, it is called an unloosing or 'analysis'. Either of these is termed 'recollection'.

An example of 'collecting-together' in humans is when they visualise building [something] according to an order [of events][13] from the material to the form of the house they propose bringing into being; for the mind-picture moves from the material to its transportation and from there to the foundations, thence to the walls and from there to the roof. An example of similar discourse in brute [creatures] is birds' nest-building. An instance of 'unloosing' among men is when the thought advances from the form of a house to that of a site where it is to be built; then of the material that has been brought together in that place; then of the actual transportation [of the material]; and then of the place it is got from. In beasts we have an example of the same 'discourse' in the same birds when, reasoning back from their young to [the production of] eggs, a nest, and [its] material, they return to a destination [previously] designated.

FOL. 342

As existing in man, these faculties differ from the same ones in beasts only in the degree and the speed of thinking. As often as we think of the end, this same 'recollection' of the means to an end would run through the means in the same [given] sequence. When it reasons from cause to effect, the imagination might be called an 'art'; conversely, its proceeding from effect to effect might signalise it as 'the science of causes'.[14]

11. Now, just as there is memory of the things we have perceived, so also there is memory of the succession from one thing to another, or of event to event, or from antecedent to consequent. This is usually called 'experience'. He who has noted and remembered many experiences, i.e. many consequences of things, is said to have much experience of things. Again, someone may have already experienced many conse-

---

[13] 'Ordo'. Hobbes intends to convey more than the idea of 'sequence'; he may intend a connection with '*ordinatus/inordinatus*' above. Cf. fols 111v, 348.

[14] For 'means to an end', perhaps: 'means to this end'; and for 'an art', perhaps: 'the art (of causes)'.

FOL. 342v

quences similar [to one another]. On seeing an event [*eventus*] resembling a past event, he thinks that the present event will be succeeded by another event which resembles the consequences of the past event. On the other hand, say he witnesses some event *like one resulting from* some previous event. He will also think that [the event which he witnesses has an] antecedent which resembles a past antecedent.[15]

Let me illustrate. Someone who notices a dense cloud expects rain because he has seen that previously it often or always happened that [the appearance of] a cloud was followed by rain. Likewise when he sees that rain has fallen, the same person will consider that there has been a cloud [over that place previously], because he recalls that this [example of cause and effect] has occurred before. The explanation must be that a concept [*imaginatio*] of the future is identical with a concept of the past. But we believe or suppose that this order[16] is related to the present. This we do by assuming that similar occurrences [*eventus*], namely a past and a present one, are not [merely] similar but are as though the same thing in number.[17] Hence we believe or suppose, [wrongly,] that an occurrence which in fact has preceded a previous one seems to follow [it]. Whatever is put after the present, however, we call 'the future'.

12. But as the memory of what we have undergone is called 'experience' in respect of the past, so the same 'memory' with respect to the future is called 'expectation'. Therefore 'much experience' is the same as 'prudence' or 'foreknowledge of the future', which cannot exist without experience. Hence, other things being equal, old men are always more prudent than younger because they have more experience. Likewise, other things being equal, the quick-witted are more prudent than the slow because they notice more and therefore also have more experience.

FOL. 343

13. Now, some have observed and committed to memory a resemblance [occurring] within a sequence of events. Such persons call the antecedent of a consequence and the conse-

---

[15] Cf. fols 344, 432v.
[16] See note at fol. 341v.
[17] 'Number'. Probably, 'the property of things according to which they can be counted or enumerated'; but cf. fols. 108f. and 248.

370

quent of an antecedent, 'a sign'. To those who recall that things resembling the antecedent have always or nearly always followed those resembling the consequent, 'a sign' is both the antecedent of a consequent and the consequent of an antecedent. For example, a cloud is a sign of future rain, and for those who have found that the usual sequel to a cloud has been rain, rain is the 'sign' of a cloud that preceded it. In this way a conjecture of [what] both the future and the past [hold] is made only through signs.

From this it is also gathered that, to those trained to see it, causes and effects are 'signs' reciprocally, owing to one thing's following upon another. To this extent, therefore, the faculties of men and those of the other animals do not differ. The other animals feel, think and imagine, and from imagining one thing they are led to imagine another (which is 'to discourse'); they also dream, as men do, and men in the grip of disease rave as they. When not afflicted by sickness, animals remember things; and they collect and unloose their 'impressions' [*phantasmata*] as men do; that is, just as men['s minds] range from a building through all the means [of constructing it] to the place where the building-material can be got, and thence back again to the building, so also other animals let their thoughts wander from their nests and homes to the material of these and thence back to their homes. Many animals, apart from man, possess also a memory of the past and a foreknowledge of the future, namely experience and prudence. For this reason they love those who look after them and they fear others, i.e. they expect future evil from those at whose hands they have suffered ill-treatment before. By signs, too, they judge what is to come, just as much as men do. Lest, by [our] recalling the known skill, craft and premonitions of many beasts, their prudence be esteemed greater than [that] of man, however, none of the aforesaid faculties is to be termed either 'intellect' or 'reason', for these rightly befit man alone. Therefore we must see, step by step, in what the human mind is understood to surpass that of the beasts.

FOL. 343v

14. Sometimes future events resemble those of the past; sometimes they do not. Conversely, past events are sometimes found to have been such as we have conjectured them to be from present events, but sometimes to differ from these. We must therefore realise that the cause [why past events sometimes

FOL. 344

differ from present ones] lies in the animals themselves, which think that past events resemble present ones when they do not. Animals either do not notice all the occurrences that preceded some event; or, having noticed them, do not remember them; or else they do not correctly compare, [each thing against each,] the things they do remember. The result is this: the consequences of present events do not resemble the consequences supposedly attendant on the past counterparts of present events.[18] Now, 'comparison' is a discourse of the mind consisting in three acts, i.e. three mind-pictures that are continued among themselves. Of these acts the first relates to one thing, the second to another, and the third is the mind-picture which refers to the difference between both [the others]. It is not, in fact, difficult to compare, each with each, the things we experience directly one after another, because both things compared are seen at about the same instant.[19] Hence the difference between them must also be apparent. So the beasts also set one present event against another; but it is impossible for them, lacking the aid of artifice, to compare past with present events owing to the dimness and wavering of the concept.[20]

FOL. 344v

The limits of any magnitude cannot be recalled without a permanent and perceptible means of measurement, nor can colours be recalled without a perceptible and permanent criterion; and without some perceptible and permanent 'note' no image or fancy can be conjured up that resembles a past one. By 'note' I mean 'a thing that is perceptible by a sense and is permanent. It may be the thing itself–or at the least [a thing] like it brought forward by our own choice such that, on our noticing it, there is aroused a fancy similar to that fancy we wish [to revive] of the past'. For example, an asterisk and characters inserted in the margins of books are 'notes' of a place we want to remember, and stones are 'notes' of the ends of paths or fields because they indicate the length of a path or field–limits previously marked out. Hence letters are also the 'notes' of words, and names are the 'notes' of the things they are assigned to.

[18] Cf. fols 342–342v, 432v. Text: 'Present events are not succeeded by events resembling those previously thought to have followed the counterparts of present things'.
[19] For 'compare' (*comparare*), perhaps 'match, pair off'.
[20] Hobbes probably means the image or 'fancy' aroused within the imagining faculty, not the latter itself. See also fol. 8v.

15. Now, the nature of man is seen to surpass the common nature of animals in two ways.

(1) It has been able to find 'notes', of whatever kind they be, in order to aid the memory. That any animal other than man does so has not yet been recorded. Although it is quite feasible that birds in returning to their nests and beasts to their lairs are guided in their passage by a recognition of the perceptible things they have noticed [earlier] in it, and that in respect of animals those perceptible things may therefore be called 'notes', yet beasts seem never to have made for themselves any note by design. We see that certain animals often lay by for themselves what remains after [satisfying] their present hunger, and this shows prudence; yet no-one has ever seen them noting a place in order to find it again (which alone would show artifice). The reason for this seems to be that some inquisitive persons have found no satisfaction in enjoying nature unless they have scrutinised her closely and known the causes of everything. They have seen, however, that this can be done only by comparing things, but that a comparison is drawn not from among [in] things themselves but in their 'impressions', and that the latter can be compared only by being recalled. So they have devised 'notes', especially *names* which, in the place of impressions that had disappeared, would suggest ones like them.[21] It is probable that beasts cannot do this because, owing to their physical constitution, they possess no pleasure other than the carnal, by which they would be able to be concerned with their impressions, or to distinguish their impressions from the actual things these are the impressions of, or to remember them [i.e. the impressions].

(2) The nature of man surpasses that of all the other creatures in every faculty dependent on the use of names; what such faculties are like I shall show next.

16. To proceed,–a name or appellation is a human sound. Say a person has something in mind, of which he retains some mind-picture. He applies to, or imposes on, the thing the human

---

[21] It is not clear whether Hobbes intends his adjectival clause as continuative or whether as restrictive, i.e. whether he means *all* names or only those performing the function mentioned. If the former, then a comma is needed after *names*.

sound as a 'note' enabling him to conjure up a similar mind-picture.²² The consequences are three:

(*a*) All things that in any way resemble one another have a certain single name in common. For instance, given any number of things resembling one another as regards colour, and one of them is named 'white' because of its colour, all the remainder will be said to be white. That name 'white' is assigned because this is the colour of all the things severally that are, as to colour [*eatenus*], alike. 'A common name' is therefore the same as a universal,²³ because things given names are individual, but a single name is called 'universal' because it is assigned to universals.

(*b*) It follows that the same thing has names almost beyond number. Anything that is compared with countless [other] things will resemble some in one respect, others in another. So the thing will have a name in common with each thing resembling it in every several comparison; it will therefore have as many names as there are ways in which it may be compared. Thus a man will have the name 'man' in common with the rest of men; 'father' in common with others who have offspring; and 'body' in common with everything that has three dimensions, etc.

(*c*) Therefore when we add the negative particle 'non' to any assigned name (termed 'a positive name') we form a 'note' of dissimilarity or of diversity. Such names are called 'nonfinite names', such as 'non-man' or 'non-animal'.

17. When two names are coupled by means of a verb we make a proposition, by which we note that the latter or consequent name matches the same thing as the first or antecedent name does. So when we say 'Man is an animal' this is the same as saying 'What is called "man" is also called "animal"'. Now if those names are indeed so chosen that the name 'animal' is assigned to some single thing called man, then we say that that proposition is true; otherwise it is false; for truth and falsehood are the same as a true or false proposition.

²² Literally: 'A name or appellation is a human sound which a person applies to, or imposes on, any thing he is visualising, because of some mind-picture he retains of this thing, as a "note" by which he can conjure up a similar mind-picture'. Whether 'as a "note"...mind-picture' here goes with 'applies' and 'imposes' or with 'retains' is uncertain.
²³ 'A universal thing' or 'a universal name'? Cf. 10v–11v, 109v.

18. Again, when two propositions are joined, having one name in common to form a link between the names not common [to both propositions], we form a syllogism. So by joining the two propositions 'Man is an animal' and 'An animal is body' we create a syllogism or a union of the gist of both: 'Man is body'. This syllogism tells us that the third name, 'body', fits everything the first name, 'man', does, and for that reason the second name, 'animal', fits the same things the first name, 'man', does, and the third name, 'body', fits the same as the second, 'animal', does. If they *do* agree, and for the reason given, then the syllogism is true; otherwise it is a paralogism.

19. Names are found in the way described. Then the propositions taken from the names, and the syllogisms taken from the propositions, are put together. By this means a ranging of the entire mind (consisting in the innumerable acts of thinking about individual things) is changed, when things are compared, into a discourse of language or speech, and is reduced to fewer but universal theorems. So long as the comparisons are correctly drawn right from the start, and the names required in such comparisons are accurately assigned to things and are consistently used throughout an entire narrative as applied, this is a most useful economy. When, however, the same name is given to dissimilar things, or is employed in a narrative sometimes in one sense and sometimes in another, the result is not a short cut to truth but rather error and endless indecision. Not only is the waverer inferior to all other creatures, but also his condition is worse [than theirs], to the degree that error is more reprehensible than ignorance, or falsehood than silence.

FOL. 347

20. We are not the first to devise names; we received them from our nurses, our teachers, our friends, and our associates; and the majority of names have been applied neither accurately not in a constant and fixed sense, but have been used figuratively and catachrestically. Owing to a certain habit of the tongue we utter common speech, a thing compounded of these [names], quickly. This is so marked that, given a point of departure,[24] there is no longer any need for mental effort to produce discourse, nor can there be any thought-process swift enough to accompany our words. So it must happen that people who let

FOL. 347v

[24] Perhaps, for the last phrase: 'after acquiring the rudiments'.

their tongues use the names of things before those [names] have been most carefully pondered, although they may say a few true things, though by chance, [and] even if they themselves do not make mistakes, their speech is at most an incoherent jumble.[25] Hence the speech of those who profess no knowledge–a speech based on things and on practical affairs they know well–is clear and luminous, but the speech of metaphysicians is a perpetual and feverish rambling. To this degree those who reason should make it their business to ensure, before they admit them into speech, that the names [they will employ] have definite mind-pictures attached to them.

21. Next, we are said to understand universals only. 'A universal' is nothing but 'a name',[26] because understanding is not of things themselves but of names, and of language consisting of names.[27] So we are said to understand a name when, on hearing or reading it, we recall the mind-picture governing the [particular] name applied. We are said to understand a [philosophical] proposition when, on listening to it, we call to mind that its subject or antecedent name is contained in its predicate or consequent; or when we remember that the later name fits everything the first name does. In the same way, we are said to understand a longish speech made up of propositions when we call to mind a succession of consequents derived from the very nature of [these] names,[28] or we connect an order of words with an order of mind-pictures that have determined the choice of such names. We are indeed said to imagine and remember a thing, but not (unless catachrestically) to understand it. When I hear someone speaking about goats and sheep, I understand, not [real] goats and sheep, but the speaker's *words*; and someone who is said to understand 'book' understands, not a book, but the topics it deals with, which is to understand the book's words.

FOL. 348

Now it is clear that all animals except man lack [the faculty of] intellect because they lack [the knowledge of] names and speech. Suppose the intellect to be a faculty, as our author believes it is, which contains [i] heaven and earth and every single thing

[25] Hobbes's syntactical tangle is here retained.
[26] Or, 'There is nothing universal but a name'.
[27] Reading the contraction *orēs* as *orationis*.
[28] Or, '...when we call to mind, through the very nature of names, a succession of consequents'.

they contain, and [ii] a series or succession of all happenings from the time when the earth began. In that case I do not see how even the rest of the animals will not possess intellect; for a horse has within it everything [necessary for the functioning of the intellect] as a man has, i.e. by means of 'impressions'.

22. Again, the 'reason' is nothing but the faculty of syllogising, reasoning being merely a continuous linking of propositions, or their gathering under one head, or (to put it more briefly) 'the calculating of names'. If these names are the names of numbers, one becomes knowledgeable in the part of philosophy called arithmetic; if of magnitudes, in geometry; if of other things, in other parts of philosophy. It must also be understood that there is 'right' reasoning. 'Right reasoning' is that reasoning which, commencing with an accurate explanation of names, proceeds by means of the syllogism, or through an unbroken linking of propositions that are true. 'Right reason' (if such exist) is the potential to do, or the faculty of doing, this as often as we please: and herein lies the infallibility of reasoning. Now, as I have said, it is doubtful whether the reasoning of any man can always be right, but everyone thinks that his alone is. Certainly those who live in a body-politic and are placed under its laws by virtue of right reason must, in the interests of peace and the public good, consider the civil laws as concurrent with right reason. This is all that need be said on the intellect and the reason, i.e. on the final goal of the critical faculties.

FOL. 348v

23. If we are to know what 'the will' is, a few things must also be said on the faculties that move us [to act]. I refer, therefore, to what [White] has laid down [as I report it] in Article 3: 'Every motion propagated as far as the heart by the action of objects either aids or impedes the vital motion'.[29] Now, those objects by whose action the vital motion is assisted are called delightful and are said to charm us, but those by whose action the vital motion is hindered are called unpleasant. Each kind [of motion] is given a different name, and for different reasons. That motion which constitutes delight is the principle of the animal motion [directed] towards an object that influences [the

FOL. 349

---

[29] 'Vital motion': Article 3 begins (fol. 337v) with a reference to White concerning *intelligere*; but the material concerning the action of objects on sense, heart, and vital motions seems entirely Hobbes's.

first named motion] and is therefore termed 'appetite'. The disgust which consists in that motion which is the principle of revulsion from an object is called aversion and repugnance. When these same things, namely delight and disgust, are considered as being present, but the urge [*conatus*] to approach [to,] or withdraw [from, the object] is not, they are called 'love' and 'hate'.

Further, delight and displeasure are found sometimes in a perception [*sensio*], sometimes in the imagination. When in the latter they constitute a recollection or fiction; those in the memory, however, are either merely recollections and vestiges of past delights and displeasures, or even expectations of future ones.[30] As was said above [at fol. 342v], this is because expectation of the future is the same as remembrance of the past, but sometimes [the two] are in a perception and in the imagination at the same time.[31] Moreover, [of delight and displeasure,] the one breaks into the other very frequently; they are produced in turn, and with so quick an interchange that the reciprocation seems rather to fuse into something midway between the two than [to be] an alternation. From this two-way motion arise those passions called perturbations of the mind, such as hope, fear, anger, envy, emulation, and remorse; the feelings of those laughing or crying; and other countless ones, most of which lack names.[32]

FOL. 349v

24. Properly, 'good' and 'bad' are applied to objects, [not to persons]; for something that pleases or delights a person or is yearned for by him is said to be 'good' for him. On the other hand, something unpleasant is evil; and we call 'beautiful' that which contains the 'signs' of good, and 'shameful' that which lacks them. 'Good', therefore, and 'evil' are applied relatively and *ad personam*.[33] Hence, because the same things do not please or displease everyone, and because they are not the same

---

[30] Hobbes seems to be using the word *memoria* in the respective senses of 'a recollection' and 'the recollective faculty'.

[31] Perhaps by *imaginatio* in this and the previous sentence Hobbes means 'a mind-picture' rather than the image-making faculty.

[32] Because his use of pronouns does not unequivocally identify the several feminine nouns under consideration—*sensio, imaginatio, memoria, delectatio, molestia*—Hobbes does not make himself sufficiently clear in parts of this passage. 'Names': in Hobbes's own specific sense or in general?

[33] I.e. they are not absolutes, but are secondary qualities present, not in the objects themselves, but in the observer.

thing to the same person on every occasion, the same things are not good or bad, beautiful or repellent, to all.[34] People say, therefore, that such and such is good or evil for themselves, for this person or that, or for the body-politic. They do so, however, *simpliciter* only if (without defining [our terms]) we call 'good' that which pleases the commonalty, such as obeying the laws or fighting for one's country. But this dictum is understood, not *simpliciter* but as the most important particular case,[35] inasfar as the state is a compact by which the citizens may procure, as far as possible, the well-being of everyone.

25. Now, the things that please us are connected with those that are unpleasant in a series so long that we cannot see forward as far as the end of the chain at one glance. Indeed, so tight is the linking that pleasant and disagreeable things (i.e. the good and evil ones jointly) must be accepted or left alone. Then if, in this series, there is more good than evil, a thing is good as a whole, and it is good to lay hold on the thing and wrong to let it go. If, on the other hand, there is more evil than good, the thing is evil as a whole, and it is wrong to seize it and good to leave it alone. We take up, then, and we forsake, according to the distance we see along some determined [*determinatus*] expanse. In the latter there may be more good than evil, but in the part we have not surveyed there may be so much evil that the whole contains more evil than good. (This is because, owing to the connection [described], we have to take the whole.) If so, we seize on not what is really good but what has appeared good. This is what is called seeming good, which is sometimes evil and displeasing.

FOL. 350

26. The result is that someone longs for a thing or is moved to perform some task that seems good to him, but he is repelled and turns tail when he sees ahead what he had not seen before,

---

[34] Perhaps Hobbes discussed his opposition to ethical absolutes with his friend John Selden, in whose view 'We measure the excellency of other men by some excellence we conceive to be in ourselves' (*Table talk*, s.v. 'Measure of things').

[35] The meaning is probably that 'good', defined with reference to the state, is merely the most pre-eminent of all the definitions of the term; but perhaps we are to take Hobbes as meaning '...not *simpliciter*, but as relating to honour'.

a future evil connected with it.[36] On perceiving, however, in the same series [as above] a greater good, he approaches again, and so [the approach and recoil continue] in turn–according as the foreseen good or the evil predominates–until he does, or does not, perform what he first wanted to do. This is called 'deliberation'.

FOL. 350v

Deliberation is therefore nothing but alternate desire and revulsion. So it seems that the term 'deliberation' has been given to this to-and-fro motion because the freedom of a person that wavers in this fashion is limited [*determinatur*] or taken away as a result of his doing it.[37] [Yet the two] do not cease following one another by turns until 'a desire and an unwillingness, either by acting [*faciendo*] or by letting the opportunity slip by' no longer means 'freedom to do or not to do'. Hence the purpose[38] of deliberation is the laying aside of liberty.

27. Furthermore, it is clear that someone who deliberates does not desire both to act and to lose the chance of acting, [these being separately considered], but desires [only] what will seem good at the final reckoning, when the whole set of good and evil things is taken into account. When deliberating, therefore, he does not wish at all; he often advances, it is true, but he turns back equally often, since *he does not yet want to act*, for, owing to his frequent withdrawals, it could equally well be said that he wants *not to act*. When, therefore, *does* he will? What *is* the will?

The will is 'the final act of him who deliberates',[39] and this very last act, if an eager longing [*appetitus*],[40] is 'the wish to do [something]'; if a disinclination [to take action], it is 'the wish not to act'. And just as volition is understood as being absent from someone who hitherto is deliberating, so deliberation is no longer present in someone who has exercised his will. Yet again, though men's testaments are called their wills, they are not really so until they are 'the last', i.e. when the testator has died, whom we understand as deliberating a whole lifetime over the appointment of his heir.

FOL. 351

[36] Or, '...When he sees a future evil joined with something he had not seen before'.
[37] Either 'deliberation' or 'this to-and-fro motion'.
[38] Perhaps, 'end, result'.
[39] This definition is pursued in Chapter XXVII below.
[40] See below, fol. 410v.

28. Hence it is certain that volition may reasonably be sought in the other living creatures no less than in man, for they too experience alternate yearnings and revulsions and at length act, or do not act, in accordance with the urge [*appetitus*] that comes last, i.e. the will.[41]

Some people say that the will is a rational appetite, because they have considered every deliberation to involve reasoning. This is not universally true, because reasoning is based upon universals, i.e. on names, and its task is the collecting of separate truths under general rules. But one can deliberate, without using words, through one's experience of past events, i.e. through a foreknowledge of future ones; and a man does not always need to use speech at all in order to deliberate fully or very fully within himself—but only when he has to calculate something. Speech is needed, however, when he has to deliberate with others or to give advice to those who are deliberating, for I distinguish between 'someone who deliberates' and 'such a person's advisers'. The latter do not deliberate, but to someone who does so they point out that series I have mentioned of good and evil results which depend on what is being deliberated. On seeing this series a person deliberates; and he to whom it is pointed out, that person wills.

29. 'Voluntary' is a name applied to actions only. A voluntary action is one that, to be brought about, requires the [exercise of the] will of its performer. So the wish [to do something] is itself not voluntary, because it is not an action. And no-one can say, without being nonsensical and absurd, 'I wish that tomorrow my desire is so and so', or 'I wish that tomorrow I have the desire to do this or that'; for if I can say: 'I wish to desire something' [I can certainly say] 'I wish to want to desire', and so *ad infinitum*.

FOL. 351v

30. As regards all the motions other than animal motion, something whose movement is unhindered is termed 'free', i.e. it is free on that path [to action] where it encounters no obstacle. A river, for instance, runs freely where unchecked by its banks, though its freedom is not due to any [particular]

---

[41] In the preceding sentence, for 'are called their wills', perhaps, 'are said to represent their wishes or intentions'. Is Hobbes indulging in some deliberate word-play ('will' being equivocal both in the English and in the Latin), or did he prepare his initial draft in English, it being good fortune that the word-identity was reproducible?

spot on them.⁴² But as regards animal motion, a thing is 'free' whose motion suffers no check save that arising from its own wish. Though, then, some person may see it as something fuller and wider, 'liberty' is 'doing what we want to', provided that I say [it is also] 'our desiring that which, in order to safeguard the freedom of the will, we do desire'.

Let us now turn to our author's observations on the intellect and on the human will. Through these he is able to assert that the Divine Nature possesses such faculties.

FOL. 352

31. He declares that the other animals neither speak nor confer together; that they establish neither states and armies nor any mechanical contrivances for any purpose; and that they neither teach nor learn skills, but that all these things are proper to men. After examining more closely (he says) by what faculty men surpass the animals, he sees that, by comparing one thing with another, he infers from these [two] things a third, and so on; but that though this advance is only gradual it is without limit. [Such] a comparison, however, can, [he declares,] be drawn only if both the things compared are near at hand to the person examining them. White's finding is that the ideas of many things, their resemblances or their natures rest, collectively, in man–on whether this applies to other animals he has not, he avers, sufficiently reflected [as to be able to say]. Then he adds that there cannot be more than one body in the same place at the same time, nor more than one colour, nor heat and cold; and that two different images cannot be present in a man together. On the other hand, [he says,] when something is to do with other things as a group, man accepts it as he would the behaviour of one of them towards another; and he adapts one thing to another.⁴³

It will be clear that we can go further [than this], for a third point follows.⁴⁴ To be able [to hold] more than one thing in view at the same time–this, says White, is 'to understand', or is at least connected with a reasoning thing as such. So if he has proved that a number of things as a group (according, that is, to their differences and individual characteristics) can be present

FOL. 352v

---

⁴² Perhaps: '...though it is not free at a particular spot', or 'It is not free because of its banks'.
⁴³ Of dubious meaning.
⁴⁴ Mersenne changes 'It will...further' to: 'Who can go further than this?' The third point follows from 'On the other hand' a few lines above.

in *ens a se*, i.e. in God, it will follow that God possesses understanding.

32. In the above statement [of our author's] some things are true, but most are false. It is true that the other animals do not speak or converse together, or teach and learn skills; but it is false to deprive them of armies and communities.[45] It is also false that, according to him, they do not build machines, for what are spiders' webs but devices for snaring their prey? Next, his claim that by comparing one thing with a second we may infer a third, is untrue,[46] for there emerges [only] the awareness of a third, i.e. [an awareness] of a difference or excess. No difference is created; [a difference] is found by comparison [alone].[47]

It is true that two things are compared only if they lie side by side such that one of them can be examined [a] immediately,[48] or quickly after the other, and [b] before the image fades of the first [of the two things] looked at. What he infers, however, 'So I see that the ideas of many things, or their resemblances, or their natures, rest collectively, in man [alone]' neither follows nor is true. Because several things can be compared one with another only if they are close to hand, he ought to have inferred as follows: 'That several things can be taken collectively means that each can be near the others',[49] so that a mind-picture can flit from one to another—unless by 'being close to hand' he understands that the things themselves which we are comparing, say towers, mountains, sky and earth, are comprehended and contained within a man's skull! What he does infer: that 'several ideas, or resemblances, or natures, rest collectively in man', is untrue; for as regards ideas (if by 'ideas' he understands 'names'), I do not know how 'the names of things that a man compares one with another' correspond with 'man himself', since if a man compares the sky with the earth, the man will not be called 'sky' or 'earth'. If White takes 'ideas' as meaning 'impressions' he himself will assert that in a man there

FOL. 353

---

[45] The original has: 'It is false that (White) *does* deprive them', etc. *EW* III.156 (*Lev.* II.17).
[46] Text: 'Next,...with a second is the third false thing he says'.
[47] Hobbes seems to mean: 'We see only that there *is* a difference', and not that 'we see in what such a difference consists'. By 'excess' he may suggest the fact that x is greater than y, or the amount by which x exceeds y.
[48] '*Immediatè*' may mean, as previously, 'at once' or 'directly, with nothing intervening'.
[49] 'Each...others': 'physically near' or 'of similar properties'.

can be no more than one impression at a time, and this [White does] almost in the very next words, where he says that in our body the mind-pictures follow one another and are not coincident.

Again, as regards the natures of things, I do not think he believes that, if someone compares a bench with a table, such a person has the characteristics [*naturae*] of a bench and a table so that he himself becomes a bench and a table, or that the bench and table are contained in the brain or body of the person drawing the comparison. So it is untrue that the intellect consists in there being present in a man more than one thing at a time, i.e. in our discourse that is communicated from our mind to our tongue. White has been deceived [through taking] the universality of names as [equivalent to] the multitude of things, [which it is not]. The beasts, also, compare things they have seen,[50] and perceive the differences between them; hence the differences themselves[51] were also the several things present in [the beasts] at the same time, i.e. [the beasts] possessed intellect. But for the sake of brevity I leave [discussion of] this out.[52]

33. Having stipulated that the human intellect (for to this point he has discussed the human one [and not God's]) consists in there being present in man more than one thing simultaneously, he then shows step by step that there are more things than one present in God, and this simultaneously.[53] That is to say, God has produced many things simultaneously and directly [*immediatè*]. From this, declares our author, it follows that, in accordance with 'His own nature and distinction',[54] more than one thing is present in God.

First I ask: 'How, [from this,] derives the sequel: "He produced them; therefore they are within Him" (as though the things He created are within Him more than before the crea-

---

[50] The text does not make it clear whether a beast x compares a thing with another thing, or whether beast x and beast y use the same thing, z, in different ways.
[51] Or, 'the things themselves'.
[52] Re-worked by Mersenne: 'If the beasts, also, compare things they have seen, and perceive the differences between them, it would follow from this that the beasts possess more than one thing within them at a time, i.e. that they possess intellect'. Hobbes's tenses are here retained.
[53] For 'simultaneously', perhaps 'as a group' (*simul*).
[54] By this phrase White seems to mean God's *Ichheit*: the qualities that are peculiar to Him and that mark Him as different from man.

tion)?' Perhaps our author will say: 'Everything was within God even before the Creation'. Then, [I reply,] the world and everything it contains existed before the Creation–and this according to their[55] own properties and distinctions, for whatever was in God indeed existed. Hence the Creation is not the passage of things from not-being to being, but their transfer from the place where they lay hidden in God to a place removed from Him where they could be seen.

FOL. 354

From what White says it also seems that men, and perhaps the beasts also, have within them more than one thing at once. So, [we infer,] they lack no faculty necessary for creating the world except the ability to lead forth or to create outside themselves the visible world which, as a whole, they possess within them.

Lastly, if we accept his whole argument,[56] the result is that God understands not only *simpliciter* but also as man does, i.e. through the action of objects[57] and through the meanings of names. Hence God is acted upon by objects–a thing that simply must not be said. Let us say, rather, that God understands [only] absolutely, i.e. that nothing is hidden from Him. If we are required to demonstrate this, let us declare that it is quite indemonstrable, because the way in which God understands passes *our* understanding. Yet we must believe [in it] as faithfully as we believe that He exists.

34. In the remainder of this Problem the disputation is about the way to answer those mockers who ask: 'What compelled the Eternal God to establish the world after so many centuries, He having lain idle for an infinite time before [then]?' The reply which is alone the true, reverent and Christian one, namely that God did it of His own free-will, our author rejects as untrue, and he supplies another one which is metaphysical and false. It is that before the Creation there existed neither any centuries nor any time, and that God is older than the world not in terms of centuries but through the necessity of existing.

FOL. 354v

The reason why, in [White's] opinion, the beginning of the world did not depend on the free-will of God is as follows. Before the Creation of the world, although it is conceded that

---
[55] Perhaps, 'God's'.
[56] Or, 'his argument as a whole'.
[57] Probably in the scholastic sense: *materiae circa quas*.

time and sequence did exist, in no respect were the instants of that time distinguishable (for the variety of things was not yet in being) or capable of being considered as differing from one another. So it made absolutely no difference from what moment of time the world took on [its] beginning. Therefore there was no reason why God should create the world at that time rather than at another. Hence if He of His own free-will had preferred one moment for creating the world to all the others throughout eternity (there being no reason why He should prefer one to any other one), He would have done so without reason. This, [White states,] is against the nature of the will, which he defines as 'Motion, or the principle of motion, in association with reason, not without it'. Hence God could not, from the freedom of His will, have imposed any beginning at all upon the world.

If, however, the Divine Will were similar to the human, i.e. if they both fell within the one definition of 'will', it seems to me that nothing can be more useful in denying the possibility of the Creation than such an argument [as the preceding]. Human volition is a motion, as our author himself admits, and, together with deliberation (he says, 'together with the reason'), it originates in external objects. The action of these creates those impressions that, by causing pleasure or displeasure, form the will and aversion. Hence a man cannot will unless there already exist created things other than himself which make him will. But surely, [I answer,] we shall not say the same of the Divine Will? Was there [present] before the Creation a motion within God, and this generated from perceptible things not yet made? The differences of things antecede in time our will and choice, of which they are the efficient causes. If, then, God created the world of His own free-will, shall we say that the will of God has an efficient cause, and that the differences of things preceded the Divine Will, which is eternal? [No!]

Again, [I say that] when men want to do anything, they want to because it is good;[58] what God does is good because He has wanted to do it; hence there is no reason why the cause of the world's having had a sure and definite beginning could not have been the free-will of God. Next, nothing prevents our saying that the Divine Will itself, being eternal, has totally

---

[58] Whether it is the thing that is good, or the motive in doing the thing is not clear.

lacked a cause. So suppose we say: 'The will of God is without cause. In a man, the desire to do something is not the reason why he does it, but in God the two are the same'. Here we shall make no unreasonable claim–unless it be thought ridiculous, [which it is not] to assert that the Divine Will differs from the human or that the nature of the Divine Will is incomprehensible.

35. That argument by which our author wishes to answer the mockers is a metaphysical one, and is this: 'Let us realise (he says) that [the existence of] God preceded [that of] the world, not by a profusion of empty centuries, but by the necessity of existing'. How are we to understand these words? What about that 'pre' in 'preceded'? Does it not imply both precedence and consequence, i.e. that God came first and the world afterwards? Does not the particle 'pre' denote time *and* sequence? But nothing is said to be 'pre' something else except according to some order and sequence either of things themselves or our knowledge of them; and 'the necessity of existing' does not mean 'a sequence', at least in Latin. That the above words of his are meaningless and incomprehensible he concedes when he adds: 'By means of which apprehensions', (i.e., I think, 'which comprehensions'), 'and in a manner of speaking, this must be adapted to our tongues,[59] to that extent let us dismiss', etc.

FOL. 356

If, however, one allows that before the Creation of the world there was no time, namely that even God Himself recognised no sequence of time (this, in my view, must not be said), how does this bear on the mockers' question? They ask: Why did God not create Himself ten thousand years ago rather than create time about six thousand years ago? Or this: God, who created for us a series of six thousand years of time, why did He not prefer to create a longer or a shorter series? Or again: If the world could not have been created more quickly owing to lack of time,[60] how (they ask) could it have been created then [?when it was created]? For, [they argue,] if God chose to create the world at the instant He created time, He could also have done so for a hundred thousand years before. [In reply I say:] It is absurd to attribute to the Divine Majesty an intellect or will

FOL. 356v

---

[59] 'This': the matter under discussion? See the beginning of the article and fol. 347v.
[60] Is not this wording paradoxical?

of a kind suitable for man, because in men these are indeed effects, but in God they are the causes of everything else.

36. Feeling that his opinion will be inexplicable, [our author] has chosen to ascribe the above theory [of his] to the difficulty of metaphysical speculation rather than to his own ignorance of the nature of the intellect and the will. He says, therefore, that three abstractions in particular make metaphysics difficult: those arising from [a] place, [b] continuous parts, and [c] sequence, or time. The last he puts far in advance of the others in difficulty when brought forward for examination.

FOL. 357

My own view, however, is that the greatest difficulty in philosophising correctly is that the speech of the greater number of those who philosophise is accompanied by no pondering on things. Words accepted without due consideration they commingle and collect into propositions in different ways, until at length these syllogisms seem to mean something carefully thought out. They are totally lacking in meaning, however; so when the philosophers want to explain and demonstrate them to others they do not know by what mode [figura] of speech they must be cast into a form suitable for delivery.[61] This is the difficulty complained about. But White himself creates it, so long as he chooses to speak of abstractions from place, from continuous parts, and from time, and also of things like these, rather than saying clearly: 'Let us consider place, parts, and time'.

FOL. 357v

It is not so difficult to consider place apart from a consideration of body, for any yokel understands that 'place', whether something is actually in it or whether it is empty, is the space that a body can occupy. But if someone tells [him] to think of the same space as abstract,[62] the rustic will think that this person is going crazy. Similarly anyone can examine time–whether something is being done or moved at that [particular] time, or whether everything is at rest–because something could be done and moved at this time; yet one cannot think in non-concrete terms what time is.[63] As regards a continuum, every-

---

[61] Perhaps, as at fol. 356, rather than 'suitable for delivery', Hobbes means: 'to fit the vocabulary of the metaphysics that is disputed publicly'.
[62] Or, for 'to think...abstract', 'to tear a space apart'; see a few lines below. In this article Hobbes seems to use *abstraho* and its cognates both in the primary and in the secondary senses. See fol. 393v.
[63] Or, 'one cannot dissect time, tear it apart'.

one knows that, even though it is finite, there is always some portion that may be considered as cut off from it, and also another piece cut off from that, but not that he may break off, or wrench and tear, part from part for ever.

For his attributing the wildness of his words about the intellect and the will to the difficulty of grasping what time is, or place, or the parts of a continuum, our author gives no reason—and I say 'his words', not 'his mind'. We may the more justifiably ascribe it to the rashness of his way of speaking, for which he has the metaphysical writers to thank.

FOL. 358

CHAPTER
THIRTY-ONE

❧❧

On Problem 6 of the Third Dialogue:

That the existing world is the

best of those creatable[1]

1. The question set out. 2. That it ought not to be disputed. 3. An examination of our author's argument. 4. That 'the most perfect'[2] is different from 'the best'. 5. That apparently it was not an inspection of the archetype [*idea*] of every creatable world that induced God to decree the creation of the present one. 6. Our author seems to feel that God's purpose in creating this world rather than another one was for men to demonstrate His attributes. 7. That our author's reason why finite things cannot vary in ways that are numerically infinite is unsound. 8, [9]. How, in his opinion, there originated 'capacity'[3] in things not yet created.

FOL. 358v

1. 'Creatable' is the term applied, not to a thing already created, but to one hitherto uncreated yet capable of being created. 'Creatable' denotes some potential present in a given subject even before there existed any material at all for [making] the world. Therefore there is no 'creatability' in the world itself either after the Creation or before–not after, because what has been created cannot be created; and not before, because

---

[1] The word 'creatable', as in the *Oxford English Dictionary*, Hobbes uses in a restricted sense here.
[2] *Perfectissimus*. Perhaps, 'completely made'. In Article 4, below, the word is not always translated in the same way but is inserted in brackets when it occurs.
[3] *Possibilitas*, i.e. a power or ability to act, or be influenced, in a given way. Cf. fols 7v, 363.

non-being has no potential [to become]. It remains, then, that 'creatability' is none other than the potential to create, which was in God always [*ab aeterno*]. So the question thus set for disputation is merely this: Could God have created a better world than the present one? Our author believes that He could not.

It is certainly foolhardy to declare of God that He could not have performed something that was not the best.[4] What, however, does the word 'best' mean? That which pleases someone it does please, everyone calls 'good'; i.e. what pleases him is, [people say,] good for him, and what pleases the community is good for the community. Now, God undoubtedly said that everything He had created was good.[5] We must therefore conclude that He called all things good because they pleased Him as God, or because He created them to His own liking and in the way that He wished them to be made—not because they had to please the men who were to conduct disputations concerning them. Before man was made, all things were good; and just as something 'good' is 'that which pleases', so also something is 'best' that pleases most. The question is therefore the same as asking: 'Could God have created a world that, of all the worlds He could create, would please Him the most?'

FOL. 359

2. In this question I freely confess my ignorance. As I can understand 'to please' and 'to be sought after' only as a kind of motion within the parts of a sensible[6] body, I dare not ascribe to an incorporeal God those terms in a sense in which men can understand them. It suffices, therefore, that God created this world and not another one.[7] From this we gather that no other world has pleased Him more. If anyone philosophises beyond this point, he speculates beyond his own powers of comprehension, and this is reckless.

3. Our author's argument by which he tries to prove that God could not have created a better world is as follows (his pages 315 and 316): 'It is certain that God does not swerve

---

[4] Of all things absolutely? Or, 'of all that He could do'? For 'performed best' the text has: 'He could not have committed anything other than a sin'!
[5] An inaccurate reminiscence of *Genesis* 1.10?
[6] *Sensitivus*, i.e. 'of a body that can perceive or be perceived'.
[7] Perhaps: 'It suffices that God created this world, and that no-one else did'.

FOL. 359v

from His own virtue [*virtus*] or reason, for the very essence of these exists essentially.[8] It is also clear that, just as it prefers good things to bad, so, if it is permitted to, reason chooses from among several good things the best of them.[9] Now, a reason would not be a right reason if it did not select, from the things that could be done, the best one to do. Likewise an action not in accordance with what the reason has repeatedly said should be done, or is the best thing to do, is not a reasoned action. Certainly the action that did not create a most perfect world must proceed from a principle other than the reason.' Undoubtedly, therefore, however learnedly they are expressed, we must examine the topics briefly, individually, and in the order in which they occur.

We call 'virtue' that disposition by which we act in accordance with the laws.[10] Our author wishes it to be certain that God acts subject to the laws *He* must comply with. Or if, as some people would have it, virtue consists in a kind of middle state of the passions (or the mind's perturbations), then White will say that it is certain that God acts according to some mean within His passions.[11] But the [faculty of] reason is the ranging of the mind or a sequence of mind-pictures whereby, according to our author, God perceives everything: not (as most people

FOL. 360

say) intuitively, i.e. by a single intuition,[12] but by reasoning, as men are accustomed to do. But why does God hold rigidly to virtue and to reason? Because, says our author, God is 'essentially the essence of virtue and reason'. So God, it seems to him, is a mean between two extremes,[13] or, at the least, *is* the quality of obeying the laws (for it is correct that the essence of virtue is either that mean, or the quality of obeying the laws). Likewise the reason being a discourse [of the mind], God (who White says is reason itself) will be a discourse!

What our author asserts: that the reason chooses good in

---

[8] Or, 'as an essence'.
[9] Or, 'God's reason chooses...' Hobbes here omits a sentence from *DM*: 'The same thing is best and more to be done if the decision [to do it] arises from all the factors connected with the action [of doing it]'.
[10] Or, 'with laws'.
[11] White's book on the middle state of souls (1659) aroused opposition within his own Church.
[12] Probably in the scholastic usage: an instantaneous vision and knowledge of an object. In particular, the word was used of angels and of spiritual creatures; it is translated as 'contemplation' at fol. 411v.
[13] Cf. fol. 435v.

preference to evil, is true; but it is not the reason alone [that does so], for even the beasts act thus, and they in fact are denied reason. At last [he concludes] [a] either that God has created the most finished [*perfectissimus*] world of all He could create, or that God's action proceeded from a principle other than the reason; and [b] that it is absolutely disgraceful to assert that God acted contrary to reason, or without right reason, in anything He did.

But we shall declare–correctly, I think–that the principle of God's actions is the Divine Will. In God the will does not follow the reason, as in men:[14] that men will something because the reason keeps saying[15] that [this thing] is good does not cause God to will something because it is good. This must be stated the other way round, however, i.e. that the action is good because He willed it. The principle of the human will, namely a decision of the reason or of the senses, is laid down; but an eternal and unchangeable will, such as is the Divine Will, can have no principle.

FOL. 360v

4. We must not pass over the fact that he here changes the word 'best' into 'most perfect';[16] for if the world were not very perfect [*sic*] God would not have completed what He had set out to do. All the things that are done, [says our author,] are performed step by step; at whatever stage you stop [to view it], a thing is said to have been 'made' up to that point, but not yet 'completed' [*perfectus*] because it has not yet been made according to its maker's idea and will. [I answer:] Any work is said to be 'completed' when it has been made according to its author's opinion and according to the form preconceived in his mind.[17] Therefore we call those animals 'not fully made' [*imperfecta*] whose body (or its faculties) is incomplete in some aspects that are very rarely absent in other animals of the same species;[18] for it is generally held that nature has been as wil-

---

[14] 'Follow' (*sequitur*) means 'is a consequent of' or 'conforms to': which is intended here is not clear.
[15] Or, 'the reason recommends'.   [16] Cf. fol. 81.
[17] 'Up to that point': i.e. at the stage when one looks, rather than 'to that degree (of construction)'. 'According to...in his mind': either, 'according to the opinion of the author of the action, i.e. of the person performing it', or, 'according to our author's (i.e. White's) opinion', i.e. that the work is 'completed' when it is made fully in accordance with its maker's views.
[18] 'Body': singular also in the text. Hobbes's general sense is here translated; his expression is syntactically and grammatically suspect.

FOL. 361

ling to endow any individual animal with limbs and faculties as much as she has been accustomed to do for most creatures. The only person, then, who can deny that the world is made completely [*perfectissimus*] is he who at the same time wishes to deny that it was made as God wished it made. But could He have made a greater habitation, resplendent in more stars [than actually are], or inhabited by more prudent and better animals, or resplendent in wiser people, or in more religious persons who would have pleased God Himself more? That is, could He have made a better world? All this is open to doubt,[19] unless the world is to be considered the best because He made this one rather than another.

5. The ideas of all the worlds He could have created are within God, says our author, for these *entia* he calls 'the offsprings of the Divine Ideas'. An examination of these ideas led God, he thinks, to choose this world He made as being the best of all worlds. This, it seems to me, denies the Divine Omnipotence, because if the ideas (in God) of the worlds were indeed finite in number, the number of creatable worlds was also finite. It follows that God could not have created another world beyond this finite number. To do so would evidence a limited potential, because if the number of ideas were [not finite but] infinite, and some were better than others, [such a world] could not in consequence be called 'the best', and the existing world would not be the best of those creatable. As, when we are dealing with infinite quantities, none is the greatest, so, in the gradations of good and evil, if the gradations are infinite in number no gradation is 'the best'.

FOL. 361v

6. Could God, from a scrutiny of infinite ideas, have created a better world? This, [our author] says, must be examined in the light of the purpose [in creating a better world], and not in relation to the number of things creatable. These, being infinite in number, are not—even if you were to compact everything into a single entity—equivalent to God. (Here we should note, by the way, that [our author] does not see that things infinite in number cannot possibly be reduced into one mass [*corpus*].) Now, suppose someone says: 'The compacting of these into a

---

[19] Mersenne adds: 'or, rather, ought to be denied'. In the sentence preceding 'wiser' may refer to 'animals', not to 'people'.

CHAPTER XXXI: 7

single thing evinces [the existence of] a God more than the present world does.'[20] White declares: 'Someone [else] will say in answer: "To what purpose? For the benefit of men who live this mortal life? To such men, [this world] which we see is for the most part beyond them [to explain], as we cannot expound by demonstration the least part of the Divine Works. Then [is it] for the benefit of those who are separated [from this world],[21] i.e. for the benefit of spirits? But to these is promised the very sight of God, which surpasses all expression. Hence to no purpose will there be some world better [than this one]"'. Therefore the world around us is the best of all the worlds God could have made as regards His purpose. Though this be granted, we must still ask with what intention He created this world.

FOL. 362

The intention was, in fact, for men to make demonstrations about it. Imagine that there is no purpose in creating a better world, on the grounds that [even] the present world contains *divine works* which we cannot demonstrate in this life and shall not need in the next. Then clearly the present world has been created in order that we may demonstrate *its own qualities* and those of its parts.[22] And would we were satisfied with this, and did not also proceed to demonstrate the properties of God Himself! If the purpose of creating the world were to give us freedom to frame such demonstrations, I would like to know whether that world were not better which was the more demonstrable; but let us press on to other matters.

7. One thing only, says our author, prevents him from demonstrating more fully the matters he has here argued. It is

---

[20] Mersenne inserts at the end of the sentence: 'in order that it may do so'.
[21] Perhaps, 'from the flesh'.
[22] Translator's italics; and these two sentences have been re-phrased. A closer translation would be on the lines of: 'Now if indeed no purpose exists why a better world might be created' (or, 'that the present world might be improved'), 'because the present world contains divine works which we cannot demonstrate in this life and shall not need in the next, it is clear that the world has been created with the intention that we might demonstrate', etc. It is unclear whether this clause of reason attaches to what precedes or to what follows. '[Even]' tries to account for the curious(?) two negatives. For 'demonstrate', perhaps 'understand, describe'. 'We shall not need in the next': because we are there promised the contemplation, not of His works, but of God Himself. Presumably a contrast is intended between *divine works* and the world's *own qualities*, the latter being the inferior.

395

that he does not understand in how many ways the elements may be combined into [making] the world. He speaks as follows (his page 321): 'Now, you claim that what I was pleading consisted in conjectures, and I do the same with you.[23] Far from me to imagine I have understood all the ways in which the elements can be combined into [fashioning] the world. But I am not rash in claiming that these ways are finite, for the following is taken to be a universal truth: wide differences are never found save by design [sine fine], because, they being [self-] contradictory, every one of them is opposed to every other'.

In these words he alters the question, for initially what was asked was: 'Is the existing world the best of those creatable?' but now it is: 'Can the elements have been better combined into the world?' This is another matter. The first question was about the creation of the elements, that is, 'Could better ones have been created or not?' And (not to limit His potential) 'if God had wished it He could have [done so], and [made] ones infinite in number'. The question now, however, is on the varying in composition of the elements already made; but the ways in which things finite in number can vary are numerically finite. The reason brought forward by our author does not, however, prove those ways finite. What he does is to distinguish names into a class [*categoria*] whereby all things are divided into substance and non-substance. Substance is then divided into corporeal and non-corporeal, and corporeal substance into animated and non-animated, etc., until man is divided into 'Peter' and 'not Peter'. [But I say:] Categories of this kind could consist in the infinity of the world and in the infinite variety of things. There could be substances, non-substances, men and 'not-Peters' all infinite in number, yet without altering to any extent their categories or their use in the sciences.

8. Lastly he sets out in the margin [the heading]: 'That the creatures owe to God their ability to do things [*possibilitas*]'.[24] Well, who has any doubts about this? The way [in which they do so] may be inquired into, however. God did not [first] grant a creature the ability to exist, and then existence itself. If He had done, He would have granted potential to a creature not yet in existence, i.e. to being [*ens*] that was not yet *ens*, which is

---

[23] Or, 'Perhaps you bring my argument back into the realm of conjecture'.
[24] The term is defined at fol. 412v.

ridiculous and is only too obvious a contradiction. But that ability [*possibilitas*] by which something could be created was an ability belonging to the Creator; i.e. it was the Divine power. So it will indeed be rightly said of God: 'He was able to create', for He had the power; but to say that a creature 'can be created' before it is *ens*, and when it is merely nothing, is to speak catachrestically.

FOL. 363v

On his pages 321 and 322 he makes known and supports his own opinion, its import being as follows:

(1) Because God possesses understanding, White infers that all things collectively wait upon Him, with their attributes and distinctions. For this is how our author interprets 'to possess understanding': to consider present things [a] not one by one, but all at once; and [b] many things not by several [acts] but by one act only. 'And this', he says, 'applies also to God, from the very nature of existing existence[25]–a belief the metaphysicians teach to many.' He infers that in God 'to understand' is the same as 'to be'. It follows, [he goes on,] that all things are present at once in the essence of God. Thence he gathers that, just as the creative abilities within the intellect of artists are the archetypes of [their] works, so also they depend upon the intellect of the Godhead.

(2) Our author supposes that it is He Himself whom God loves, and he concludes as follows: 'It is necessary that God love Himself. Therefore the reasons governing these (i.e. governing the creative abilities [just referred to]) also derive their ability-to-exist from the Divine Will. Just as (he continues) –to make [the matter under discussion] easier to understand [by giving an illustration]–a workman chisels the veins of marble to the required shape, not through any manual skill he may have, but because they were of this shape within the marble itself, so also a man, a horse, a lion and a sheep are what they are because they are such in that existing existence, and had been such before this very possibility [of existing] was given them'.[26]

FOL. 364

---

[25] Or, '...of the nature of one that exists'.
[26] The wording and the illustration here may be awkward, but the meaning is clearly that, just as the shape of the statue is the same when it is prefigured in the block of marble and when the artist brings it into 'existing existence', so a man, a horse, etc., are the same when they are virtually, and actually, existing. On 'possibility' (*possibilitas*) cf. fol. 358.

9.[27] This reasoning is faulty because (a) one can show, by the same [argument], that creatures derive their ability to act from man, [not from God]. Say that, for the possibility-of-existing of a thing to be established, it is enough that a person possess understanding and that he esteem himself. Then take a man who has understanding and who, according to our author, considers that several things are simultaneously at his service, and that in addition he very much esteems himself.[28] Such a man will endow with the possibility-of-existing the things he understands.

(b) [The reasoning is faulty] in that White says: 'Of necessity God loves Himself'. Neither Holy Scripture nor, I think, the Church proposes that we are to believe this, and we have no right to attribute to God those things (self-esteem among them) that in man usually become vice. Further, if God loves Himself, it is unworthy to say He does so 'of necessity', for 'necessity' is 'a lack of potential'.[29] We may even come of necessity to think that God loves Himself because arguments persuade us so; but certainly philosophy does not submit to us arguments like these.

FOL. 364v

Again, what *is* 'to love' if not 'to wish well of'? Or, how can someone 'wish well' of anyone except of a person for whom all is not well, or can be not well? Neither condition applies to God. Moreover, White's declaring that, in order to make [it] easier to understand, [by giving an illustration,] that a man, a horse, a lion, etc., are such because, in that existing existence, they were such, even before [they possessed] the very possibility-of-existence, is madness. Whatever God has been able to create, has He not had through all eternity the ability to make it? So could not a man, a horse, etc., have existed from all time? If, then, a man, a horse, etc., were prior to their very power to be made so [*possibilitas*], they were so before eternity! And 'before eternity'—what is this but the language of madmen, i.e. metaphysical? Furthermore, if a man,

---

[27] Article 8 begins on fol. 363, and the present article, number 9, is not separately listed in the table of articles at fol. 358.
[28] Hobbes ridicules White's demonstration by applying to man what White applies to God. 'Present in his mind' (*praestare*) corresponds to 'at His command' on 363v: the things are not present in time and space, but may be given the possibility of existing in time and space. In the case of man the possibility is imaginary: the fact of his having several things in his mind does not mean that he can bring them into existence (at least in the sense in which God can).
[29] Cf. fols 371v, 387ff.

a horse, etc., have existed prior to their very possibility-of-existing, then they were such when such creatures could not exist! Who doesn't see here a very obvious contradiction? We have decided to pass over the obscurity of the language where [White] says that a man, etc., 'are in existing existence', because this obscurity is found only–and frequently–in the metaphysicians. They need it in order to seem to be saying something when they are saying nothing.

FOL. 365

CHAPTER THIRTY-TWO

❦

On Problem 7 of the Third Dialogue:
How it was out of His own goodness
that God established the world

1. 'Purpose'[1] does not mean the same to those talking about man as it does to those talking about God. 2. That philosophy does not teach us that 'God founded the world for a purpose'. 3. An example based on [magnanimity in] humans which our author uses to explain how God, through His goodness, could have created the world, is not applicable to God. 4. That the argument is invalid by which he tries to prove that God established the world through His own goodness. 5. An objection which alleges the contrary, and which he claims to remove, he has not removed.

FOL. 365v

1. Here it is asked: 'To what end has God established the world?' Now, whatever men do they do with the desire of securing something pleasant; and the 'end' they always take to be that which, through the mind-picture [that it generates], moves or urges them to secure it. Yet as soon as they have obtained what they sought, then what was once their goal is no longer so, but they press forward to other things, because in his lifetime no-one is without the wish to acquire things [*appetitus*]. No desire exists except that of reaching a goal, or self-benefit, which people think they can gain through their own efforts. But if anyone ascribed such a purpose to God

---

[1] *Finis*, here rendered also by 'end' or 'goal'. In articles 4, 5 and 6 of Chapter XL a distinction is understood between *terminus* ('end') and *finis* or *scopus*.

CHAPTER XXXII: 3

when He established the universe, clearly such a person has claimed that He has not been the Most Blessed from all time, and that He has appetite and greed.[2] If such a person wishes to interpret 'purpose in God' differently from 'purpose in animals,' i.e. as something analogous[3] and above human understanding, then the present disputation does not pertain to philosophy, or to any natural theology; it is to do with religion, in which case the argument should have been conducted not according to man's reasons but according to Holy Scripture and the decrees of the Church.

2. Every cause, philosophy has ordained, precedes an effect; hence the goal which is the cause of desiring precedes the desire. God's wish or desire to establish the world was timeless; so if He established the world because this was His purpose, then such a purpose preceded eternity, which is impossible. Next, if what our author has tried to prove in the chapter immediately preceding be true–that God was unable not to establish the present world–then God cannot be said to have founded it intentionally. What emerges from nature takes place by necessity, not by design: it is not because a purpose [i.e. of shining] has been held out before it that the sun shines, and if a man falls from some high tower it will not be said he chooses to fall perpendicularly. We believe, it is true, that from the deeds [related in] Scripture everything God performs is for His own glory, but not that we must ascribe to Him a glory such as men pride themselves on,[4] or that we must think that the Creator has to strive to be praised and honoured by His creatures; for that is not to His good but to ours. Lastly, God established the world because He wanted to, and the Divine Will can have no cause. Therefore, properly and philosophically speaking, the reason why God created the world was not 'That His purpose was to do so'.

FOL. 366

3. Our author next tries hard to prove philosophically that God established the world 'out of His own goodness'. Had he used these words in order to show the Divine Goodness to

FOL. 366v

---

[2] Or, '...and [suffers] need'.
[3] I.e. arguing that what applies to God applies also to man, and vice-versa. See below, fols 369, 369v, 381v.
[4] I.e. vain-glory; or is Hobbes taking *gloriantur homines* passively: 'that by which men are glorified'? Cf. fol. 367.

be, not the end [for which] the world [was made], but its efficient [cause], I would have agreed. God's goodness is His benign [*bonus*] will, and 'the will of God' may be taken as being, not the end to which the acts of God are performed, but an *efficient* cause. But by this expression 'out of His goodness' White understands '*a final* cause'. How that may be interpreted forms the subject of the whole Problem itself:

(i) He first brings forward an instance from human compassion. 'Don't you see', he says, 'that we also pity the poor (from which we may expect neither advantage nor reputation), and this for no reason other than that to do so befits a rational nature?[5] Imagine you have with you a man anxious to train his soul in those acts which are in accordance with nature. Don't you perceive that, from the fact alone of his seeking to possess a soul ordered by reason, it follows that he displays much kindness to those very close to him? In like manner I apply this to God, etc.'

FOL. 367   By these words he wishes it to be understood that man's pitying man is due to a final cause; for this reason White uses them in order to prove that God's goodness is the final cause of His creating the universe. The words themselves evince the opposite, however, because if indeed someone who wishes to, and does, help a poor man does so neither for advantage, nor for reputation, nor for any joy [he may derive], he does not do it for the sake of [receiving] any good. Every good thing falls within those three classes, and is therefore quite independent of any overruling purpose [*finis*].

(ii) Whoever thinks that men pity the poor out of a desire to act in accordance with a rational nature is insufficiently aware of the spirit in which men do so. Some do it out of compassion, that is, in order to lessen the sorrow they feel when they imagine a similar wretchedness in their own persons, for the nature of pity is this: it is a sorrow arising from the contemplation of someone else's misfortune. Others do it in order to avoid the solicitings of those who ask. Both these groups, therefore, do it in order to avoid sorrow, i.e. they do it from pleasure. Others, again, do it for vain-glory,[6] namely so as to appear merciful or generous, i.e. for reputation's [*honor*] sake. Finally,

---

[5] Hobbes later stretches the word 'pity' into meaning 'to assist, to relieve'. See also the opening of fol. 367v below.
[6] Cf. fol. 438.

others do it so as to merit the Divine Mercy themselves, this being to God the best motive [*finis*] but nonetheless a useful one.

FOL. 367v

Someone who [pities the poor] in order to exercise his soul rationally has in view another, ulterior motive: either that very useful one of being rewarded by God, or the worthy one of being praised by men. To live by right reason is not the ultimate good which we seek after; it is a means [to achieving this], and it [exists] because of something else which, to the religious, is without doubt the enjoyment of the Kingdom of Heaven, but to others the brief and empty triumph of this world.[7]

Now if, as White tells us to, we must ascribe these motives to God, it will follow that He created the world either to free Himself of vexation [*molestia*] or to secure honour or some other agreeable thing He lacked. To say this in accordance with the correct and philosophical significance of terms, as our author purposes in the present work, is most ill-advised, however.

4. His argument to prove that 'it was out of His own Goodness that God created the world' (his page 324) White sets out as follows: 'Of necessity God loves Himself. It follows that He wishes to have the will to perform things worthy of Him. Since the first of these is the establishing of the world, His desire to do this will stem from the good inherent, not in a world [already] founded, but in a desire to establish the world'.

FOL. 368

We have shown above that philosophy does not tell us: 'Of necessity, God loves Himself'. Even if we concede on this occasion that He does, the sequel speaks nonsensically: 'God wishes to have the desire to perform things worthy of Him'; for what is 'to wish to have the desire' if not 'to wish to wish' or 'to have the desire of having the desire'? But let us take the consequence of 'God loves Himself' to be 'He wishes to perform worthy deeds'. So whoever loves himself wishes of necessity to perform deeds worthy of himself; hence (for there is no-one without self-esteem) all men do things worthy of themselves, which is false. Lastly, as he says that to found the world is worthy more of God than of all others, I ask if his knowledge

---

[7] Perhaps: '...to others of this century (generation, age, spirit of the times) it is a brief and empty triumph'. Probably 'this world' is seen as opposed to the next.

FOL. 368v

of, and his passing judgment on, things worthy or unworthy of God is a philosopher's task. If not, he ought, in this work, to refrain from declaring, as a philosopher, that the world is worthy of God.

FOL. 369

5. Next he sets against his own opinion the following: Creating a world brings no benefit to God. God's will is in accordance with right reason, and as God wishes good for Himself, nothing prevents the will of God from being able to be in accordance with right reason even if God did not create the world. In order to remove this objection White distinguishes the 'good' into two kinds: the useful and the honourable. Here surely everyone would expect him to declare, 'God created the world because to do so was either useful, or honourable, or both'. Why White decided otherwise I would signalise thus: he had said that something honourable is good of itself but that something useful is good, not of itself, but because it is honourable. Therefore, by changing the word 'honourable' into 'perfect' he is asserting that perfection is what we call goodness. Why he does this I do not know. He now postulates not two but three kinds of good: the useful, the honourable, and the perfect. It is to be expected, therefore, that he would affirm: 'God created the world with the intention of gaining, from doing so, either some usefulness, or honour, or perfection'. But our author says nothing of these. Why, then, does he declare: 'To something imperfect, we call useful that which is the cause of improvement [and which], rightly, [is said to be something] good to that thing'? 'Because' [he continues] 'the consequent of perfection is (say) for something to create a thing resembling itself,[8] this consequent we describe as good inasfar as it is the effect of good; therefore, whereas God is indeed perfect in every respect, we must not seek in Him "goodness" in the first sense [i.e. "utility"]. That which exists does not need a cause of coming into existence. But in the second sense it is observed that the term "good" may occur by analogy in respect of God, not as "such a good as appeals, as such, to the will" (for this corresponds only to that which is good *per se*) but as "of such a kind as attracts the will through some other factor".

[8] 'For something to create a thing resembling itself': perhaps the meaning is not that a thing is created but rather that it is brought to resemble itself—in other words, to improve or to strive towards perfection.

CHAPTER XXXII: 5

Thus from [His] pleasure at a will in conformity with His nature God is derived [*derivatur*] to establish the world.'

If, having taken care, someone wanted to guard against his words' seeming to make some sense, I do not think he could find a more ridiculous statement than this, or one more suited to the metaphysicians, who are to be scorned.⁹ For I understand neither what he means by 'good in the first sense [*modus*] or in the second sense', nor what 'to assume a term analogously' is; nor what the good that 'does not appeal to the will' is, and yet the will is drawn to it by something else; nor the meaning of 'pleasure at a will in conformity with its nature'; nor of 'God is derived'; nor how 'the world' can be accepted for 'the heaven, the earth, and the universal world'. I am therefore unable to say whether White has or has not met the objection he himself has raised.

FOL. 369v

Reading *irridendis metaphysicis* for *irridentis metaphysicis*.

FOL. 370

CHAPTER THIRTY-THREE

❧

On Problem 8 of the Third Dialogue:
That God established the world
of His own free-will

1. A definition of 'free' in common use. 2. The true definition of 'free'. 3. On coercion. 4. In animals, freedom is the changing or alternation of appetite, and yet is correctly termed 'freedom'. 5. God acts freely. 6. That God's freedom does not consist in a consideration of several creatable things. 7. That the philosophers' objection concerning the eternity of the world has not been answered, and how it should be.

FOL. 370v

1. Who first defined 'something free' as 'that which, all requisites to action having been laid down, can act or not act' I do not know, but whoever it was, he negatived all freedom; for there is nothing that does not act immediately everything requisite to action is present. It is clear that, in agents that have no will, each one acts with all the potential it has, as fire warms and a heavy body falls as far as it can. Organisms possessing volition [*voluntaria*], however, if they do not do the same thing, do not because they are unwilling to.[1] In this instance volition, one of the requisites to action, is absent. When it has arrived, though, they are unable not to act, since someone who can, and who wishes to, act cannot prevent himself from acting. There are, in fact, many things which a man or any other volun-

---

[1] By 'as far as it can' it is not clear whether Hobbes means 'as near as it can penetrate to the centre of the earth'. 'The same thing' is ambiguous also. It may mean (*a*) 'all do the same thing as one another' or (*b*) 'to warm or to fall'.

tary agent would wish to, but cannot, perform, and which he can do but does not wish to do; but whenever both–the ability, and the desire–are applicable, i.e. everything is requisite, an action necessarily follows. Now, of the things we do not lack the strength to perform, we carry out some but not others; and we perform the same things now but not at another time, so that we appear governed by a kind of chance. The reason is this: we can will some things but not others, and we can will the same things now but not on another occasion. Actions alone are voluntary; passions and faculties, such as feeling, understanding, loving, fearing, wishing, and not wishing, are not voluntary.

FOL. 371

2. Among animals, the one is properly said to be free which, though possessing the remaining power [cetera potestas] to carry out every action, does not yet possess the wish [to act]. Among inanimates, that is called free which is in no degree hindered from doing whatever, from its own nature, it can do. Hence a river runs freely that is prevented from flowing neither by its banks nor by any external obstacle. Nor are those who speak thus unaware that water runs down by force of its own gravity, i.e. by natural necessity; they acknowledge, therefore, that liberty is opposed not to internal necessity but to an external impediment. Likewise also those who allow a man[2] liberty of action are well aware that there are some things that he cannot will (such as those that seem the worst to him) and some that he cannot avoid desiring (such as those things which seem the best that have been done for him); they do not deny, however, that he acts freely and by free-choice. But of free-choice, as of everything else, there exists some cause, and this a necessary one; yet he who necessarily chooses nonetheless chooses because of that fact, unless we also say that a stone, because it falls of necessity, does not fall.[3]

FOL. 371v

3. The necessary causes why someone chooses one thing rather than another are two. One is called coercion (or fear); the other may be termed inducement (or love). There is coercion to act when the reason for acting is the expectancy of greater evil by not acting than of good by acting. There is coercion not to act when the reason for not acting is the expect-

---

[2] Plural throughout the sentence in the text.
[3] So fols 366, 387ff.

ancy of a greater evil by doing than by not doing. So someone who throws his possessions from a ship into the sea is said to be coerced by his expectancy that the ship will sink unless he does this; and he who wishes to take someone else's life, but does not, is said to be coerced by the expectancy of being hanged if he does. Such coercions are none other than 'the reasons for choosing'. The latter are therefore necessary because the ill anticipated on the one hand outweighs the good anticipated on the other. This imposes limits on [*determinat*] a will that initially (namely when the outcome on either side hung in the balance) was free or that swung up and down as scales do when you drop a further weight into either pan alternately. An inducement to act is [the situation] when the reason to act is the expectancy of a greater benefit by acting than of a disadvantage by not acting; and an inducement to not acting is [the situation] when the reason for not acting is the anticipation of greater good by not acting than by acting. Hence those attracted by [the hope of] reward to treason or to other crimes commit them when tempted. This limits [*determinat*] the volition springing from love of reward. Now, when the advantages of an action were being weighed against the inconveniences, such a volition was free, i.e. it swung to and fro. In fact, it makes little difference whether one is driven on by fear, or encouraged by hope, to perform some action; the two differ [only] as propulsion and traction, which produce their effects by an equal necessity. Of the good things experienced by men, however, none can outweigh the greatest of the evil ones, namely sudden death.

FOL. 372

FOL. 372v

4. The freedom of animals is, then, merely the exchange or reciprocation of appetite and aversion; and the reason for this is that appetite and revulsion, and the will of all animals, have their causes. The said alternation is correctly called freedom, because the impediment [to action] works not through external factors but through internal, i.e. through the intellect and through the mind-picture of things to be chosen; for if the will lacked a cause, [the will itself] would be freedom.

5. The Divine Will and the several volitions of God being eternal, it follows that there has never been any reason for God to will one thing rather than another. Cause must precede effect; but 'earlier than eternity' is inadmissible; hence there is nothing that has ever been able to impose necessity upon the

CHAPTER XXXIII: 7

Divine Will, let alone coerce or inveigle it. Hence God acts absolutely without constraint, be the converse of freedom necessity or an impediment [to freedom]. Yet this freedom of God's is not, properly speaking, choice (choice is the limiting of things previously unlimited,[4] and this cannot be said of the Divine Will) but a consonancy between things and the eternal will of God.

FOL. 373

6. The divine freedom in founding this world our author epitomises thus: God had considered worlds without number that He could have created, from which He chose the present one. Now, [White continues,] what He fixed upon as requiring doing God performed freely. That He was unable not to choose this world, namely the best one, in no way deprives Him of liberty. It is indeed true, [I say,] that a cause–though necessary–of choice does not remove choice. We cannot, however, say that God first considered several infinite worlds that He could have made and then decided He must create this one. If before He chose to do it God pondered what He would do, then He could not be said to have chosen from all time. Yet all those who acknowledge that God is eternal must say and believe this.[5]

FOL. 373v

7. From our author's stating that there must be a creation not only of a world, but of the existing world, two difficulties may[6] arise. The first is that the world must be coeval with its creator. The second is that God could have wished to make another world, i.e. He could have had a will other than the one He possessed from all time; that is to say, God's will could have been changeable. White therefore turns to remove these difficulties, the first one in the present Problem, the second in the next.[7]

The first difficulty is: before the world was fashioned, the only cause that might[8] create it was the Divine Power. Therefore everything requisite for creation existed from all time. Therefore it seems necessarily to follow that the world also must have existed from all time. This difficulty White replies to as follows: 'Philosophers', he says, 'are very prone to the error of

FOL. 374

---

[4] Or, '...determining...determined...'
[5] For 'this', perhaps '[that He *did* so choose]'.
[6] Or, 'can'.
[7] I.e. Problem 9, which forms the subject of the following chapter.
[8] Or, 'could'.

409

believing that the world may be equated with an offspring of the Divine Power. They see it not as one world chosen from many worlds as being the best, but as the only one that can be made. Hence from what they say you would believe that the world was born rather than made and was also co-eternal and consubstantial with God. You see how monstrously this argument departs from our own rules both in the question itself and over terms. *They* ridicule the world's freedom, free choice, and temporality, *we* the ignorance of those who compare the finite with the infinite.'[9]

FOL. 374v

The above words perhaps reflect, but do not confute, the philosophers' view. Our author had to show that the opinion under scrutiny not only was false but also was wrongly inferred from 'that necessity of creating the world' ([a statement] which he defends). He has not shown this, and does not state the philosophers' opinion correctly, for no philosopher has been stupid enough to believe that the world etc., was [both] born and eternal. 'Born' and 'eternal' are conflicting terms. Aristotle, who denies that the world has always existed [*aeternitas mundi*], denies also that it had a birth. 'Whatever has been generated', he says, 'will also pass away'; but he denies that the world is corruptible. Moreover, if it be indeed true, [I say,] that an act and a deed of God are performed in the same way as the acts and deeds of created things we can understand, then clearly the world must be eternal. For acts and deeds are conceivable only if motion be a pre-requisite [to action]. They presuppose two different bodies, one of which is the agent, or the mover, or the efficient [cause], and the other the patient, or the moved (which receives the effect). But this state of affairs is inconceivable [as existing] before the creation of the world. How, then, are the philosophers to be answered? Not by philosophy, but from the teachings of the Christian Faith: that God is incomprehensible, as is His way of acting, and that it needs neither motion [nor effect] to...[10]

---

[9] 'Temporality': 'material possessions' or 'ephemerality'?
[10] There seems to be a hiatus here. The words above conclude a leaf, and fols 375 r and v are blank. It is possible, however, that the article is complete and ends thus: 'God's way of acting does not require [the agency of] motion [or effect]', but this seems rather an abrupt conclusion.

CHAPTER THIRTY-FOUR

※

On Problem 9 of the Third Dialogue:
That God's ability to have established another
world does not show Him to be changeable[1]

1. Two conflicting opinions of our author's. Explanations of (2) 'act', (3) 'potential', (4) 'mutability'. 5. The ability to change is inferred from the potentiality to do several things. 6. An examination of [White's] reply [to this]. 7. That a potential which never becomes action serves no purpose, and that this is universally true. 8. Our author's answer to an objection concerning the way in which the will of God can be restricted. 9. This answer scrutinised.

1. In the Third Problem of the First Dialogue our author has tried to show that not more than one world can have been made, his argument being based on the absence of a place where the worlds [already] made could be stored. In the Sixth Problem of this, the Third Dialogue, on the other hand, he declares not that God formed the present world as the only one that could be created, but that He chose it as being the best that He must create of the countless ones He could. Therefore in the two Problems our author sustains two arguments that are contradictory and conflicting, namely that several worlds both could and could not have been made–or, which is the same thing– that God could have done what simply could not be done. Assuredly those who use words with no conception of what

---

[1] I.e. God cannot be changed; He suffers no vicissitude; not, 'that God is capricious'.

they mean but solely out of familiarity with them in speech, must fall into incongruities of this kind. If White had really understood what is meant by the terms 'potential', 'act', 'action', 'effect', and by other names relevant to these questions, the fact that one of the above views conflicts with the other could not possibly have escaped him. But before we go further, and to ensure that we do not fall into the same error, we shall define the terms one must know if the present Problem is to be examined, namely 'potential', 'act', and 'change'.

2. We shall first say what an 'act' is. Take note, therefore, that, as regards names, there are two main kinds: the positive and the negative. The positive is that used in order to recall a thought, or an impression [*phantasma*], of some definite and delimited thing. For example, the name 'man' is used to arouse the thought of a definite thing, namely of some animal that in its shape and movements is distinguished from everything else not of a similar shape and motion. A negative name is the one used to arouse a definite thought, which can be of any kind you like other than that aroused by its positive. The negative is therefore formed from the positive by adding the negative particle 'non', as in 'non-human', and this term signifies anything—a stone, a tree, or what you please—except man. A negative name has a significance that is indeterminate, and [the former] is usually called 'an indefinite name'.[2] Positive names, on the other hand, denote two kinds of things, i.e. *ens* and *esse*. *Ens* includes things that exist, have existed, or will exist; *esse* includes the [philosophical] modes in which we understand *entia*, and the latter are usually called 'accidents'. Therefore under *ens* are numbered bodies such as the heavens, earth, a star, air, water,[3] stone, metal, animal, movable, at rest, thin, thick, white, black, and the like (which are the material of things), and anything *per se*. But under *esse* are listed the natures of *entia*: either [to be] entities, or essences, or the things we say happen to, or are present in, *entia*, such as 'to be body', or

---

[2] The text has *nomen infinitum*, which could be correct if *indeterminatus* means 'unlimited' rather than indeterminate; but nowhere in this work does Hobbes speak of 'an infinite name'.

[3] The word *terra* is here repeated in the original. For '...*entia*, and the latter...' perhaps: '...[those] *entia* which...' So the text; but what follows a little later suggests that 'the latter...accidents' should refer to 'modes' or to *esse*.

'corporeity' (the essence of body); 'to be the heaven', or 'the celestial nature'; to be air, water, earth (or the airy, watery, earthy natures that some people would probably call 'heavenness,' 'airness,' 'wateriness,' 'earthness'); to be stone or of a stony nature; to be metal or of a metallic nature; to be animal, or 'animalness'; to be mobile, or mobility; to be at rest, or to be quiescent, or rest; to be thin, thick, white, black; or thinness, thickness, whiteness, blackness, etc.

Such a division of things is the same as that attributed to Aristotle in those short verses:

> Whatever lay within the world
> (So as to scrutinise the ends of things)
> The greatest Aristotle classified
> Into two types.[4]

[That is,] he distinguishes *that which exists* from *existence*; nor could it have been done otherwise, for [what applies to] the one cannot be said of the other. If indeed someone were to say 'Body is corporeity', or '*Ens* is *esse*', or 'A white thing is whiteness', or 'Something that runs is "to run"', he would be speaking nonsense, nor could anything at all be deduced from such words. What we call *esse* (or the nature, property, essence, or accident of *ens*) is what is called 'an act'. To say 'Colour is *esse* or *accidens* or nature, or the nature of a coloured object' is the same as saying 'Colour is the act of a coloured object', except that we do not always use the term 'act', but this only when we are speaking of the present. For example, suppose we say, concerning the present, 'Socrates is one who walks'. Then 'to walk' or 'to be a person who is walking' will be said to be 'an act' of Socrates. When we are talking about the future, however, as in 'Socrates will walk or can walk', an 'act' of Socrates will be said to be, not 'to walk' or 'to be a person walking', but 'to be able to walk'. An act is therefore the same as essence [*esse*], as is conveyed [either] through the words 'to be', as in 'to be a person running', or through an infinitive, such as 'to run', or through an abstract noun such as 'running' or 'coursing'. This seems a sufficient explanation of what an 'act' is.

FOL. 378

FOL. 378v

3. Given that the word '*esse*' is associated with the future, as

[4] Source unidentified.

when we say, 'Socrates *can* walk or is able to walk'. Then that act, 'to be able to walk', is called potential. Suppose we [now] say this of the future, thus: 'Socrates *will* walk or will be one who walks'. This is allowed as equivalent to saying 'He can walk', but not as equivalent to calling such an act 'potential', since although the act is to be, we do not *know* whether it is to be or not. But 'potential' and *esse* have the same meaning, and in talk it does not matter which of them we use, unless on occasion for euphony or clarity in our conversation we prefer now one, now the other.

To make clear what 'potential' means, let us take some example. Suppose two bodies are in contact, and that both are at rest. As long as both are motionless one will not move the other or be moved by it. Hence (let us say) the first can move the other or has the potential to do so, but the second can [only] be moved or has the potential to take up motion. It is therefore obvious that the first can move the second because the first can itself be moved. For, if the first is moved against the second, with which it was already in contact, the second will necessarily be moved. So the first's potential to move the second is a potential that is none other than the motion of the first itself. Therefore just as the first's motion is the potential of the first to initiate motion in the second, so in every other transfer-of-action the potential to produce any act not yet produced is some act already produced in the agent. Likewise the potential of the second body to receive motion, or any other act, from the first is some act previously produced within the body itself[5] by which it becomes capable of producing acts subsequently. From this argument a definition of potential may be inferred, thus: 'Potential is the act by which an act not yet produced will be produced later'. Hence 'to be able to be' is exactly the same as 'to be afterwards', or 'a future act'. 'Potential' being explained, there is no need to explain 'lack of potential'; it is either the impossible or the necessary, for that is called 'necessary' which cannot possibly be other than it is.

4. What 'change' is has been stated several times,[6] so we

FOL. 379

FOL. 379v

---

[5] But which of the two bodies? Or does Hobbes mean 'a body'?
[6] At fols 45vf., 327v, 329; but if Hobbes means 'by other writers' Cicero is an instance (e.g. *De Natura deorum*, 3.12.30). One avoids using the word 'mutability' in the text, since it implies impermanence—that a body *will*, at some time or other, change. 'Changeability' (380) does not well express

understand what 'the faculty to change' or 'to be able to be changed' is. 'Change' is adequately defined [by White] in this very problem at his page 337 in the following words: 'The faculty to change is the possibility that an act [will occur] in something said to be capable of change. Hence the only thing capable of change is the one which can receive another act that it did not possess before'. This definition of our author's I receive and applaud as true, concise and clear.

5. These points being agreed, let us return to his purpose in the present problem. As he shows, this is to demonstrate that, if we grant that God could have founded other worlds, it does not follow that God can be changed. It seems possible, [he says,] to formulate an objection, thus: If God could have created a world other than the present one, then He could have wished to do so; for if He could not have wished to create, and if creation is voluntary, it follows that He could not have created. If, then, He could have wished to create another world, He could have had a will other than the one He possessed from all time. He could therefore have had within Him another 'act' (the will being an act). But 'the ability to have another act' *is* 'power to change' through the definition actually stated by White; if, then, He could have created another world, it will follow that God *can* change or be changed.

FOL. 380

6. In reply, White first of all denies that God can be said to change, on the grounds that God can perform a number of things only as long as these are in potential [to be done. Now, I reply,] it is indeed true that, because a person can do different things, then, so long as he does not do them, he is not said to have been changed; but 'to have been changed' is one thing; 'to be able to be changed' is another. Just as someone who *does* things actually different from those he has decided to do *has in fact been* changed, so someone who *can do* things different from what he purposed doing *can be* changed. For someone to be able to change, it is enough if he can be changed even if he is not yet changed. The objection states God's ability to have made another world. From this it clearly emerges that He could have had another wish, i.e. another act; and in the latter, as our

---

the idea that a body *may* change. Hobbes has previously discussed *change*, but not 'the ability to change'.

author himself admits, consists the nature of changeability. So this part of the reply in no way diminishes the difficulty of the objection.

FOL. 380v

Second, on his page 336 he says that the potential of God is such 'as is satisfied by a single act'; which he thinks he explains by adding: '...just as you see to be the case with the whole class of electives; for as they are constituted by nature to choose one thing from several, it is required both that they can perform several things and that they be satiated by one thing'. To the words (they are metaphorical) 'the potential of God is satiated by a single act' the only meaning I can assign is: 'A single act performed by God represents all He could do'; for unless 'is satiated' means 'is filled' or 'is equalled', i.e. that there is as much in the act as there was in the potential, [White's] expression is quite meaningless. But if he meant that [in God] there is as much *actually* by this creation of a single world as there was *potentially* in Him before the world was created, he patently contradicts himself, who previously had declared that God possessed sufficient potential to create other worlds innumerable.

Furthermore, what White adds about electives is untrue. A person, [he says,] who from several things was able to choose any one that he had to do, but has not done that one thing immediately, has lost such potential [or power to do it], so that later the person will not be able to do other things. [I reply that] someone who can do several things does not necessarily have to do them all simultaneously. Now, of the things they can do, men choose to do different ones on different occasions. Why, then, shall we deny to God, who could have created several worlds, the potential to create a world different [from the present one], either now or when He pleases? At least, if we suppose that, one world having been made, remaining things cannot and should not be made, then [the making of] these latter ought not to have been undertaken. Further, being omnipotent even now after the world was created, God can not only do many things from among these[7] but can do even now all the things He could have done before the Creation. Man has

FOL. 381

---

[7] The things just referred to? Or the things God can do in the world He has created? This difficulty relates to the sentence preceding, where 'one world having been made' should perhaps be 'one thing having been done', and 'remaining things...should not be made, then [the making of] these', etc., could be 'remaining things...should not be done, then [the doing of] these', etc.

CHAPTER XXXIV: 7

no right to prejudge what God ought to do; hence the objection remains unresolved so far.

7. However, having assigned to God a potential sufficient to perform many things, White claims that He was able to do one thing only. (This is a contradiction.) Next he considers that he must answer those who would cite the old saw: 'A potential not consummated in an act is useless'. (This [adage] means that no potential exists save with reference to the act it produces. Therefore if God had the potential to create more than one world, such worlds had necessarily to be created at some time or other.) 'That axiom', he says, 'is true for potentials that relate to potentials we know as potentials [*quae veram rationem potentiae habent*] and that are potentials because their acts make them so, however much the axiom is misleading if applied to electives. Suppose that the reason why we always make the best choice is that potential has been involved to the greatest extent, and not otherwise (except where potential is absent). Then potential is used to the best end by choice, and the ensuing acts will never take place. But potential is not therefore vain; so we term God a potential, in accordance with that analogy which people call "deliberate equivocation"; but this is not said of any act external to Him.'

FOL. 381v

In this rejoinder several points are not only correctly made, but also remove the objection completely. Less justifiably, however, others are mingled with them. It is true that it is only 'by analogy' or by 'deliberate equivocation' that God is said to be 'in potential to' [perform] any act external to Himself.[8] It is also true that God's potential is not understood in the usual meaning of 'potential', i.e. that when we apply that term to God we neither understand nor come to know anything. Because God Himself is incomprehensible,[9] everything within His nature is incomprehensible too. Yet we assign that term ['potential'] to God equivocally, it being an honourable one to us, for it is our purpose not to philosophise about God but to honour Him. Hence [to use] such a word is rightly said to be 'deliberate equivocation', namely [when we do so] with the intention of honouring God; for if anyone intentionally used equivocation in philosophy, no reason could be given for such

---

[8] Cf. Galileo, *Two chief World-systems* (Drake), p. 103.
[9] Cf. fol. 291v.

an intention, except, perhaps, in order that he should seem to know what [in fact] he did not.

The objection [at folio 381] is removed, however, if we say: 'It is indeed true that the potential to choose one thing from many things does constitute an argument for changeability', where the term 'potential' is employed properly and *rei ipsius causa*. Suppose that, with the intention of speaking in terms of honour, we use the word 'potential' of God. (The only *act* we can conceive of in God is that He exists *simpliciter* and has as His own title none other than that which is, is.) Then it does not follow that, from a potential thus attributed, God is either mutable or immutable. Moreover, that 'potential has been involved to the greatest extent because the best choice is made', said at the start [folios 376, 381v], is not said correctly, because a potential attributed to God is eternal. What is eternal has not been made; and to deny that that axiom, 'A potential not consummated in an act is useless' concerns electives is incorrect. If someone who has chosen one of two things could at the same time have chosen something else that he possessed later, he could have removed the impediments to his *not* choosing the thing he had subsequently, but he could not have removed the impediments themselves. What he chose, therefore, he chose of necessity, i.e. he was unable not to choose. Hence the axiom 'A potential never consummated in action is worthless' is true even of electives.

FOL. 382

8. [White] raises a further objection against himself: It does not seem, [he says,] that we can restrict the will of God to [the making of] this world without at the same time understanding God to be imperfect. In reply he demonstrates the manner in which the Divine Will is restricted to [doing] one thing only. 'Just as', he says on his page 340, 'we can understand one thing out of many, without considering another one, so we can understand that a person may desire a certain thing, whatever it be, ignoring anything else he desires at the same time. We therefore see that God wills Himself before we see that He desires other things *because of* Himself. Here we accept that to this extent God is unrestricted; and we declare this of God, not of Himself, but according to our conception of Him. Looking further, we see that to this volition, which arises from His reason, it is no matter whether or not there is more than one world. We also see that, from other adjuncts we perceive in God, He is to be

FOL. 382v

limited by no object other than the world we inhabit. And since, through a scrutiny of our own nature, we diligently search out this step towards God, it is necessary that we adopt this way of understanding and speculation. And so it[10] is true; that is, it is of a kind that suits us. And this solution I think is everywhere clear: it is more than adequate to silence the most stubborn, whoever he be, if only he understands it.'

9. Personally, I do not find this solution 'everywhere clear', try as I may to grasp it; but as far as I understand it, the drift of his words is this: 'God wills many things at the same time, whereas we understand or examine His wishes only one by one. First,[11] we consider that God wills Himself before we consider that, at the same time, He has desired other things because of Himself. That is why we say that the will of God has not yet been circumscribed. This is so, not because it is [in fact] unlimited within Him, but because we who ponder one thing *after* another have not yet conceived of a restriction of that kind. Looking further, we see that it makes no matter to the will of God in itself whether there is more than one world. We also see, from a consideration of other acts, that within God His will has no object other than the creation of this present world alone. Hence it is true to say all this; for we have no means of understanding God's will unless it resembles our own. We examine God's will exactly as we examine the human will; such discourse as suits us is true discourse'.

FOL. 383

The gist of the above long-winded solution is this: we attribute the same things to the Divine Will that we do to our own. Although the two have nothing in common, and although we say things of God that are unbecoming Him, what we do say is still true, because it is in conformity with our nature, and we cannot reason otherwise. In this solution four things are false:

(i) The validity of the claim that whatever we shall say of others is true if it is similar to what we apprehend within ourselves. [If this claim were true] we would be able correctly to ascribe even to inanimates the will, the intellect, and all the functions of our own mind; for any objection advocating the contrary could be resolved by saying that it suits our nature to

---

[10] I.e. the method ('way') of proceeding towards God.
[11] 'First'. No 'second' or 'third' is specified in the resulting argument.

FOL. 383v

grasp and discuss the will, the intellect, etc. But truth, [I say,] consists in speech's matching not what we feel, but the nature of the things themselves being talked about; so such speech is merely the conjunction of several appellations for the same thing, according to their received meaning.

(*ii*) Also false is White's assertion that such a manner of apprehending and discussing is a necessary one. [I say] that although we ascribe to God [the faculties] to will, to understand, to know, to see, to hear, to do, and similar attributes, these expressions need to be interpreted only as evidence of the reverence and the religious feeling of those who speak in such terms; for we cannot comprehend how God wills, understands, etc. Far less is it necessary to discuss them philosophically; for who compels us to?

(*iii*) False yet again is our author's declaring that the will to create a world is indifferent to the possibility of creating several worlds, even if we speak in terms of our own will; because there is no volition that does not seek one definite and clear-cut thing.

(*iv*) Finally, the words 'Before, therefore, God realises[12] that He desires other things because of Himself, we see that He wills Himself' are both false and words that seem to originate with someone who has not the slightest inkling of what 'the will' is. The will and every appetite is (*sic*) of the future, for what we will does not yet exist. If, therefore, God willed that He existed, then, because that was His wish, He was not in existence! Not even the metaphysicians could say anything more ridiculous than this.

[12] Perhaps, 'we realise'.

CHAPTER
THIRTY-FIVE

※

On Problems 10 and 11 of the Third Dialogue:
The basis of the Stoics' 'fate'; and that
God is not the cause of any evil

Definitions: (1) agent and patient; (2) integral cause; effect; and (3) cause *sine qua non*, or 'hypothetically necessary'; efficient cause; material cause. 4. That 'potential' is the same as 'cause'; 'active potential' as 'efficient cause'; 'passive potential' as 'material cause'; and 'full potential' as 'integral cause'. 5. An effect or an act that must be produced cannot, at the same instant, follow upon an integral cause or full potential. 6. Definitions: 'necessary' and 'fate'. 7. That, before they were brought about by the force of those preceding them, all past events were necessary. 8. All future events will necessarily take place because of the force of preceding ones. 9. The necessity of the truth [demonstrated] in propositions is the same as the necessity of the outcomes [contained] within causes. 10. Propositions are not to be termed 'true' and 'necessary' before we know that they are so, and it is rash to assert anything concerning the future. 11. Of [two] contradicting propositions about the future, one is true, but we do not know which; yet a disjunctive proposition is always true. 12. We cannot find out what the basis of the Stoics' 'fate' was like. 13. The Stoics' argument has not been correctly answered. 14. A repudiating of the argument by which our author tries to prove that propositions about a future contingent[1] are neither true nor false. 15. The distinction he uses to reconcile the necessity of things with the con-

FOL. 384

FOL. 384v

---

[1] A thing which may or may not happen. Text: 'are not "neither...false"'.

tingency of propositions is not proved. 16. 'God is not the cause of evil'–in what sense we must say this.

1. If the question 'On fate' or 'On the necessity of things' is to be rightly understood, we must define and explain the terms on which the question turns, i.e. agent, patient, cause, effect, necessary, contingent. There being two classes of things, one of which Aristotle called *that which exists*, i.e. *ens*; and the other, existence, [*esse*],[2] we cannot conceive of a passage from the first, i.e. *ens*, to not-*ens*. Hence, as every philosopher believes, *entia* do not, by reason of their natural or ordinary virtue, decay absolutely.

FOL. 385

There is, however, a natural passage from '*ens* of a certain kind' to '*ens* that is not such', e.g. from white to non-white, from hot to non-hot, or from man to non-man, but these are not the decayings, but rather the changes, of *entia*. In such metamorphoses that which perishes completely is [not *ens* but] accident, or some quality, or some determinate existence, as whiteness, or 'to be white'; heat, or 'to be hot'; humanity, or 'to be man'. We may say of coming-to-be what we say of passing-away: *ens* is not created simply and naturally from non-*ens*, or [created] according to the usual working of God, but is merely changed. In this change, '*ens* of a certain kind' is created out of '*ens* not of such a kind', as white from non-white, man from non-man. Hence there can be created and there can perish completely, not the *entia* themselves but the acts, forms and accidents which distinguished them [the *entia* under consideration] from other *entia*. But it is unimaginable for *ens* to be changed, i.e. for some act to be created within *ens*[3] or to be destroyed without [the intervention of] some agent. We therefore say that 'to act' means the same as 'to produce or destroy some act within some subject that is *ens*'. Hence 'the agent' is 'that *ens* which produces or destroys some act'; and 'the patient' is 'that *ens* in which some act is produced or destroyed'.

FOL. 385v

2. That potential, therefore, by which the agent produces or destroys an act within the patient is some special act. For the production or annulment of some act it is not enough that the agent, or the patient, or both of them, *is* [or *are*], i.e. that it

[2] Cf. fols 293, 293v, 377v.
[3] Or, 'within itself'.

exists; the agent and the patient must be of particular kinds. [That is,] some *ens* burns, not because it is or exists, but because it exists in a certain way, i.e. as something hot; likewise the patient's potential by which it can receive an act *is* an act or a definite quality and is not simply 'to be', or 'to exist'.[4] Hence there are, for example, different 'acts' in wax or clay by which the one is melted, and the other hardened, by the same fire. Therefore if any act is to be produced or annulled, there are required definite accidents or acts or forms or natures, both in the agent and in the patient. All these acts, as much of the agent as of the patient, are concurrent, and this is called the 'integral cause' of the act produced. So–to define it–an integral cause is 'all acts, taken together, as much in the agent or agents as in the patient, necessarily requisite to produce any new act', and the act they produce is called the effect.

FOL. 386

3. Any one of the above acts, of itself, is usually termed a cause *sine qua non*. Now, since such an act is necessarily required if an effect is to be produced, it is clear that no effect can be produced without [the existence of] that act. The latter is therefore correctly named a cause *sine qua non*, and it is part of an integral cause. It [a cause *sine qua non*] is also called, not more truly but more learnedly, 'something hypothetically necessary', because we suppose that without it there can be no effect. For instance, if a sick person cannot get well without medicine, the medicine will be necessary to him on the hypothesis, or supposition, that he wants to get better. Of the causes 'hypothetically necessary' or the causes *sine qua non* some are in the agent, some in the subject. Those in the agent are called 'an efficient cause'; those in the subject,[5] 'a natural cause'. Those in every agent and in a patient constitute, if taken as a whole and as I have just said, integral causes.

FOL. 386v

4. Cause, then, and potential are the same. Because all *ens* produces an effect,[6] 'potential' is 'some act without which [the *ens*] cannot produce any effect'. The agent's potential, therefore, which is also termed 'active potential', will be the

---

[4] I.e. passive potential is not merely the being or existence of the patient.

[5] For 'or' understand 'and'? Causes *per hypothesim* are not alternative to causes *sine qua non*. 'Subject...subject': the original's 'patient' is here avoided because of the immediate medical context.

[6] Or, 'Because every entity produces an effect'.

same as causes *sine qua non* or as acts 'hypothetically necessary' within the *agent*, i.e. the same as an efficient cause. The potential of the patient, or 'passive potential', will be the same as acts 'hypothetically necessary' within the *patient*, i.e. the same as a material cause.[7] Full potential, consisting in active and passive potential jointly, will be the same as an integral cause.

FOL. 387

5. The above terms being thus explained, two things are clear. (i) An integral cause cannot but produce its effect. If it does not, then some factor necessary to produce the effect is missing; hence the cause is not, as was supposed, integral.[8] (ii) At the instant when all the conditions (as much in the agents as in the patients) necessary to produce the effect are fulfilled and a cause becomes integral, the effect is produced. If it is not then produced, there was not present everything 'hypothetically necessary', and that cause was not an integral one, as it was believed to be. Because an integral cause is merely the full potential produced from the active potential of every agent, and from the passive potential of the patient, it also follows that (since, [for] as often as the potential [to perform] some act is full, the act must be produced) the potential therefore becomes full, and the act commences, at one and the same instant.

FOL. 387v

6. That is called 'necessary' which is unable not to be; that which 'necessarily occurred' is that which was unable not to happen; and 'a necessary event' that which could not but occur. A proposition is said 'necessarily to be true' when it cannot be false. So why something cannot be otherwise [than it is]; [and] why it is unable not to occur; and why some proposition cannot be false, depend on what goes before, i.e. on causes. Therefore necessity, or (as the Stoics called it) fate, is nothing but the consequence of a result upon what precedes it.

7. These points being understood, it is easily shown, first of all,[9] that every past event has been necessary not only by virtue of the proposition 'Whatever is, necessarily is', but also because of a cause that anteceded it. Now, to whatever has been performed has been attached, or has not been attached, an

---
[7] Cf. fol. 296v.
[8] On necessity see also fols 320, 342v–346v and 371v.
[9] 'First of all'–1°. Cf. note to fol. 382v.

integral cause compounded of all the causes *sine qua non* (or of all causes 'hypothetically necessary'). If not, then [the act] lacked some cause essential were it to be performed. Consequently it has not been done. This is contrary to what is supposed [by White], namely that the deed was done; for if what was done had an integral cause, then it could not have been left undone, as we showed just now in Article 5. What, therefore, both was done and could not *not* be done, necessarily was done, according to the definition of 'necessary' given in the previous article and universally accepted. Even our author confirms the truth of the theorem–in the present Problem, and in the following words (his page 359): 'We are not, therefore, afraid of fate, if by "fate" we wish to understand "a series of causes that is controlled by God and that stretches firmly and evenly from one end to [another] end which He has preordained"'. And again (his page 361): 'Since men are neither able nor accustomed to look upon all causes [at once], they absolutely deny that there are [both] necessary effects and causes [at the same time]–but this, however, not in such a way that, if we prefix the words "with respect to all causes or to the First Cause," they [these?] cannot be called "necessary effects"'.[10]

FOL. 388

8. As past deeds derive [their] necessity from the causes preceding them, in like manner will future deeds also be necessary, for past deeds were once future ones, and yet were necessary. The necessity of the latter is proved by the same argument by which we prove the necessity of past deeds, since that which is 'future' is 'future' because at some time or other it will have an integral cause.[11] Those acts which will form an integral cause when collected are, individually, future acts, because each of

FOL. 388v

---

[10] Obscure and dubious: the printed text of the *DM* is here preferred to Hobbes's slightly different version. Perhaps we are to understand '...they absolutely deny that there are necessary effects and [necessary] causes', or '...they absolutely deny that necessary effects are necessary causes', or, indeed, '...they deny that there are necessary effects and causes [operating] absolutely'. 'But this, however': attaches to 'deny' or to its noun object-clause–one cannot tell. Possibly White's meaning is that to set the Greek article before a noun (or concept) changes the latter into the class or quality to which it belongs–here, 'causeness', 'First Causeness'; at fol. 16v Hobbes adopts this procedure. If such an interpretation is correct, then the issue of necessity is (White claims) avoided.

[11] There is perhaps some play on the word 'future'. As noun and participle (in the Latin)? With emphasis on 'the future' and 'what must be' or 'what must come to pass'?

them has its own integral cause. So, by extension, subsequent acts depend necessarily on what has preceded them, right as far as present acts.[12] At the moment, therefore, those acts exist which at some time will produce that effect we declare to be a future one.

9. The 'necessity of propositions', by which we say that such and such an event will take place, follows from the necessity [which demands that] events [result] from causes. The necessity governing the truth of the proposition 'That which is a future act will be done' is the same necessity governing the performance of a future deed; and the truth of the proposition 'Caesar defeated Pompey' depends on the same necessity by which it came to pass that Caesar once did defeat Pompey.

10. Again, observe that we must at once declare to be true, not every proposition which is true, but only those propositions we know to be true; and as for a proposition which is necessary, we must not at once say 'It is necessary' before we know whether it is necessary or not. It is on record that a certain Roman was cast into prison for declaring, before anyone knew it and on the very day when the battle was fought, that Paulus Aemilius had beaten Perses in battle in Macedonia. Why, unless because he had asserted to be true what he did not know to be true? and it was no thanks to him that he had not told a lie, to the injury of the state. Fortune-tellers, too, whether what they predict does or does not happen later, are to be considered liars and should be punished—because they have not guarded against lying themselves, but the Author of Future Things has, and they do not know it. Hence those who speak correctly will call nothing in past events 'necessary' except what they know to have already happened. In future events [they call 'necessary' only] that whose necessity they consider to have been explained already, i.e. its following upon, and its connection with, present events. No proposition is either necessary or true, they say, whose predicate is not a consequent of its subject, [and this] through [the use of] an agreed and received meaning of words. Though it later appears true and necessary, a proposition concerning the future must not

FOL. 389

FOL. 389v

---

[12] 'Right as far...present acts': probably, 'later acts depend on earlier ones right up to the time that the former take place, here and now'. The 'them' in 'preceded them' is ambiguous.

at this moment be declared true or false, or necessary or not necessary. If, however, we see later a result derive from it, then we may say not only that [the proposition] *is* true and necessary but also that it *was* true and necessary. No-one is sure of the future; that the sun will rise tomorrow we know only by supposition, i.e. if it continue in its usual [apparent] motion. Therefore we call 'necessary' and 'true' only those propositions whose predicates we already know to include their subjects. For example, that 'Man is an animal' we say is 'necessary' because we already know that the term 'animal' includes the term 'man'. But[13] a proposition about the future is true by supposition only: someone might say, for instance, 'If tomorrow something will exist called "man", the same thing will also be "animal"'.

When a proposition is called a 'contingent' one, 'contingent' means the same as 'unknown'. 'Such a proposition is contingent' is equivalent to saying 'I do not know whether it is true or false'.[14] But to say: 'This proposition is neither true nor false because we do not know whether it is true or not' is irrational and contrary to the definition of 'proposition', which is distinguishable from other kinds of speech in that it alone is always either true or false.

FOL. 390

11. Those propositions are contradictory [one to another] in which the subject of each is the same thing, but in the one the predicate is a positive name, and in the other it is the negative of the same positive. A positive name combined with its negative creates a contradiction. For example, 'future' and 'not-future' are contradictory, and propositions [utilising these] are also contradictory, [as] 'Some given thing is a future one'[15] [contradicts] 'It is not a future one',[16] [or] 'Socrates will die tomorrow' [contradicts] 'Socrates will not die tomorrow'. There are similar illustrations. Such propositions are so constituted that one is always true and the other false. Now, say that both [their] predicates are combined into one by means of a disjunctive particle, such as 'or'. Then we know that a proposition having [such] a predicate is true and necessary (e.g. 'Tomorrow Socrates will die', or 'Tomorrow he will not die').

FOL. 390v

[13] Text: 'And'.
[14] Cf. fol. 289.
[15] Or, '...will happen'.
[16] Or, '...It will not happen'.

From the very meaning of the terms used we know that this proposition is true and necessary, so we say and know that such propositions, though to do with the future, are yet both true and necessary. Let us now turn to our author.

12. First of all, he fixes the basis of the Stoics' 'fate' as follows: Two propositions, an affirmative and a negative, to do with the future are contradictory; therefore one of them is necessarily true.[17] Beginning with this basis, [the Stoics] argued (he says) [the existence of] fate, or the necessity of things. How he can know this defeats me. That some [others] have similarly argued is no reason for taking it as the conceptual basis upon which the originator of the [Stoic] sect depended. They seem to me, rather, to have arrived at the above opinion through studying the definitions of 'cause', 'effect' and 'necessary'. Neither of us has any means of knowing what principles and what methods were used by the teacher who first explained to his pupils the above argument used by the Stoics.[18] It seems sound enough to me, however.

FOL. 391

13. We may set it out thus: Given two contradictory propositions relating to the future, for example 'It will rain tomorrow' and 'It will not rain tomorrow'. Therefore if it rains, the first must be true and the second false; but if it does not rain tomorrow, the second must be true and the first false. Yet the proposition that 'tomorrow either it will rain or it will not' is true and necessary. Therefore one of the propositions is necessarily true and the other false.

Our author answers, in the first place, that neither proposition is true and neither false.[19] This is contrary to the definition of 'proposition', for a proposition is an utterance indicating true or false—unless he says that 'It will rain tomorrow' and 'It will not rain tomorrow' are not propositions. The reason why he denies that they are either true or false is as follows: 'For myself', he says, 'I seem to see that men cannot perceive future things within themselves in the way that they comprehend present events by means of the sense, and past

FOL. 391v

---

[17] 'First of all,' cf. note to fols 382v, 387v.
[18] Alternatively, '...used by him who was the first of the Stoics to explain to his pupils the above argument'.
[19] 'In the first place' (the text has merely '1°'): cf. fol. 382v, 387v and the end of fol. 392.

428

events by means of the memory. Now note this: I see that a doctor examines the pulse and urine in order to say whether the patient will die or live; that a sailor notes the clouds and stars and winds in order to foretell the weather; and that in every prognostication similar things are looked into. From this I am assured that these people are speaking of "futurity" only inasfar as [things] lie within causes. So at length I see that it is innate in the human species to utter nothing without having, or without seeming to itself[20] to have, some kind of reason, however absurd, for what is said. Hence, if experience is to be consulted, assuredly I must commence with that [kind of "experience"] by which the Aristotelians and Thomists prepared the way. So propositions to do with a future contingent are not, in my opinion, "determinately" true'.[21]

FOL. 392

These words say only that because men cannot be certain about the truth and falsity of such propositions, and are not in the habit of labelling as true or false what they do not yet know to be one or the other, such propositions are therefore neither true nor false—as if, [I add,] the fact of a proposition's being known was to do with its being true. For the same reason someone who has not yet applied himself to [the study of] geometry will say that the theorems of Euclid and Archimedes are neither true nor false, because he does not yet see whether they are true or false, and says [merely]: 'The propositions are not "determinately" true or false'. What this word ['determinately'] means in this context I do not know; but I suspect that White wants it to be understood as 'they are not known [to be true or false]'. Hence a proposition said to be 'determinately true' means one that we know to be true; but whether

---

[20] Perhaps, 'one's self'.
[21] 'If...consulted': presumably: 'if experience (of the past) is the criterion (when we consider things of the future)'. Hobbes omits White's words 'which is the mistress of fools' after 'experience'. In 'that [kind of "experience"] by which...the way' the word 'experience' (translator's quotation-marks) means either 'experience considered in the abstract' or (the quotation-marks being omitted) the Aristotelians' and Thomists' own experience by which they prepared (their own? or others'?) way (to the truths of philosophy?). Though Aristotle and Aquinas themselves both consider experience (e.g. the opening of the *Physics* and the *Summa theologica*, Pt II, 40.5.1), that their followers did so is less easily demonstrated.

'So propositions...': the clause may be final, not consecutive: '...if I am to believe that propositions to do...are not determinately true'. The rather abrupt turn from the consideration of experience to that of contingents is White's; a parallel occurs at p. 207 of the DM.

he will say: 'A proposition "determinately" false is one that we do not know to be false' I cannot tell.

Next,[22] as he has had to concede that of [two] contradictory propositions one is true and one false, he answers: 'To call these propositions by names to do with contradictory propositions is one thing; to attribute to them the nature of a contradiction is another'. I do not fully understand this, for I do not see how one can ascribe to propositions 'the nature of a contradiction' except by calling them contradictory. If one is incorrect in terming them contradictions, one is also incorrect in ascribing to them 'the nature of contradictory propositions'; likewise if one is correct in the first, one is correct in the second.[23] Hence one must show that they are not contradictory. White therefore inserts an addition.

'Propositions', he says, 'such as: "Someone is a man", "Someone is not a man", and such as "A man runs and sits down", and "He does not sit down" or "He does not run", and the like, may be called contradictory—as far as their wording goes—but they cannot have the force of contradictory propositions.' Who fails to see that the propositions he advances as illustrations are not contradictory, even as far as their wording goes? For the 'some man' in the one proposition is different from the 'some man' in the other. [Surely] in the one proposition 'the man' is understood to be one [particular] man, and in the other proposition as being a different man? But in the propositions now under examination, such as 'Socrates will die tomorrow', 'Socrates will not die tomorrow', the contradiction is obvious, not only as far as their wording goes but also as regards the truth of the fact. In no way, therefore, has White undermined the Stoics' argument which he believes to be the foundation of their 'fate'.

14. There follows our author's argument by which he proves the above propositions to be neither 'determinately true' nor 'determinately false'. 'Are contingent effects determinately

---

[22] 'Next': perhaps this relates to the 'First' at the end of fol. 391; but it may follow directly from what Hobbes has just said in the preceding paragraph.

[23] This is far from clear. 'In terming them contradictions': the original has merely 'in so terming them'. The verbal expressions 'is incorrect' and 'is correct' have no subject in the original: whether they are predicates to 'one', as translated above, or whether one or both are predicate to 'White' understood, is not apparent.

within their causes', he asks, 'or are they not? For instance, is rain so unquestionably in a cloud that the cloud cannot be dispelled by a wind blowing more strongly than was expected, and rain avoided?'[24] 'No', he answers, 'The existence of an effect within such a cause is completely indeterminate. A proposition put forward about such an effect cannot be determinate either in conformity, or not in conformity, with the effect.'

Here he cites the cloud as a sufficient cause to produce rain, but this in no way advances the subject under review. The Stoics did not consider that rain must follow whenever and wherever a cloud was seen. Every effect, they thought, lay [a] within an integral cause composed of everything needed to bring about the necessary effect, but not [b] in a part of the cause, because if a cloud is to be dissolved in rain many factors are involved apart from the cloud itself. If all these conditions are simultaneously fulfilled, the cloud must dissolve. How a proposition derived relative to an effect not yet determinate can be neither 'in conformity' nor (as he says) 'not in conformity' he shows by means of an analogy.

FOL. 393v

'Just as', he writes, 'a white wall is neither similar nor dissimilar to a wall that is neither white nor not-white, so a mental proposition can *abstrahere* and be *indifferens*, just as a cause that contains its effect before [the latter happens] is *indifferens* so long as it is equally disposed towards any effect of the many that lie within it.' In what sense, indeed, a proposition is said to *abstrahere* or to be *indifferens* I fail to understand.[25] I see, however, that he so manages things that, if a proposition is not stated but remains, as though suspended, within the mind, there results an effect in conformity with that proposition. Then and only then is such a proposition true. But 'if it happens that the effect is "in conformity" because of time', the proposition, he says, was true.[26] This is no different from saying: 'After the event, we know that the proposition was true before the event as well'. A little earlier, White as-

FOL. 394

---

[24] White (and Hobbes) seem to mean: 'Is a cloud's being so thick that even an abnormal wind cannot disperse it (and hence remove the threat of rain) evidence that rain must follow?'

[25] For the term '*abstrahere*' and its adjective see fols 77v, 357. A definition and a discussion are offered by J. C. Scaliger as article 342 of his *Exotericarum exercitationum de subtilitate* (1557).

[26] 'Because of time'–a reference to 'simultaneously fulfilled' (17 lines above), or 'for the time' (*pro tem.*)?

serted that the proposition was neither true nor false, but could be either. So even though he has failed to see that such ambivalence lay in our not knowing [whether the proposition was true or false], he is nevertheless forced to admit this, but in a confused manner. Almost in the very next words, therefore, he concedes that 'someone who has seen all the causes that can prevent an effect [from happening] can pronounce with certainty whether there will be an effect or not'. From this it follows that, if he forecast that such an effect would occur, this same person would be about to speak the truth, and hence that the proposition about the future contingent was true. This contradicts what White was pleading previously.

15. He next voices an opinion that is in sympathy with the Stoic point of view. 'We do not', he says, 'fear fate, if by "fate" we understand "a series of causes"', etc.[27] True, the Stoics did understand 'fate' in this way;[28] but he immediately contradicts this same opinion [of his]. Someone asks him: 'So you acknowledge that the wisdom of God imposes necessity upon chance causes and effects by means of a series of causes?' 'Not at all', he replies. 'You are afraid, it seems, that you are offending the divinity of fortune.' Now either, it seems to me, providence does not exist, or fortune is nothing except our own ignorance.[29] In order to reconcile the two, he distinguishes [the question]. He says that every future event is indeed necessary and that in the nature of things there is nothing haphazard—but that we are right to call future events adventitious. 'The question must be divided', he declares. 'We must discriminate between what is to do with *res* and *natura* and what is to do with terms in common use. It has been established beyond dispute that men adapt their words to fit their thoughts, etc. Therefore man contemplates a cause *per se* (which is succeeded by an effect when the latter does follow) and some few things as well. It is on the basis of these last that he announces [his decision]. They bring about, however, not a clear-cut emergence of an action settled upon beforehand, but a vague one. It is therefore necessary that an effect should appear

FOL. 394v

---

[27] For Hobbes's ' &c.' White has the words (*DM*, p. 359) cited at fol. 388.
[28] But Cicero (*De fato*, 9.20) says that such a definition deprives the mind of the exercise of free-will.
[29] Perhaps: '...exist, or else (leaving out of consideration our own ignorance) fortune profits us nothing'.

contingent to a man, that it be called contingent by him, and that it be divorced from necessity.'

According to these words, the reason why future events are to be termed 'fortuitous' is that men pay no attention to the reasons for necessity. Next he adds: 'But if all causes were to be examined, the outcome might be, on the other hand,[30] that every effect smacked of necessity from contingency. Hence no discussion [of what *might* happen] would remain [for examination]'. Here he clearly asserts that everything happens by necessity. I do not see how these last words can be made to fit the previous ones. Perhaps he intends that he, and the philosophers who agree that the connection between causes is an unbroken one, must speak as if future happenings were necessary, but that others must take them to be [merely] contingent; and because a man's words cannot outstrip his mental capacity, [he claims that] both parties are speaking the truth.

FOL. 395

With all due respect to our author, I deny that this is a reconciliation [of the two views]. Although the meaning of words depends on the way men use them,[31] the truth of propositions, i.e. the linking of names in accordance with their meaning once established, depends not on the same things but on reason. Hence what he says a little later is false: 'The aptness of words depends on how the populace uses them. Therefore effects and their causes are truly called "contingent"'. If I owe to the common people my understanding of 'fortuitous' and of 'future', I do not consider that the words of the commonality must govern my decision whether this proposition: 'Future events are adventitious', is true or false. It seems to me, therefore, that our author's opinion is the same as mine. Through a desire to pronounce it before he got it into proper order, however, he has entangled himself in many contradictions and unusual words, such as 'futurition', and obscure ones like 'determinately true' and 'an indeterminately true proposition in the mind can *abstrahere*' and the like, and has involved himself and others in a fog [of obscurity].

FOL. 395v

16. The above opinion of his [fol. 394] would seem to imply that God is the cause[32] of evil. Therefore in the next Problem he

---

[30] 'On the other hand': perhaps, 'the reverse to what one might expect in the circumstances'.
[31] Perhaps: 'on current usage' or 'social convention'.
[32] Or, 'a cause'; and so in what follows.

tries to show that, given that every action must derive from God by a series of second causes, God Himself is not the cause of ill. Our author's reasoning to this end derives from the fact that evil or sin is *non-ens* and hence is not an effect. Therefore, [says White] although He is the cause of every effect, God cannot be said to be the cause of evil, because evil is not an effect but is a defect.

This is a pleading quite often practised by writers, but is one by which we can equally well prove that *man* is not the cause of evil. Personally, while I hold that the nature of God is unfathomable, and that propositions are a kind of language by which we express our concepts of the natures of things, I incline to the view that no proposition about the nature of God can be true save this one: God exists, and that no title correctly describes the nature of God other than the word 'being' [*ens*]. Everything else, I say, pertains not to the explanation of philosophical truth, but to proclaiming the states of mind that govern our wish to praise, magnify and honour God. Hence those words 'God sees, understands, wishes, acts, brings to pass' and other similar propositions which have only one meaning for us–'motion'–display, not the Divine Nature, but our own piety–[a piety of us] who desire to ascribe to Him the names most worthy of honour among us. Therefore the [words cited] are rather oblations than propositions, and the names [listed], if we were to apply them to God as we understand them, would be called blasphemies and sins against God's ordinance (which forbids us to take His name in vain) rather than true propositions. [Again] neither propositions nor notions about His nature are to be argued over, but are a part of our worship and are evidences of a mind that honours God. Propositions that confer honour are correctly enunciated about God, but the opposite ones irreligiously; we may reverently and as Christians say of God that He is the author of every act, because it is honourable to do so, but to say 'God is the author of sin' is sacrilegious and profane. There is no *contradiction* in this matter, however, for, as I said, the words under discussion are not the propositions of people philosophising but the actions of those who pay homage. A contradiction is found, [wherever one is found,] in propositions alone.

CHAPTER THIRTY-SIX

On Problem 12 of the Third Dialogue:
That decisions based on astrology
are worthless[2]

1. That future events exist previously within the stars as [their] causes is one thing; that future events may be foretold from the stars is another. 2. 'Influence' defined. 3. That every influence of every star is concerned in every effect. 4. The sum total of causes within the stars is the total number of causes in the universe. 5. That all future happenings depend on the present influence of the stars. 6. That if the aspects change, this alters the effect. 7. Some less correct allegations against astrology. 8. That the weather in the heavens cannot be known beforehand. 9. Predictions about the characters and fortunes of men have no basis. 10. That the work *Tetrabiblos* is unworthy of Ptolemy.

1. As regards the judgments passed by astrologers, two questions must be asked: (*a*) However random they seem to be, do future events nonetheless so depend on the present influence of the stars that they cannot fail to be brought to pass?[3] (Here we call 'stars' those large bodies which, together with that transparent fluid which flows between them and which we call the ether or the air, constitute the world, and in this sense we also include the earth's sphere under the name of 'star'.)

---

[1] Fol. 397rv is blank.
[2] MS. Chatsworth E.2, fol. 33, lists a number of books on astrology in Hobbes's personal library.
[3] Andabata (*DM*, pp. 2, 6) answers in the negative.

(b) From what men have learnt about the nature and motion of the stars, can those manifold happenings we call 'accidental' be known and stated before they occur? That which will bring about future events is one cause and one necessity; that which makes us suspect that they *are* future happenings is a different one; and the causes of those future events may be in the stars, though [these causes] are not therefore clear to us [*non inde nobis conspicuae*].

2. On the first query: whether chance events depend on the stars, something may usefully be set down here on the way in which bringing-to-pass [*operatio*] can extend to remote things. Now, as has often been said [by me, e.g. at folio 338], every action is motion, namely local motion,[4] for rest alters nothing, i.e. it achieves nothing. When we see some effect being produced out of motion, we all agree that a bringing-to-pass can well be equatable with motion. Most people, however, when they see the effect but not the motion, say that in this instance there is no motion, and to the bringing-to-pass by which the effects have been produced they assign, according to the differences that men perceive within themselves, different names at different times: heat, cold, moisture, dryness, light. When the effects follow immediately on [the cause] they call the bringing-to-pass sympathy or antipathy or an occult quality, or, indeed, an influence, but never motion—as though nature's qualities and the potentials of bodies were infused into bodies in the same way as water or another fluid is poured or flows into a vessel!

We must therefore understand that motion is propagated from body to body, across a distance however large, by the continuous thrust of a neighbouring and adjacent body. Hence, if an effect is to be produced, it is unnecessary for any body, or any part of a star, to fly towards the earth; it is enough that the part of the star facing the earth press forward or be advanced, even by a very small amount. In such forward motion the nearest part of the air or ether is forced onward; it forces forward another part closest to it; and likewise motion will be propagated to any body on the earth by means of a continued thrust.[5] Such motion will be undetectable by the sense, how-

---

[4] Cf. fol. 161v.
[5] Cf. fols 66vff.

ever. In the way described a star is said to heat, to cool, to dry, to give off dampness [or] to shine; and whatever is attracted is attracted in this way, as iron by a magnet [or] heavy bodies by the earth: such motion is what, in the stars, is commonly called 'influence'. To define the term, influence is just this: 'that motion in parts of the stars (the senses cannot detect it) by which a motion is propagated to a distance by a continuous thrust of the medium'.[6]

FOL. 400

3. But the influence of the stars, or their motion, extends in the way described from the fixed stars as far as the earth, and for the same reason it stretches also from one star to another. Again, like any other star,[7] the earth also has its own force or influence by which it can act on bodies far removed from it. Hence everything is brought to act on everything else at the same time, i.e. if any effect whatever is brought about, the influences of all the stars combine to produce it.

4. Again, either the space intervening between the stars serves to propagate their influences, each upon each, or if there are corpuscles in that ether or liquid medium, of their nature such that they operate on a distant object,[8] these are nothing but astral particles, e.g. vapours and exhalations. Therefore any effect within any body must be wholly ascribed to the virtue of the stars. Hence the sum total of all the causes within the stars is the sum total of all the causes within the universe.

FOL. 400v

5. Now, in the tenth Problem of this Dialogue at his pages 359 and 361 our author concedes that the necessity governing all events derives from the sum total of all causes acting jointly. His opinion also makes it clear that every future happening is necessary by reason of the sum total of all causes that lie within the stars. Hence whenever it appears to be due to chance, every future event 'still depends necessarily on the present virtue of the stars'–and this was the question being asked.

6. It is also obvious that, if any two stars have one effect when diametrically opposite [one another], they have another

FOL. 401

[6] This does not make clear whether the medium impels the motion or is squeezed by a body in motion.
[7] So fol. 398v.
[8] Cf. fol. 403.

effect when in conjunction, and yet another effect when in a different aspect, because opposite motions weaken and retard one another, whereas combined motions increase and reinforce each other. The degree to which all aspects other [than those just named] increase or diminish their force is greater or lesser according to whether they are more in conjunction or more in opposition.

7. So he is not correct when he asserts on his page 371: 'How vain are the conditions of causation!–because they are brought about by opposition (whether it be at the quadratures, or sextile) or by conjunction; whether [the stations and retrogressions are] rising or sinking; whether scorched by the solar rays or not'. This is because (as I have just said) through the variety of their aspects the stars either impede or assist their own influences, each upon each. Nor is it correctly asked, a little earlier on the same page, 'Will Jupiter in no way be affected by his four satellites, unseen over so many centuries? Will Saturn be not at all influenced by [his] two? Will all the remaining planets' orbits around the sun, which were [once] thought to lie round the earth, be in no way transformed into "effects"?'[9]

FOL. 401v

Undoubtedly, even before they became visible to man, those secondary planets influenced Jupiter and Jupiter them. Yet we do not know how; for among the stars changes, unless very great and such as have never taken place so far, are imperceptible to us owing to distance. So beyond any doubt Saturn is influenced to some degree by his two companions, be they stars, or parts of himself; but no reason can be given why matters are not the same as regards the effects of the [stars'] influences, whether the sun and the planets move round the earth or the earth and the planets move round the sun. Whichever of the latter is the case, the virtue of the stars is the same, and the aspects are the same, and these two factors are the only ones that can alter an effect.

Nor is he right to ask (his page 372): 'Say a child is on the point of death because a roof is [about to be] destroyed: what motion [directed] towards the mother's blood or towards the child's body can cause the roof to fall?' Astrologers do not, I think, say that such a happening occurs only because of that

FOL. 402

[9] Perhaps: 'transformed in their effects'. Galileo (Drake), pp. 368–369.

CHAPTER XXXVI: 7

influence on the child's body, or only because of that influence upon the roof, or because of one influence out of a few instances; it is many influences that, combined into one influence, not only cause the roof to collapse at such a time but are also [and] at the same time the cause of the child's going under the roof. This means that that effect becomes necessary because of the joint action of all causes, a fact that he himself also says is true.

Again, what he writes on his page 373 is also without reason: 'For the most part, effects are such as to need a defined collection and order of causes. Further, it is clear that the cause *per se* of such an effect cannot be any dead body, but that, of necessity, we must invoke an animate body[10] or an intelligent cause'.

First of all, when he says: 'For the most part, effects are such as to need a sum total of causes', this seems insufficiently in agreement with him who had declared earlier, not that most, but that all effects need a collection of causes.

Next, White says that the 'cause *per se*' of 'a defined and ordered effect' cannot be a dead body. Here he puts 'a cause *per se*' for 'any one agent to which the effect is mainly ascribed', and for 'an inanimate body' he puts 'a dead body'. He denies that an inanimate body can be the cause of 'an ordered and defined effect', i.e. [he says] that there is no cause of voluntary actions that lacks the power of understanding. Now, as regards the First Cause of all, God the Omnipotent, we readily concede this. But God performs all natural actions, the voluntary no less than the involuntary, by means of second causes, namely the bodies constituting the universe. Therefore all actions should equally pre-exist in second causes, and hence all the appetites of animals, i.e. the wishes of men, should also possess their own causes composed of the gathering together of everything requisite to volition. That is, appetites should necessarily originate in second causes, be these animate or not-animate. So, to [produce] voluntary actions or those ordered by a kind of purpose leading towards a defined end, it is unnecessary for every agent to be animate, much less [for it] to possess intelligence. Many species of animals originate from putrefying matter, yet their actions are marked out and

FOL. 402v

FOL. 403

---

[10] For 'an animate body' the MS. has 'any body'; the reading is inserted from the White printed text.

ordered by design towards a specific end;[11] but neither the sun (that generates them) nor the decaying material is on that account to be considered as possessing intellect.

Lastly, his question: 'Finally, since they shine in common upon the rest of the animals, why do the stars work no effects on them?' seems to mean that the stars do not act otherwise than by illumination.[12] We must not accept this, for we see that a magnet acts upon iron and that other things bring about effects upon one another at a distance,[13] but not that they are shining. Why, then, shall we deny to the stars every influence other than light? But no matter of what kind the stars' influence be, someone will say: 'It is common to every animal and affects it equally; hence its effect on everything ought to be the same'. He is easily answered: the wide range of effects arises for two reasons. (*a*) Two bodies cannot be in the same position relative to any one star. Hence the diversity of an effect is due to the diversity of the aspect. (*b*) But succeeding effects are further varied because of the already diverse nature of some from [that of] others.[14]

FOL. 403v

These arguments that our author cites against astrology are therefore invalid, if what he has postulated–that the necessity of things arises from the grouping together of causes–be viable.

8. The second question is: 'Can chance events or voluntary ones[15] be foretold by what men know from the natures and influences of the stars?' Here we must examine what we said earlier, in Article 3, i.e. that in order to produce every effect all the influences of all the stars combine. Owing to the countless number and the imperceptibility of the stars' movements it is impossible to comprehend these influences; therefore the effect deriving from them is certainly not predictable. Some of the things that depend not on the influence but on the circular course of the stars can be foretold by conjecture and from

FOL. 404

---

[11] 'Ordered by design' seems to mean 'ordained by providence towards a specific end'. It is not excluded that Hobbes attributes this power of foresight and decision to the animals themselves: decaying matter does not possess intellect, but animals may.

[12] Or, 'that the stars do nothing other than illuminate'.

[13] Cf. fols 281v, 400–400v. There and here the words are *ad distans* ('at a distant object'). Here Hobbes seems to mean *ad distantiam*.

[14] Or, '...of the nature, now diverse from others, that some have received'.

[15] I.e. deeds whose performance or non-performance depends on someone's decision. See the next chapter.

signs,[16] however, as the positions and eclipses of the planets are predicted by observing past periods; but such predictions are the province of the astronomer, not of the astrologer.

Perhaps some advance estimates could be made concerning the changing of the air if someone had accurate commentaries in which were set forth, from a millennium ago to the present day, what kinds of weather occurred every day in every place on earth, i.e. whether the sky was cloudy or clear; where and when there was thunder, rain, hail, etc.; in what position any of the planets was on any single day, and at what aspect the planets were, both each to each and in relation to the place whose climate was being investigated. Perhaps the weather could be forecast, I say, for if it were constantly and frequently recorded that similar storms occurred at a similar position of the stars and place, surely by the same token a conjecture concerning a similar event in the future would be quite plausible. Since, however, these commentaries neither exist nor will exist, such predictions must be worthless. The reason for this our author has most amply furnished in the few words: 'Conjecture proceeds from a few things, an effect from many'.

FOL. 404v

9. But for what reasons or by what experiments[17] can we know how the twelfth part of the heavens, rising, gives hope of life, and the next one, of riches;[18] or in which sign which planet increases in strength; which planet is dominant all year or throughout a day; which signs are fruitful, which sterile; or, of the planets, as opposed to the sun,[19] which one is hot, which cold, or which male, which female, which beneficent, which noxious? Those who ascribe to the signs of the *primum mobile* properties different from[20] those of the signs of the eighth sphere are glaringly at fault, however, for in the *primum mobile* there are no signs at all. Owing to the precession of the equinoxes an order of signs in the *primum mobile* is reckoned from the intersection of the [celestial] equator and the ecliptic, whereas in the eighth sphere the signs are reckoned commencing

FOL. 405

---

[16] Either, 'evidence in general' (but see fol. 343) or 'the zodiacal signs'. Cf. 'token' 14, lines below.
[17] Or, 'experience'.
[18] Of the twelve houses of the astrological horoscope the first two concern life and riches respectively.
[19] The meaning of *praeter solem* is not clear.
[20] Perhaps, 'varied because of'.

FOL. 405v

with the same first star in Aries. Hence it is not the signs that are different, but the mode of reckoning that is; and the latter can in no way contribute towards producing an effect.

10. Our author has correctly said much that bears on this matter. In fact, I wonder whether Ptolemy was the writer of the book called *Tetrabiblos*.[21] I am loath to believe he was; I think its author was some gipsy of the band of those who wander through every nation and who, being beggars, even nowadays answer you when you question them about the life, ways, fortunes, and death of men—if you give them a trifle.

---

[21] At *EW* VII.75 Hobbes refers to this work under the title *Judiciae astrologicae*, paraphrasing 'De astrorum judiciis', which is part of the title some editors give to the *Quadripartitum*.

CHAPTER THIRTY-SEVEN

※

On Problems 13, 14 and 15 of the Third Dialogue:
That the earth was created for the benefit of man;
that men rejoice in free-will; and that
liberty does not conflict with providence

1. That the earth was formed for the uses of man is not established by reasons taken wholly from nature. 2. An inspection of the argument by which [our author] proves [that it was]. 3. What constitutes a free being. 4. On decision-making [*deliberatio*], on the will, and on the liberty of those who deliberate. 5. That 'possible' is spoken of in two senses: absolutely and conditionally. 6. That there arise from this [a] a double meaning for [the word] 'liberty' and [b] different interpretations of free-choice. 7. That our author seems to confirm [the validity of] both meanings [of 'liberty']. 8. An examination of the description of deliberating put forward by him. 9. That the soul is not moved of itself–or at all. 10. An exploration of the reason why liberty does not conflict with providence. 11. A review of the objection raised by our author against liberty, and of his resolution of the difficulty. 12, 13. That [a view of] liberty based on the statement 'Volition has no cause' conflicts with [the existence of] a First Cause no more than it would were it based on 'The soul moves itself'. 14. The origin of the opinion held by those who say that free-choice consists in there being no cause whatever of volition.

1. In speaking of our *intention* in doing something or failing to do it, people can take the word to mean only 'the cause

why we do or do not *want* to perform the thing'. Philosophers, therefore, in order to bring any matter they have in mind under, and make it conform to, an appellation [in philosophy], are always in the habit of using a technical vocabulary ['appellations']. So they always employ the term ['intention'] in the sense of 'the cause for wanting or not wanting'. Thus our author should have taken it in the same sense here, for at the beginning of his book he declared that he decided from 'reasons taken wholly from nature' everything offered for consideration in these Dialogues.

FOL. 407

He now states that the earth was founded with the intention that it should serve the uses of men. It follows, then, that the will of God does have a cause and that, because a cause precedes an effect, something existed that preceded the Divine Will, i.e. anteceded the Eternal. This is a tricky point. However, if he is saying that when applied to God 'cause' and 'effect' do not have the same meaning as they have if applied to created things, by exploring the question further either [*a*] he ought to have made us picture to ourselves a cause that does *not* precede its effect, or [*b*] he ought to have admitted that this proposition [under scrutiny] is not known *ex natura rerum*, but was believed in from Holy Scripture.

2. Yet the argument by which he thinks that he is demonstrating the proposition derives from the fact that men actually utilise the earth and those things within it, and that the latter all have, through the will of God, that property of being useful to us. This, [he says,] is because, if there exist in the world things not of any use to us, this is due to our indolence and our sin—because, in the event, man rules over all these things; and because of [his] reason and knowledge, man alone is worthy to take account of [that] end [i.e. of creating the world]. But the other creatures, [White continues,] must be numbered as mere instruments, beings that experience neither good nor evil and have no free-will.

FOL. 407v

First, anyone can see that the argument, 'Man utilises the earth; therefore the benefit of men is the sole purpose for which the earth was established', is ineffectual. By the same reasoning one could say: 'Lions, bears, wolves and other fierce animals feed on human flesh; they kill men, horses, oxen, etc.; therefore God created men, horses, oxen, and so on, with the intention that these be devoured by lions, bears, wolves, etc.'

Moreover, if animals were created for the uses of man alone, by virtue of their having been given, by the will of God, a nature suitable [alike] for our sustenance and our delight, it would follow that man was created on account of lions, bears, etc., for it is also the will of God that man be fit nourishment for *them*. Likewise an argument based on man's sovereignty over the other animals can be turned round to prove that God created the world, or a great part of it, with the intention that it should come under the rule of a Caesar (for it was once actually subject to his command), because man does not rule over lions, bears, etc., as our author believes, but wages war against them: with equal justice man kills bear and bear kills man.   FOL. 408

We must not pass over White's declaration that man is worthy to be aware, in the Divine Counsel, of [God's] purpose [in creating the world]. Our author has not proved this, and has not tried to prove it, 'through reasons taken wholly from nature' —nor can it be [?so] proved. In whatever is worthy of anything [else] the possession of merit must be prior to [the existence of] the object deserved. So if man were worthy of being the end for which God created the world man possessed that merit before God settled upon him as His reason for creating the world, i.e. before he possessed any merit. [So our author contradicts himself.]   FOL. 408v

It may also be asked: If man were the end for which the earth was made, what then was the purpose of creating man? Say we reply, as we should, 'The purpose for which man was formed was his worshipping God from a contemplation of His works, i.e. from a contemplation of the world'. It will follow that God's intention to create the world preceded His intention to create man, if only (properly speaking) the principle, the means and the intention all have, initially and latterly, the same meaning in God as in us. But what does this worthiness of man consist in? In reason, in knowledge and in free-will, says White. Why not, rather, in morality and in the Christian faith? For as regards sense-perception,[1] and if we leave aside the faculty to reason, man is far inferior to the beasts. Yet just as we surpass the other creatures in reasoning-power as often as we reason correctly, so whenever we reason falsely they are to be considered superior to us; error is worse than ignorance.   FOL. 409

Further, the rest of the animals are, for the most part, the

---

[1] Or, the faculties of acquiring knowledge generally.

instruments of man, for the same reason that makes man the instrument of man, i.e. the defeated the instrument of the victor: [animals] do not serve man, because they do not feel good and evil.[2] Indeed, if we are talking of natural good and [of] natural evil, this is not true; if of moral good and of evil, it is the same as if he had said: 'Beasts cannot sin'—yet I do not believe this to be a fair reason for condemning them to subservience. Last, he says that this dignity of man has its foundation in free-will—overlooking the fact that all animals possess the same freedom of decision-making as man does. But about this liberty of choice more is to be said in the next article, in which he contends that such liberty befits man alone.

FOL. 409v

3. As regards the question of free-choice, we must first of all accept that, as our author says on his page 399, liberty consists in motion.[3] That of which the motion is unobstructed, [I say,] is free; liberty is the absence of impediments to motion. For as long as its motion in a given direction is not hindered, for so long, and with reference to the same direction, is something said to be free, as a stream has the freedom to flow between its banks but not across them. What nullifies or hinders any motion, therefore, negatives or restricts a corresponding amount of freedom. Suppose that, on the one side of a balance, a weight descends freely and some other weight is dropped into the pan opposite. The first weight sinks less readily, or not at all, according to whether the weight thrown in is heavier or lighter.

Second, we must be clear that whatever is moved or is movable is body or substance possessed of three dimensions. All philosophers believe with Aristotle that this is so, and it can easily be proved. Next, motion is a passing from place to place. What is moved, therefore, must be able to assume one position after another, i.e. it must be able to be in a place, i.e. to fill a space consisting in three dimensions.

FOL. 410

---

[2] I.e. if animals did serve man, they *would* feel or experience good and evil (at his hands)? In the text no comma precedes the *quia* clause; but to follow this would imply that animals do indeed not serve man, but for some reason other than an insensitivity to good and evil. In what follows, perhaps: 'of good, and [of] natural evil'.

[3] White's examination of the question (*DM*, 399–400) takes the form of an inspection of the statement rather than, as Hobbes claims, an assertion that it is true.

Third, we must be clear that whatever moves is moved[4] (save God the First Mover); for a moving body either drives forward what lies ahead of an 'endeavouring' [?body] or drags what is behind it; of such 'endeavours' [*conatus*] both are motion.

Fourth, nothing can move itself. This our author not only concedes but has also proved in Problem 1 of the present Dialogue, and I have demonstrated the same thing somewhere above.[5]

Lastly, every motion is predetermined [to tend] in some direction; for it is not motion unless it is directed towards some definite and specified place and proceeds along a predetermined line.

Relevant to the question of free-choice, the above facts, I say, must be known about what constitutes a free being. But there are others, concerning the nature of the judgment or the will, that must be known first.

FOL. 410v

4. It must be realised, therefore, that [*i*] the will is an appetite; no-one, as far as I am aware, will deny this. It must also be realised that [*ii*] appetite is the first[6] *conatus*, i.e. an invisible motion of animals' nerves or spirits towards an object they perceive or imagine; for there is no motion save of corporeal substance. Likewise an aversion [felt by] the mind is the first *conatus* or motion towards the [bodily] parts that feel distaste for[7] the object felt or imagined: aversion is also commonly called avoidance [or withdrawal].

[*iii*] The cause both of appetite and of avoidance is the mind-picture of the good or ill that will result from the object.

[*iv*] From the experience and the recollection of there being good and evil consequences of a similar object,[8] the mind-picture is modified through our expecting, in turn, good or evil. Likewise there arise both a *conatus* towards an object, i.e. appetite, and a *conatus* in the opposite direction, i.e. revulsion; and these alternately. This oscillation is called deliberation.

FOL. 411

---

[4] Cf. 325v, 417. 'Whatever moves' cannot mean 'whatever moves itself' (see next sentence), so it must mean 'whatever moves something else'.
[5] Particularly in Chapter XXVII.
[6] Or, 'main'.
[7] Or, 'that move in directions away from'.
[8] Presumably, 'Because the same (corporeal) body can affect us both for good and for evil (on different occasions)...'

In a balance, now one pan is depressed, now the opposite one, as fresh weights are thrown in from one side or the other; likewise deliberation is a *conatus*, sometimes to action, sometimes to inaction, until at length the action being pondered is either carried out or is so far from being carried out that it cannot be carried out afterwards.

[*v*] In this series of appetites and aversions the last appetite, which is in direct relation to the action being pondered, is called 'the will, or the wish to do something', and the last aversion, in direct relation to not-doing, is called 'the will or wish not to do something'. Hence when we put 'will' for 'the act of volition' the former must be defined thus: 'Will is the final [mental] act of him who deliberates',[9] whether that action be appetite or revulsion. Those who define the will as 'a rational appetite', i.e. the one that follows ratiocination,[10] do so in the belief that every deliberation is [an exercise in] reasoning, which is not true; for future events, over which alone one can deliberate, are decided not by syllogising but by the contemplation of similar events of the past that are stored in the memory. So beasts deliberate; and from alternating recollections of good and evil on similar occasions they are spurred now to act, now to refrain, until either they do act or else they leave the scene where the action would have to take place.

[*vi*] An attraction towards something cannot exist at the same time as an aversion from it. We cannot think good and evil simultaneously; these are two [separate] mind-pictures; and two mind-pictures cannot be contemporaneous. Our author admits this in the Fifth Problem of the present Dialogue at his page 308.[11]

Point (*vii*) follows from the last. When someone deliberates, none of the above acts is a mixture of appetite and revulsion, but is appetite or revulsion pure and simple. Any one of the [actions] except the last is called 'inclination' or 'propensity', but never 'the will'.

[*viii*] As long as the deliberation lasts and the action being pondered has not yet been performed, nor has the opportunity for [performing] it been completely missed, that which deliberates is called 'free'–not simply because it can act or not

---

[9] The more terse and familiar version occurs in *Leviathan*, 'The last appetite is deliberating' (I.6).
[10] Text: 'follows after ratiocination (*sequitur post ratiocinationem*)'.
[11] On the grounds that two concepts must be successive. So fols 348v*ff*.

act but because it can act conditionally, i.e. [it can act] if it wants to. The freedom of one who deliberates, therefore, consists in his being able both to act if he wishes and not to act if he wants not to act. So 'free-choice' is the same as 'the freedom of him who deliberates'. But when either an action takes place or the opportunity to act is irrevocably cast aside, clearly this is necessity, and there is no liberty beyond this—a man or other animal is said to have chosen and to have deliberated, or to have laid down his freedom of choice.

[ix] In actions dependent upon volition, he who cannot will [anything] cannot do [it]. Whether anything must be willed is not subject to deliberation, but whether it must be done is; for the cause of [our] future voluntary actions lies in ourselves, but the cause of volition itself lies outside us [and] in those objects that arouse a mind-picture of good or of evil.

FOL. 412v

5. But to most people liberty is someone's ability to act or not to act, i.e. to possess both faculties simultaneously, even though he cannot at the same time act and not act. So we must distinguish the use of the terms 'possible' and 'impossible'. Sometimes these are associated with future events, sometimes with propositions, i.e. with affirmations or denials, insofar as [these last two] are true or false. 'Impossible', when applied to occurrences, is used without qualification, and this is so because of the necessity deriving from the nature of things and from causes. (We said this in Chapter 35 when we were examining the word 'fate' as used by the Stoics.) In our author's opinion this necessity, if regarded as being the collecting[12] of all causes taken as a whole, makes every result a necessary one, since nothing can possibly take place except from a sufficient cause (which lacks none of the causes called *sine qua non*). It is impossible, on the other hand, for any of the last-named causes to arise except from other[,] integral causes. Thus by going back [from a standpoint in the future] to the present time, in which must be contained the causes of all causes to follow, [we see that] a result that is thus impossible is impossible *simpliciter*; and a result that is possible will necessarily be a future one.[13] However, the 'impossibility' that is applied to

FOL. 413

---

[12] *Collectio* may mean merely 'a collection.'
[13] An obscure passage. Hobbes apparently intends *simpliciter* to balance 'necessarily', and perhaps also: 'a result that is impossible is simply impos-

FOL. 413v philosophical propositions is conditional, i.e. it is subject to the condition and to the hypothesis 'if there is or is not such a thing'. We shall clarify this by means of two examples:

(i) It is said to be 'impossible' for man not to be an animal, or that man is a stone. This signifies either (a), that he cannot possibly not be an animal provided that he is a man, or (b), that he cannot be a stone. If he is a man, i.e. so long as he remains a man, then he cannot be a stone. This is 'conditional impossibility', since no-one doubts that someone who is now a man may soon be earth, and that what is now earth may later become a stone. Properly speaking, because it applies here and now, this kind of the 'impossible' is not impossible; for 'potential' and 'lack of potential' are properly understood as applying to the future only, whereas the term 'impossible' is used in statements of the same kind for 'false' or 'absurd', the causes of which lie in the past.[14]

FOL. 414 (ii) Another instance among future events is: 'Socrates may be rich', a statement which implies only that the words 'Socrates' and 'rich' contain nothing that would render the statement impossible. Therefore Socrates can become rich, provided that there is someone to make him rich. This is conditional possibility.

6. The meaning of the word 'liberty' is equivocal, because people take the terms 'potential', 'possible', and 'impossible' in a twofold sense. Some persons restrict 'liberty' to meaning 'potential in both directions', i.e. an absolute power, unlimited by any condition, of doing or of not doing; others take 'liberty' as conditional potential, in which, however, we are not aware of the [existence of a] condition. Those who interpret 'liberty' in the first sense say that free choice cannot subsist in the presence of any efficient cause of necessary volition; but those who say that [to have] liberty is 'to be able if they wish to be', i.e. that it is conditional potential, profess that whatever we do willingly is performed of our own free choice. Hence there arise

---

sible' and 'a result that is necessarily possible will be a future one'. The words following the phrase 'on the other hand' represent his own views, not those of White.

[14] As the text stands, the meaning of these closing words is that instances of falsity or absurdity are due to previous causes. But Hobbes may intend: '"Impossible" can replace "false" or "absurd" in statements dependent upon past causes'.

dissension and controversy: the second party declares that humans do possess freedom of choice; the first says that they do not.[15]

FOL. 414v

7. But there is a third opinion. Certain people say that in fact men possess free choice as regards evil-doing (so that they have absolute power of doing, or of not doing, a mischief suggested to them), but not as regards doing good. This is because, when it is a question of choosing the right, men behave as they would were they dead; since the dead have no choice, i.e. no free-will, whatever.

That opinion is irrational. If people are free to choose evil, they do not necessarily do so; therefore they can opt for the good; therefore they can make both choices. This last point of view is held by few, however; and is such as perhaps is useful [for proving] their opinions [on] other [matters]. But for the fact that I considered it to be a wavering between two opinions rather than his own view, I would say that our author holds a fourth opinion: that men do not have freedom of choice. This [belief of his] stems from his imagining that at one and the same time a man not only is able to do and able not to do, but also is able and unable to act. We shall see this from what comes next.

FOL. 415

8. So on his page 390 he proves as follows that men, and they alone, possess free-will: 'There is present in man', he says, 'the power to let the mind range, and to transfer[16] from one thing to another, and from some known things to others, his own certainty [concerning them], or his ability-to-define. And since the achievement of the sciences, books, and almost everything which we see performed by men beyond the industry of the other animals depends upon this operation, we do not argue grossly [when we say] that this operation is not to be found in the other animals.'[17]

Here we must note first that the power of allowing the mind to range and the power to advance from one thought to another are present in all animals. That beasts do not derive from their

---

[15] A point taken up at the end of Article 7.
[16] A few lines later, Hobbes takes White's verb *derivare* to mean 'derive'.
[17] It is one of his interlocutors in the Dialogue that is speaking: here White does not refer to mankind at large. Hence he means: 'On this particular issue we speak good sense; our argument is not without some finesse'.

experience of one thing their knowledge of other things I grant him freely; for this is done with the aid of words, and without them there is no proposition; without a proposition no syllogism or argument; and without an argument concept cannot develop from concept. I admit, therefore, that there resides in beasts neither knowledge nor any of the attainments that depend upon knowledge. I concede, also, that on knowledge depends the making of books and many other achievements that set man above the rest of the animals. Hence I shall also deduce, if he so wishes, that no animal except man possesses free-will in a deliberation where the matter at issue is: 'Must I write a book or not?' or 'Must I do something else of those things that only man can do?'

Continuing, he says: 'It is no less clear that usually with regard to the things he can do and before he is led, as it were, to a settled state and repose of the mind, man is beset by many things which pull a doubting mind to and fro, drawing it hither and thither until some single idea of such importance is brought to it that it[18] is powerful [enough] to turn a fully-formed inclination in a different direction [from that to which it purposed going]. So we may firmly conclude: initially, the mind to some degree lacks a determining-faculty, but later there is such a faculty. As long as the reason is pulled [out of equilibrium] by many things whose strengths are near enough equal, no decision is made; but when the reason is directed towards one, to the neglect of the others, a resolution is taken'.

Here he explains how deliberation takes place. I examined the same matter above, in Article 4, where I showed that deliberation is an alternation of appetites and aversions, i.e. of strivings, i.e. of motions. In Article 3, moreover, I showed that every motion is limited to [advancing] towards some fixed place, and by an unaltered path; and likewise that no motion exists save of corporeal things. Therefore, properly speaking, the mind cannot be 'pulled', nor, if moved, can it be moved indeterminately. It is moved determinately, now towards an object, now in [various] directions away from the object. So reason is not said to be 'pulled [out of equilibrium]'; for who can imagine how the reasoning faculty, or any other faculty, can be moved? This entire account of deliberation is meta-

---

[18] It is impossible to tell if White means the mind or the idea; probably the latter.

physical, and cannot be reduced to a correct form of speech unless such motion is attributed to corporeal organs alone; in which case the deliberating applicable will resemble in every way the throwing of weights into either side of a balance in turn. Again, because appetite and revulsion are found even in the rest of the animals, the deliberatings of men will resemble those of other creatures; therefore as far as men's judgments are, or are not, free, those of animals will be the same. If this is so, I see no reason to believe what, in a Problem above, our author affirms: that, because he possesses free-will, man is worthy to be the cause for which God founded the world.[19] Such merit, [I say,] man shares with the beasts.

FOL. 417

9. But, our author says, there is a motion of this kind in the very soul, both in the soul itself and [also deriving] from principles implanted in the soul earlier, i.e. [a motion from the soul] itself; later (he says) this motion quits the soul. Again, [he continues,] that the soul must be said to move and to delimit itself is unquestionable, and this very fact—that the soul has the power of moving itself—is called its free-will and its liberty; man, who possesses these gifts, is accounted free.[20]

(i) As opposed to Aristotle and all the philosophers,[21] White attributes motion to an incorporeal thing, for all deny that the soul is corporeal, which is illogical,[22] for what is moved passes from place to place; but what is in a place adapts itself to the dimensions of that place, and hence is body.

(ii) He who on many occasions earlier has tried to prove that nothing can move itself[23] [now] declares that the soul is self-motivated!

---

[19] This is not a quotation of White's own words, nor do they occur in Article 7; hence 'A Problem above' rather than 'the Problem immediately above'. In fact, the matter is mentioned in Article 2 of the present chapter.
The translation conveys what seems to be the intention of the text, but neither is the interpretation clear nor is the sentence unambiguous; it may also mean: (1) '...that man, for whom God founded the world, is worthy of that name because he is endowed with free-will', or (2), as (1), but for 'he is endowed' Hobbes may mean 'God is endowed'.
[20] Or, 'A man who...gifts is...'
[21] Cf. Galileo (Drake), p. 121.
[22] Hobbes's loose usage of 'which' is retained. It seems to refer to White's attributing motion to the incorporeal, rather than meaning: 'It is illogical to deny (or that some do deny) that the soul is corporeal'.
[23] E.g. fols 211v–213v; 305v–307; 325v; 410.

FOL. 417v (iii) He states that the motion of the soul proceeds 'from principles implanted earlier'. If someone now asks: 'Whom were these principles implanted by?' and White answers, 'By somebody else', then the principle of the soul's motion comes from without; so [the soul] does not begin to move of itself. If [he answers], 'By the soul itself', then that chain of causes he established earlier[24] and described as being the 'fate' that is (and rightly so) accepted is now visibly broken.

10. Next, he strives to prove that the soul's motion both can exist of itself and can also possess an external principle in a series of causes [that have existed] from all time. That is, the chain of causation is not severed but adapts itself well enough to the fact that the soul is self-motivated. He therefore heads this fifteenth Problem, 'Liberty does not conflict with Providence'.

True—if we equate 'liberty' with 'conditional potential', so that 'to be able if I want' means that liberty, the series of causes that makes me wish, does not prevent me from acting when I wish to; but if liberty be defined as 'the power of something to move itself', then I cannot act.[25] The essence of the proof lies

FOL. 418 in God's bringing it about through His Providence, i.e. through a series of secondary causes, that man (or the mind) sets bounds upon himself (or itself).[26]

'I acknowledge', he says, 'that [the soul] is delimited by someone[27] else; but to what end? Surely that it must [eventually] set its own limits? Whether [the soul] delimits itself or is delimited by something else[28] is of no relevance except in that [this other] thing itself performs two functions for itself.[29] There is no reason why these [secondary] causes should not let man—or, indeed, should cause and bring it about that he

---

[24] At fol. 394.
[25] For 'I cannot act', perhaps, 'is inconsistent with providence': text has merely *repugnat*.
[26] Here and in the succeeding sentences it is not clear whether Hobbes and White mean 'sets limits on' or 'sets forth for itself' (and hence 'makes decisions'). At the beginning of fol. 419 'defined beforehand' seems to imply determinism.
[27] Or, 'something'.
[28] Or, '...delimits itself, or whether something else delimits *itself*...' From here to 'that of the moved' Hobbes is loosely paraphrasing White; but the words in the manuscript may have been underlined by mistake and the opinion therefore be Hobbes's adaptation of White's plea.
[29] The words 'except...for itself' translate Hobbes's (not White's) *nisi quod ipsum sibi duorum munere fungitur*. But *ipsum* may be a slip for *ipse*: '...except in that [the soul] itself performs', etc.

should–select among several things. Liberty consists in this: Our nature has a double force–[that of] the mover and that of the moved. In man, the very limitation–by [something] extrinsic–to "final circumstances" stems from none other than man himself. But causes have been postulated such as will make him limit himself to "final circumstances", so much so that, when these have been fixed, he must choose them and no others. Otherwise, surely, they would be neither causes nor effects.' FOL. 418v

Is this not to make man's voluntary actions necessary and not necessary at the same time–necessary, because they must be narrowed down to 'final circumstances' by causes outside the will–not necessary, because of itself the soul narrows them down? By such reasoning cannot anything be said to be moved by itself and to set its own limits? For instance, could we not say of a stone hurled upward by the hand that it is moved by itself and that it delimits its own perpendicular motion, although it is necessarily driven by the hand in such a way that it cannot be otherwise moved or otherwise determined? On this reasoning, therefore, the actions of men will be 'free' in exactly the same way as the motion of heavy bodies in flight [is 'free'].

11. Then he cites an objection he could raise against himself, as follows: 'Man cannot do what cannot be done. But a thing other than what man does cannot be done. Therefore FOL. 419
man cannot do that thing'. His minor premise I confirm thus: 'From the very fact that it has been defined beforehand, either in a first cause or in a series of secondary causes, that one particular thing be done, it has also been preordained that *other things* shall not be done. But that which is preordained not to be cannot possibly be. Therefore it is impossible for *another thing* to be done, or, consequently, for man to do another thing. Therefore man is unable not to do what he is about to do'. Indeed, this reasoning seems–to me, at any rate–legitimate.[30] So let us see how he answers it.

'Have you not read', he asks, 'that many things are about to occur which will not occur?' This (though I have come across it [elsewhere]) is false, because [to say] 'There are many future things that will not occur' is the same as saying: 'There are many future things that are not future things', which is an open

---

[30] Or, 'To me, this reasoning seems legitimate enough'.

FOL. 419v

contradiction. And what he subjoins: 'Many things *may* come to pass that *will* not come to pass' is itself the subject of controversy. Suppose that what our author lays down be true—that 'if a series of all causes, seen as a set, is taken into account, all effects are necessary ones'. Then it clearly follows that it has never been possible for such effects not to take place.[31] Therefore what *will* never take place *has* never been able to.

Distinguishing 'possible' into 'possibility through the essence of the thing' and 'possibility as used by the logicians' does not help at all. The distinction is badly explained by him. A possibility said to exist because 'nothing in a thing itself impedes the thing' does not cancel out an impossibility originating in obstacles [that lie] outside the thing. We may say (to use his example) that an axe can cut because there is nothing in the axe that stops it from cutting. Yet there may well be, in the nature of things, causes that make it impossible for the axe—or anything else—ever to be picked up, and as a result the axe cannot cut.[32]

FOL. 420

12. After establishing freedom of choice as being such that men cannot avoid choosing what they do choose, he proceeds to consider (on his page 397) [another] viewpoint. Some interpret free-choice as meaning that 'because of the mere motion of the will, man does one thing or the other, precisely because he wants to'. This formulation White rightly impugns, both because of other shortcomings that he enumerates in it and also because it conflicts with this principle: 'There is nothing without a cause'. 'He who denies this last principle', he declares, 'does an injury to the First Cause, in that he says that this limiting of the will is so absolute that [the will] is independent of God.' And rightly indeed does he seem to censure such freedom of choice in these terms. But we would be wrong to infer that he [actually] does so; for he says: 'The will is moved and delimited by itself'. This is connected with the same shortcoming,

FOL. 420v

for whatever generates for itself the beginning of its own motion is the first cause of its motion, and nothing can be said 'to move itself' except that which supplies for itself the beginning of its own motion.[33]

---

[31] Mersenne's emendation of the original.
[32] Cf. fol. 434v.
[33] For 'beginning', perhaps: 'principle'.

13. 'But the will', he declares, 'moves itself not directly but indirectly; for example, volition moves the passion round the heart, the passion sends spirits to the brain, the spirits conjure up images [*species*], the images set in motion the judgment, and this in turn directs the motion back to the will.'

So, to pass over the folly of attributing motion to the passions, faculties and images, i.e. to the ghosts and mere fantasms of things, I ask: 'How is motion, in this instance from will to will, connected with external causes?' Clearly the motion and these last are not connected. It follows either that that motion by which the will has been acting upon the passions has commenced externally [to the will]–so the will does not activate itself, and hence the will's freedom is lost; or alternatively, the motion has not originated externally–so the series of causes is broken, with injury to the First Cause, a process he censured a little earlier.

FOL. 421

14. Further, he says that the standpoint of those who define liberty as 'mere motion of the will' originates as follows: 'Men have heard[34] condensed into a single word what has been done only through a long sustained succession of motion. For example, in "someone *built*" there is expressed the whole building-process. So they came to believe that in deliberation there was no motion, but that the whole awareness of, and reason for, the deliberation was contained in the final instant; hence, at the final instant, the will consisted[35] in its full freedom of choice, either to do or not to do [something]'. For my own part I do not believe that that opinion originates in such a cause, for few of those who uphold free choice are so metaphysical as to think that 'in building there is no progression' merely because it has been said that someone has 'built'. I believe, rather, that men consider their actions to be performed from the mere motion of the will; partly because they do not easily see the cause of volition, but partly because they foresee what Oedipus, in vindication of an act of his, once said in the *Colonos*:

FOL. 421v

>My deeds are more
>What I have suffered than what I have done.

[34] Text: 'seen'.
[35] Or, 'rested'.

That is, if the will had an external cause, [this cause] could be extended to cover all the crimes ever committed. On the other hand, those who uphold the opposite view do so because they consider not only that for the celestial and eternal chain of causation to be broken is an affront to the Divine Majesty; but also that [for this to happen] is against natural reason.

CHAPTER THIRTY-EIGHT

※⁂※

On Problem 16 of the Third Dialogue:

Why human affairs seem to be

ruled by chance

1. The true reason why human affairs seem to be ruled by chance. 2. An examination of the complaint that the good fare ill, and the wicked well; and a reply. 3. That our author's purpose in this Problem is to show that in this life the virtuous have more pleasure than the evil. 4. What constitutes a good man. 5. That happiness consists in a seeking for things that are to come. 6. What 'worldly happiness' is. 7. What constitutes joy, or the mind's delight; and the definition of 'happiness' [in Article 5] confirmed. 8. That happiness is not the enjoyment of riches, honours and outward pleasures; and that these do not really give delight to men, unless as a proof of their own prudence. 9. 'Prudence' and 'slyness' defined. 10. A part of quantity is quantity; and one substance is not, as substance, better than another. 11. An inspection of the argument by which he proves that man could not have been created better [than he is]. 12. The argument by which he proves that man could not have been given a better brain, or (13) greater prudence. 14. That prudence does not consist in the reducing of passions to a mean; and that skilled artists have always been honoured. A scrutiny (15) of the reasons advanced to prove that good people grieve less than bad; and (16) of the reasons by which he proves that the pleasures of the good are longer-lived and stronger than those of the bad; and of the arguments by which he proves (17) that the virtuous possess greater civic power; (18) are more honoured; (19) have more

friends; and (20) see greater pleasure in their food and in their love [of others] than the evil do.

1. The true and only reason why a man thinks that human affairs are ruled by chance seems to be this: he does not know their integral and necessary causes.[1] For if someone knew in advance the way in which all causes meet or combine in order to produce some future effect, [then] he would never declare: 'That effect will happen by chance', but [would say: 'It will always happen] of necessity'. If the certainty of his knowledge convinces a person that something is about to happen, he will say: 'This will happen necessarily, i.e. not by chance.' Our author seems to notice this, [as we see] in these words at the foot of his page 358: 'It is he, therefore, who has seen every cause which can prevent [a thing from happening] who will be able to say with certainty whether there will be a future effect or not'. And in quite a few places in the same Problem (where he is disputing about the Stoics' 'fate') he declares that, with regard to all causes taken as a whole, every result is a necessary one. Hence, [I say,] all results are necessary because of their causes; consequently they seem fortuitous for no reason other than this: that we do not perceive all their causes.

FOL. 423

2. Some of those who complain that evil befalls the good and that benefits come to the wicked have said that this is not design but chance; others have wondered how such a design can be just. Anyone you care to point to supposes he is good; hence if he is beset by evil he imagines either that this is because all occurrences happen by chance and irrespective of providence or that, if providence does govern events, she does not do this rationally. So [such a person] believes that the world is run rationally when everything in it is most advantageously directed to the happiness of good men, i.e. to one's own.[2] But, [I say,] we must not accept that 'the good suffer and the evil prosper' as universally or usually true. The contrary is the case: even in this life the good fare better than the bad, and there is no art of outstripping others in the gaining of wealth or honours (or anything else in this life which may be more delightful than

FOL. 423v

---

[1] 'A man', etc.: plural in the text; here altered so as to avoid awkwardness over the word 'their'.
[2] Or, 'i.e. to this person's (the one pointed to) happiness'.

these) more effective than honesty. Undeniably the complaint [that the evil prosper] is a universal one; but its cause is that no-one admits his own depravity, and that, out of jealousy, the virtues of those whom one envies for their goods one calls wickedness.[3]

FOL. 424

They[4] wish that the control of the world be changed–for the same reason that they wish for the governing of a country to be changed, i.e. through envy of those placed above them in prudence and virtue. [To prove] the seeming justice of this kind of complaint there is no need [to claim] that the dregs of the populace can provide more good men than are to be found in a like number of the rich and powerful; it is enough that one evil person prosper or that one good man meet disaster. Therefore in Psalm 72, verse 2, David says: 'My feet were nearly shifted, my footsteps near slipping away, because I grieved over the righteous when I saw the sinners enjoying their ease'.[5] But here David is not to be considered as having believed that every wicked man lives in peace; and Job pleaded with God not because the virtuous (or most of them) were suffering, but because he himself was. But if all the good, or many of them, or [just] some of them, are afflicted, this takes away neither God's providence nor [His] justice.

As regards providence, there is nothing to prevent God, by an equal necessity of causes, from being able to enrich or to raise to earthly honours either the good or the evil, or both of them indiscriminately, just as He has wished. And, as regards justice, if it be examined either according to the laws of nature or against the promises contained in the Holy Scriptures, God has an obligation to His creatures no more than the potter to his clay. On the contrary, if His promises be examined, it is enough that they will be fulfilled in a future world,[6] when everyone will gain his deserts.

FOL. 424v

3. Our author had supposed that the world was created for man, and most of all for the virtuous. The consequence of his hypothesis was, he saw, not only that this same world ought

---

[3] 'Out of jealousy' may belong after 'one envies'.
[4] Presumably those mentioned immediately at the start of the present article.
[5] Psalm 73 in A.V. Perhaps, for 'enjoying their ease' one should translate: 'in tranquillity of mind' or 'in grace and favour'. *EW* III. 347 (*Lev.* II.31); IV.293.
[6] Perhaps, 'age'. Cf. fols 426, 445v, 447.

FOL. 425

to be controlled by its creator, as was most advantageous for men, but also that the good should derive from this more benefit than the bad: that is to say, good men should have more worldly happiness than bad men. But this, [I observe,] would appear to be untrue, for, as White himself cites from the wisest of writers, [Solomon,] 'the craftsman enjoys no esteem; neither do the brave win the war nor the wise bread'; i.e. the better craftsmen are not always held in the greater honour, and the braver are not always chosen to lead armies, or the wiser always afforded the more opulence. Now according to the end [of God in creating the world, as defined above]—that the best should receive the best treatment—the best men in any one craft ought to have derived the greatest benefit therefrom. [Even] if this were so, it could still be concluded that the world is ruled not by providence and reason but by chance. So in this Problem our author has tried to demonstrate, first, that a world more adapted to the uses of men could not have been made, and second, that the good have more earthly happiness than those who are not good.

FOL. 425v

4. We shall next examine his reasoning, in which [i] the words 'good', 'prudent', 'wise', 'swift', 'skilful' and others occur indiscriminately as though they [all] meant the same thing, and [ii] worldly happiness is [said] to be found in riches, honours, and carnal delights. But first we must insert a few remarks so as to be clear about [a], who is to be called a good man, who a prudent, and who a happy, and [b] in what consists that happiness that concerns this life alone; and then we shall go on to consider our author's arguments.

You must take note, therefore, that any individual you choose calls what is pleasurable to him 'good'; likewise someone he approves of he simply calls 'a good man'. Hence person (a), whom (b) calls good, (c) calls bad; and person (d) will sometimes call (a) good, sometimes bad. There are also certain people whom the state terms good and bad. In this disagreement among private citizens as to whether someone is good or bad, those must be considered good who the state says are good and bad who it says are bad, for every private citizen has vested his individual opinion in the state. This being so, a good man is he who keeps the decrees of his forefathers[7] and

FOL. 426

[7] Or, 'rulers' (perhaps Senate or Parliament). So fol. 440.

who observes the laws and legal institutions, that is, he who takes upon himself obedience to the civil laws as much as lies within his power. Perhaps someone demands here–rightly, indeed–obedience to the natural laws, among which are found fair-dealing, self-effacement and all the moral virtues. The whole of the natural law is contained within the civil, however. This is because the state demands not only [a] that what has been written be performed *as* written, but also [b] that what is unwritten be done as natural reason dictates: for example, do not do what you do not wish to be done to you, but do as you would be done by. Now, the term 'good' includes as parts of itself both moral honesty and piety, of which the first refers to daily living and the second to faith. What then constitutes prudence, consisting as it does in foreknowledge of the road to happiness, we cannot very clearly know without first knowing what happiness is–the happiness, I say, of this life, with no regard to a future world. (This is the sense in which our author takes 'happiness' at this point in his text.)

FOL. 426v

5. First, it is clear that anyone's happiness consists in what he finds good: no-one finds something good for which he has no appetite. So he who has nothing to seek after enjoys no happiness; and, because everything we seek must be sought with an eye to the future, we must class happiness as 'the desire for good that is to come'. Does the present situation not please us, it may be asked, while we are [actually] enjoying it? Certainly it does; furthermore, happiness will perhaps seem to consist in enjoying the things we have now rather than in hankering after those of the future, and we ought therefore to reflect that the delight excited by things of the present is short-lived. We are said to enjoy these for as long as we experience them, that is, for as long as the object of delight affects our sense-organs through the very action by which it pleases us; and in every kind of enjoyment this time is very short. But those pleasures are usually termed sensual in which the object's action, initially attractive, becomes, because of its lingering and because of the change wrought in him who feels it, immediately loathsome. This is very evident in repulsive pleasures such as gorgings and evacuations of the body;[8] the same thing is also

FOL. 427

---

[8] Perhaps Hobbes intends the meaning obtained if one moves 'becomes' to a position after 'feels it'. 'When he [Hobbes] did drinke, he would drink

seen in other delights of the senses, for the same sight, or the same musical note often repeated, offends the eyes and ears rather than delighting them. Hence it is clear that whatever pleases the senses, i.e. every good thing present here and now, does so (inasmuch as it *is* here and now) not only by its own merit but because of our need of it, or, at the least, because of its novelty.

Again, if we consider 'here and now' to be, not 'time' but the instant, we shall confess that sensual delight, if it persists, is no longer in the sense but is in the memory. Things in the memory do not please inasmuch as [they are] past events, except inasfar as the latter are signs of pleasures to be expected in the future. Now, the future exists only in the imagination of us who suppose that the past is linked with the present in such a way that the former *follows* the latter. That which succeeds the present, however, we call the future; hence, in the very object we enjoy, enjoyment itself exists only as further yearning that springs from the contemplation of the object's parts. So it remains true that the grounds of good, and hence of happiness, consist in seeking. We must next examine, therefore, of what kind the good things are, the desire for which is called happiness; and we are speaking of earthly happiness.

6. (i) For us to be happy, the good things we seek must be easy to win, that is, they must be things we think can be acquired through the faculties and through the potential we now possess. For happiness is secured through the choice of the ways leading to it; we deliberate over the ways to be chosen; deliberation exists only with reference to the things that lie within the power and the choice of those deliberating; therefore the yearning for things which there seems no means of attaining is not happiness: it is torment.

(ii) The search for things easily obtained (if they are procured, and are being constantly lost, then they are again desired, and so it goes on) is not happiness; if it were, then, as one of the ancients—I do not know which—aptly said, someone would be happy who was ever itching and scratching himself in turn. Felicity therefore consists in the advance of the appetite from

---

to excesse to have the benefit of vomiting which he did easily by which benefit neither his witt was disturbt (longer than he was spuing) nor his stomach oppressed'–Aubrey.

a good thing that has been obtained to another good thing that is to be obtained.

(*iii*) There is no happiness in the desire for, or in the securing of, benefits, if the desire and hope of acquiring them be ceaselessly accompanied by a greater, or an equal, fear of losing them; for the pleasure in the hope is offset by the anxiety in the fear. To gain happiness, therefore, the hope of preserving it ought to be combined with the urge to secure it. Consequently happiness consists in an acquisitive advance such that the acquisition of new things seems to lead to a holding fast to the old. And since not only to acquire but also to retain what one has procured are potentials, happiness will be the perpetual advance of appetite and hope from a lesser to a greater potential.

(*iv*) If the advance is not so easy and free of those problems that usually lower the spirits it will be said to have been made with trouble and apprehension rather than hopefully, and a life thus lived will be considered irksome rather than joyous; the advance ought therefore to be easy and serene. So happiness is *the joy noticed in a prolonged and serene progress of searching from potential to the next potential*; and that peace of mind the moral philosophers speak of is not rest or inactivity or the deprivation of desire, but a gentle motion from a good that has been acquired to one that must be acquired.

FOL. 428v

7. This can also be confirmed if we consider that passion we call joy, or the mind's delight, which is quite alien to that pricking of the flesh and of the organs that constitutes sensual pleasures.[9] Whatever is distasteful to the mind is repellent because it lowers our high opinion of our own potential. Potential applies to the future; therefore to settle the nature of that potential, [and] by what means we are to acquire every good thing, is the same as to settle [what] hope [is]. We grieve over the loss of riches and of friends because we feel ourselves deprived of the potential and of the protection that have raised our hopes of advancement; insults pain us not only because they arouse others to contempt for us (for even private reprimands hurt) but also because we are [thereby] reminded of our own weakness. We are stung by the preferment of those who have seemed our equals, as this makes us have doubts

FOL. 429

---

[9] Perhaps: 'alien to the delight of the flesh or to the pricking of the organs that', etc.

about the potential we thought we ourselves had. For potential exists only by comparison, because if everyone's potential were the same [as everyone else's] it would not be potential, since if one member of a group of people acts upon another the potentials [of the two] are mutually annulled.[10] Likewise [the effect of] differences of opinion[11] upon those in disagreement: these people hurt one another, as do those who throw in one another's face their errors and ignorance, i.e. their weakness. We also find distressing the evils [that befall] others, [but only] if we suppose or believe that similar ones will happen to us [as well]; otherwise we do not, because mental stress consists in an awareness of our own weakness, not of someone else's. And if in fact every distress of the mind resides in the recollection or visualising of the mind's own weakness, then the mind's every happiness must consist in the recollection, or at least in the vision it has conjured up of its own potential or excellence.[12] Joy, therefore, or the mind's delight, is nothing but a kind of triumph of the mind, or an internal pride, or boasting about its own potential and excellence in comparison with another [mind's]. Now, if such pride springs from the assessing of [future] potential on the basis of previous deeds, the latter give ground for hope, because he who has done seems to have the potential to do again. Hence a self-evaluation like this gives rise to zeal, and often, through assessing our potential correctly and fairly, with success. One success is the cause of a second, because of the new potential secured at every success; and happiness is said to consist in the successes continued in this way, together with reasons to hope if they persist.

FOL. 429v

FOL. 430

If, however, someone assesses his potential by his own image of himself (for anyone can, through the activity of his fancy, ascribe to himself any actions contrived by the poets), this does not produce the hope of a happy outcome but produces an immediate and empty pleasure only. This last is the reason why [self-glorification] is also called an empty glory; and in no way does it lead to happiness. Alternatively, if someone gauges his potential not by reference to his knowledge of his own deeds but from the testimony of his flatterers, that also is

---

[10] So with contrary motions (fol. 40v).
[11] Text: 'a difference of opinions'.
[12] *Recordatio* ('recollection' or 'the recollection') may have the sense of 'heart-feeling'; and 'visualising' (*fictio*) may mean an image rather than the process of imagining. 'Vision': i.e. 'mind-picture' (*imaginatio*).

a worthless self-glorification; it arouses him to activity, but the outcome will not be happy, because potential does not square with things not begun. If, then, happiness consists in joy or in the mind's pleasure (which is not short-lived), happiness must consist in the awareness of a continued advance from a benefit already won to a benefit to be secured later—as was our definition.

8. The same point will be further strengthened by an examination of 'happiness' as understood by our author, namely (as he puts it), 'a life that is strong in potential,[13] outstanding in honours, abundant in friendships, sustained by riches, and ever flowering and prospering in all kinds of enjoyment'. Here we must note the following:

FOL. 430v

(i) Honours, friendships and riches are parts of potential, for honours and riches are its consequents, but friendships are rarely so—yet envy and hostilities are, since the most powerful men are the most hated, they being the most feared. But granted that anyone who possesses all these things has absolute power, nonetheless he is contented and pleased by it only as a means to something further. If the hope and opportunity of seizing other realms are offered them, kings (whose power is very great in this life) also seek these things, as Alexander; if they have no such hope, they turn to the arts, i.e. [sic] to public affairs, as Augustus; or to music, as Nero; or to gladiatoral shows, as Commodus; or to licentious pleasure, as Heliogabalus, Nero and many others. No more does a person with great possessions take pleasure in what he has now than does someone only moderately well off. Indeed, if happiness consisted in the pleasure of possessing, but not in the pleasure of expectation, we would call no-one happy, because everyone despises what he possesses in comparison with what he would like to possess. It is therefore clear that potential pleases only in relation to the gaining of delights not yet acquired, and that the pleasure derived from potential, however immediate it is, consists in the imagining, i.e. in the hope and expectation of [winning] a benefit that has not yet been won. Undeniably happiness consists in a life's being lived with pleasure, i.e. with the greatest delight; but the question remains: In what does this delight lie? Now, for man 'always to flourish and prosper

FOL. 431

---

[13] Perhaps, 'power'. The point is again in doubt in what follows.

through every kind of pleasure' is impossible because delights differ so much one from another—one cannot always be gorging and disburdening oneself. Perhaps by this [our author] wishes happiness to be the varying of one's pleasures, and wishes it to appeal sometimes to the eyes, sometimes to the ears, sometimes the belly, sometimes the genitals. Yet who is unaware that anything to do with the pleasure from all these senses reaches satiety, and that they please no more than insofar as they are required by natural necessity? [To believe] that lewdness pleases even more than this is [evidence] not [of] physical pleasure but [of] the mind's vain-glory; for [wantonness] consists in people's thinking that they have pleased others, i.e. that they have 'served not without distinction'. Other than this, the lecher's pleasure is of little importance.

(*ii*) How may any happiness be thought to consist in the enjoyment of those things that are common to the beasts and to ourselves?[14] It remains, then, that happiness consists not in the pleasures of the senses but in the gratifying thought of advancing from the enjoyment of one good thing to that of another. This, however, must be taken as meaning that an advance of this kind pleases not as an advance but as a proof of one's own virtue and excellence. These two are the cause of the advance, just as they are of bravery, or of [the strength of] one's intelligence—of this last especially, because those born to riches or to civil power or to honour derive no more pleasure from these things than they derive from their being men or from possessing arms and feet. Someone who has won these same things, however, glories in them as evidence of his own mental prowess, but chiefly in his knowledge of affairs and of [his] skill in dealing with them. If not employed in the public service the ambitious are offended by reason of being passed over, as though insufficiently competent;[15] yet in their own eyes they would like to appear more experienced than those who are so employed. Least of all is it the wealth and titles of those born rich and honoured that wounds the poor; rather it is the riches and titles of those who, from their own original condition [i.e. their low extraction or poverty], have *become* rich or worthy of honour, because the winning of possessions can be cited as evidence that in these men [who have risen to honours, etc.,]

---

[14] Or, 'Surely happiness cannot be thought to be *any* enjoyment of...?'
[15] Here, and with 'experienced' (later in sentence), perhaps, 'prudent'. See the argument that follows.

the mental prowess is of a higher order than is that of people who have not been able to acquire these same possessions.

(iii) As I have said, every joy of the mind consists in the opinion any individual has about some virtue of his own; but especially about [?his] prudence, as being greater than that of people who seem to be competing with others,[16] or to be about to compete, over [possession of] the same virtue.[17]

9. Prudence, then, is the knowledge of the way to happiness before one sets off on it. As, however, happiness [lies] not in the goal but in the way itself, prudence must be the path from potential to potential and the seeing in advance of the obstacles present in those potentials. The obstacles being foreseen, one can enter upon that path in which both the greater potentials and the lesser obstacles are seen before [one starts]. Now, foreknowledge of the future consists in the recollection or experience of past [events' having] consequences the like of which are expected to follow in the future.[18] Hence the most prudent are those with the greatest experience, and the most experienced are those who in the past have widely observed the consequences of good and bad things. So he who uses his experience in order to acquire property and who sets about this in a way that no law forbids is prudent, but he who advances by unlawful methods is sly. Both of these persons can be called skilful, however.

FOL. 432v

10. Now for our author. In the first place he endeavours to prove that there could not have been created [by God] a man better than the man that was created,[19] and no earth more adapted to the uses of men [than ours]. As regards man, White says this: 'Man is composed of body and soul, like parts fitted together. Do you think any of these parts could have been better made? I say "as regards their substance" and not "with regard to their accidents"'. To this I ask:

(i) Is there anything that has dimensions such as length, breadth and depth, and yet a component part of it has none of these? In other words, is there a quantity, a part of which is

FOL. 433

---

[16] Text: 'with them'.
[17] 'The same virtue': prudence, or some other virtue?
[18] Cf. fols 342, 344.
[19] Or, 'no man has succeeded in becoming a better person than he was made'.

non-quantity? If, as all philosophers believe, there is none such, and if man possesses quantity, then how, if the soul is a constituent part of him, will the soul not be quantity? Shall we say [a] that quantity is a combination of quantity and non-quantity? Or [b] that the soul has quantity? The philosophers deny [a]; everyone denies [b]. Therefore the soul does not seem to be a component part of man.

(ii) I should like to have a clearer explanation of the distinction of the word 'better' (this latter is too metaphysical) into 'better as regards substance' and 'better as regards accident'. The nature of good is the potential of a thing we call 'good' to arouse passion or desire. Such potential is essential not to body as such but to body [already] possessed of it. It seems, then, that some substances are called good and others bad according as they have the potential to act in such a way as to be sought after; they are therefore 'good and bad because of some "accident"'. Hence it is unthinkable that a substance, *qua* substance, i.e. as the *prima materia* that can be affected by all accidents equally, can be superior to any other; but let us proceed.

FOL. 433v

11. 'If the soul were better [than it is]', he states, 'it would not be the same soul; we would be, not men, but some other species of animal put here in our place. Now if the same soul were placed within a body of this other species, the soul could no more use it than could the art of playing the flute use the lyre or the guitar.' Why so? Because, he says, 'diversity of form causes diversity of type. As the art is to the instrument', so the soul to the body. That is to say, [I interpose], something cannot improve, and yet remain the same; or if Paul, from being criminal and irreligious, becomes a righteous man and godly Apostle, he is not Paul but some other man! In the same way, [I say,] the same living soul can sometimes incline more to virtue, sometimes more to vice, i.e. it can sometimes be better, sometimes worse. For in order for one soul to be better than another, they need not be dissimilar substances, but they should produce different actions. That the species, i.e. the appearance and look of the material, varies according to the diversity of the form[20] clearly argues that form consists not in substance but in a collection of accidents by which a thing that possesses form

FOL. 434

[20] Cf. Chap. XII, Art. 2.

acts on us and arouses in the [senses of the] observer its own image, or picture, or idea.[21] Now the soul, if it were better [than it is], would not have to make the human body appear in the form of a horse or other animal put in our place. Either, then, his statement that the soul is to the body as musicianship to the instrument is false, or else we must say: 'Just as art [is *accidens*], so also the soul is *accidens*'.

12. Next he proceeds, saying: 'These matters being thus settled, what is your opinion about the brain? Is it not the immediate mover of the soul and, in consequence, the soul's principal instrument? Daily happenings tell us that without the arousing of fancies we do not perceive anything new or anything we already know. It is equally agreed, however, that these fancies are, as it were, inhabitants within the brain'.
See how disastrous much of this is!

(*i*) The brain being the soul's mover, the soul will be movable; it will therefore have spatial dimensions, i.e. the soul will be body.

FOL. 434v

(*ii*) Because the brain moves the soul, he infers that the brain is the organ of the soul; when the inference can be, by a contrary [philosophical] mode, that the soul is the organ of the brain. Because the hand moves an axe, the axe is the hand's organ; likewise because the brain moves the soul, the latter ought to be the organ of the brain.

(*iii*) He says that fancies, i.e. mental pictures of external things, are the brain's 'inhabitants'. [I comment:] The brain would have to be of truly enormous size to accommodate so many 'inhabitants', some of which are no smaller than the entire visible world.

13. Because man could not be said to have a soul better [than the one he has], our author has inferred that man could not have been given a better brain to wait upon the soul.[22] By the same reasoning he now infers [that man could not have had] a better heart to serve the brain, or better lungs or liver to

---

[21] The point here at issue between Hobbes and White is a procedural one, as often elsewhere in the MS.; the argument is being conducted within the framework of then existing philosophical definitions rather than solely from the standpoint of Hobbes's own views.

[22] Or, for 'that man...given', 'that one could not postulate'.

FOL. 435

serve the heart, or better limbs, or better nourishment of the limbs, or a better earth, or better water and air (these things conduce to the nourishing of the limbs), or better plants or animals, or better objects of sense which, by means of the feelings, drive forward the thinking-process and the inclination [to act]. Hence the learning arising from these things could not have been better, or other, than it is. Now, 'men's prudence has its origin either in the temperament of the human body or in learning–even that prudence deriving from nature itself; also, wiser men could not possibly have existed unless nature were changed'.

Apart from the fact that this is not reasoning at all, the conclusion seems to suggest that the lack of prudence in individuals arises neither from vicious habits nor from a perverse temperament or from its own individual erring discipline [i.e. acquisition of knowledge], but from a plot by nature as a whole. So if God had wished a person to be more prudent than he was, not only ought He to have altered that person but also He should have established an entirely different world-system. This has not been, and cannot be, proved.

FOL. 435v

14. Given that the lack of prudence is due to nature, in the way that we have argued, then we readily grant the inference: it is no accident that men are not honoured, or do not grow rich, even when their virtues merit it. Usually unhappiness like this comes to anyone, because of his own lack of prudence; but if [his imprudence] is due to nature's ordinance, then the imprudence is necessary and quite supplants chance. But White's next point, on the way in which our lack of prudence results in men's not being esteemed for their merits, does not seem to be true in every respect; it must therefore be examined.

The force of his reasoning is as follows: All men are addicted to the passions, and cannot live without [indulging] them; imprudence originates in the passions; hence prudence [lies] in the moderation of the passions, i.e. in reducing them to the mean. Now, [he continues,] one can stray from the mean in more ways than those by which one can reach it; therefore nature disposes that there are more imprudent people than there are prudent, and we find fools everywhere; nevertheless

FOL. 436

prudence is mistress and ruler of every art and science. Consequently, while through lack of prudence he despises his own art, believing that other things are of more weight; or so long as he

CHAPTER XXXVIII: 14

worships money, power [or] pleasure, and, for the sake of money and the other things [just mentioned], prostrates himself before those by whom he should himself be petitioned, the artist [himself] is the reason why he is not valued as his craft deserves[23] and why someone else, his inferior in skill but possessed of more prudence, is advanced before him. Hence, the nature of man being what it is, craftsmen are not favoured, and expediency and causes *per accidens* rule the day.

In the above pleading there is much we must agree with. I think, however, that we must deny that prudence is the mean between contrasting passions. It would be prudence to slacken the reins upon any desire (which, the more violent it is, the more swiftly it brings us to the goal we seek), but also to check as far as possible those which will get in our way. This supposes that our passions are voluntary, as our actions are;[24] but since they are not, the prudent will be said to be those in whom is aroused that desire or those passions most needed for the serenest passage from advantage to advantage throughout life. Anger is not a mean; during battle there is [no such thing as] a moderate hope of victory; there is no such thing, [either,] as a moderate fear or a moderate shame of evil-doing, but these are such as a situation demands. If in the allotting of benefits close-fistedness and the urge to generosity were equally potent, the mean in such a case would be stupidity, for one would neither give nor withhold; likewise if, through their equal force, anger and fear were brought into equilibrium, there would result neither anger nor fear; and so of the rest; for the keeping of passions within bounds is brought about by the mutual annulment of contrary motions.

FOL. 436v

Again, his declaring that artists are not evaluated according to the deserts of [their] art is not acceptable, for that assumption seems to contain more of the satirist's indignation than of tested truth. Aristotle, Plato, Democrates[25] and Epicurus were honoured, even during their lifetimes; Homer too was revered, though not made rich; Archimedes, the consummate artist, was held in the highest honour, not only with his king but also

---

[23] Or, 'for his own skill in his art'.
[24] Hobbes uses the word *passio* in two senses: (*a*) e.g. at 349v, and here, in the present sense of 'emotion', and (*b*) as meaning 'being acted on'—the patient is acted on by the agent, e.g. 454v. Hence the present context does not imply an antithesis.
[25] An obscure sculptor. So the text; read 'Democritus'?

FOL. 437   with the enemy.[26] In sum, in every branch of art the engineers, the geometers, the philosophers and all the poets who excelled in their skills attained a fitting reputation. The reason why others, I myself included, are without honour is either [a] that we are neither craftsmen nor of the best, or [b] that if we are [excellent] we are not recognised as such. However, one must not expect those civil honours, which are connected with public business, to be conferred on philosophers: it would be unjust to the human race to withdraw from their studies for public responsibility those minds that strive for the good and the honour not only of one country alone but of the entire human species.

15. Next, he had cited against his own case the happiness of wicked people as an argument apparently proving that the world was founded for their benefit rather than for that of the good. In what follows he strives to demonstrate that even in this world (quite irrespective of the future blessedness in the Kingdom of Heaven) the happiness of the good is greater than that of the bad, because [i] they feel pain less, [ii] their delight

FOL. 437v   is both longer-lived and more intense than the pleasures of the bad, and [iii] the good have a larger share than the bad even of those pleasures that seize and hold evil people.

Under [i], that the good suffer pain less than the evil do, he ascribes sorrow to a consuming desire for things that cannot be won or kept, or at least only with difficulty. (To this class belong things that are rare, that many people fight over, and that become tainted in the use.) But these, he says, are objects of desire not for good people but for those under the sway of their passions, i.e. for the evil; hence the good feel less pain than the evil.

This argument supposes, it seems, that a good man is neither someone who yearns for food or for something else pleasant that perishes in the enjoyment; nor someone who craves riches or grandeur, for many strive over these; nor he who passionately desires knowledge because it is won with difficulty. In sum, the good man is not a person who desires anything at all; appetite is a passion; indeed, if it has been kept under control, this is because of some other passion, a contrary one.[27] Those

FOL. 438   who desire, then, are always given over to the passions, and

---

[26] I.e. the Romans besieging Sicily.
[27] Cf. fol. 429.

are therefore evil. Moreover, they who seek rare things, those hard to obtain, and those competed for, are no more the prey of the passions than are they who desire common and easily secured things and things that others do not want. (Provided only that the latter persons cherish the objects of their desire equally strongly as the former do.) If, therefore, a good man seeks with dedication to perform what the laws require; if he should try very hard to aid the wretched, and to do things of that sort; if he is very angry at those he sees cursing God or behaving like that, [such men as he] are as equally given over to the passions as are those who are obsessed with the search for riches![28]

According to this teaching, then, those who do their utmost to perform what the laws demand, etc., are evil; which is ridiculous. On the contrary, the good are to be distinguished from the bad not by the vehemence but by the objects of their passions. Hence an evil man is one who strives towards unlawful things, and a good man he who strives for things only if they are lawful, whether he do so vehemently or not.

16. [ii] That (as is seen by comparing them) the pleasures of good men are longer lived than those of the bad he proves as follows. 'If', he writes, 'you compare the extension [of the two kinds of pleasure] notice in what direction victory lies.[29] The pleasures of the zealous are: to do good to those whom you can; to have regard for your country and its citizens; to shield the needy against the insults of those stronger; to comfort the wretched; to restore the afflicted; to offer counsel to, and defend, those in want, and so on.' What the pleasures of the bad are he does not go on to say, but I think I understand them to be the acquisition of riches; the painstaking search after honours, i.e. the civil dignities that are to be won; and the

FOL. 438v

---

[28] 'I remember, [on Hobbes's] goeing in the Strand, a poor and infirme old man craved his Almes. He, beholding him with eies of pitty and compassion, putt his hand in his pocket and gave him 6d. Sayd a divine (*scilicet* Dr Jasper Mayne) that stood by, would you have done this if it [had] not been Christ's command? Yea, said he. Why, quoth the other? Because, sayd he, I was in paine to consider the miserable condition of the old man, and now my almes, giving him some reliefe, doth also ease me.'–Aubrey.

[29] This is of course meaningless outside its context. White has been contrasting the permanence or the outward-looking nature (*extensio*) of some pleasures with the inward-seeking nature (*intensio*) of others. He finds the former preferable.

having to hand of all that can gratify every sense. If this differentiation be true, then those who wish to be good and have previously paid attention to this, should be evil! Now, those who want to do good to others must previously prepare riches [for this purpose]. Riches, however, cannot be provided unless they are both strongly desired and diligently sought after. So an earlier pleasure (which characterises an evil person) precedes the pleasure of conferring benefits, which is the hallmark of the virtuous! Likewise since the natural law forbids you to attend an assembly unless summoned, it is a sign of a bad man to take thought for, i.e. to give counsel to, his country and citizens,[30] unless they summon him to that office. This is evidenced in our land's present civil wars, the sole cause of which was that certain evil men who were not called to office thought that their own wisdom was less fairly valued [than it deserved] and advised the citizens to take up arms against the King. Before a righteous man can counsel citizens, then, he must be elected to office, i.e. he must acquire rank; and without effort and industry this is impossible. Therefore if it be the sign of an evil man to crave honours, but of a good to minister counsel to his country, then someone who wants to be righteous should become bad! In the same way, to shield the needy, to raise the afflicted, and the like presuppose the setting aside for this purpose of riches and power, things that only the evil hanker after. Again, I do not see why the unrighteous cannot also take pleasure in doing good, in protecting, in advising, in giving succour, etc. Indeed, when comparing Caius Caesar with Marcus Cato, Sallust believes the latter to have been the better man. It is to the former, however, that he attributes qualities which, in the present context, our author says befit the good man. Yet Sallust ascribes to Cato a different quality, saying: 'Caesar won glory through gifts, help and forgiveness; Cato by never stooping to bribery'.

Next, White proves that the pleasures of the good are greater, i.e. more intense, than those of the evil. 'Those benefits are the greater (he says) by which men gratify themselves; that is [a], to contemplate the natural and most moral side, or those principles under which life is constantly ruled by reason and understanding, and with no need of another guide; or [b], as our philosopher writes with great feeling,

[30] Cf. fol. 426.

CHAPTER XXXVIII: 16

> [To occupy] the temples–those high raised
> Upon the placid teachings of the wise.
> From here you can set eyes on others, glimpse
> Them wandering to and fro, seeking the path
> Of life in fruitless quest.[31]

Against these words assess the actions of the imprudent and of those who seek after bodily pleasure because it always exists and can be indulged in without hindrance. Yet who envies these things in someone else? Who hastens to snatch [them from him]?'

I would not deny that the happiness which Lucretius describes in the place quoted is great and lasting, namely 'to view, from a temple raised high upon the placid teachings [of the wise]', the wanderings of others who seek in vain the happy life. Such happiness is the greater in that he who is placed at so high a peak of true knowledge can observe not only the limited wanderings of the ignorant (for people who wander about at a slow pace, and hesitantly, err less than those who leave the path) but also the wanderings of those who, believing themselves to be looking down from temples of no less height upon the wanderings of others, themselves err the most.

FOL. 440

I must observe, however, that the question [under examination] is here altered. The comparison being drawn is no longer between the joy of the good and that of the bad, but between that of philosophers and non-philosophers. But [our author] may consider 'a good man' to be the same as 'a good philosopher'. I do not think this admissible, and it will be denied by anyone who believes that a good man is he who keeps the decisions of his fathers and the laws and statutes of his country. Must one be a philosopher in order to keep the laws? Rulers will say 'no'; the ignorant will say 'no'; the very philosophers will say 'no' as they squabble among themselves.[32] Finally, it will be denied by the practices of many philosophers who cut one another about with abuse and do all the things the rabble do–the more disgracefully, because they set themselves above the common herd.

I note, further, that this pleasure of philosophers is no more

---

[31] Hobbes (see four sentences later) agrees with White's version of these lines, but most modern translators interpret differently: '[To seek] the bright places high raised upon the teachings', etc.

[32] Or, 'because they all agree on this point, though they may differ on others'.

FOL. 440v the lot of those who philosophise truly than it is of those who philosophise falsely. Whether someone has known truth or has only thought he has, [?his] happiness is the same and pleases him as much [as the happiness of true philosophy would please him]. Hence this pleasure belongs no more to the good than to the bad, or to philosophers than to those who are losing their reason.

Finally, [our author's] belief that this delight is enhanced because no-one is jealous of someone else's having it, [and] no-one hastens to seize it [for himself], seems inadequate to commend such delight; for there is most weighty proof that we envy happiness in someone else. Unless it proclaims [its existence], happiness is incomplete, for with what aim do philosophers write so many books, if not to have as many witnesses as possible to their learning? But I turn to the remainder of this Problem, in which he endeavours to show that the virtuous possess more of the external benefits than the evil do: power, respect, friendships, and physical delight.

FOL. 441 17. [iii] That [they possess] power he argues as follows: 'We grant power to those from whom we expect fair-dealing and peace; now, we expect these not of evil persons but of good; so we confer power upon good people, not upon evil'. In this reasoning it is untrue that we do not expect fair-play and probity of evil people if at any rate we do not know they *are* evil.[33] If we are discussing the evil people whom we [already] know to be so, and whom we entrust with power nonetheless, the first proposition [of the syllogism just quoted] is false, for those from whom we expect fair-dealing we do not [necessarily] entrust with power. Again, it is not universally true that we give power to those from whom we expect fair treatment: in order to save our lives, when vanquished we cede to the victors, be they good or evil, the power over us.

FOL. 441v 18. That the virtuous command respect he argues thus: 'We honour others because of their virtue and knowledge; but the evil have no virtue and knowledge; therefore not the evil but the good are honoured'. This is the same paralogism [as the last], in that we honour people because of [our] opinion

---

[33] A tortuous way of saying (presumably): 'If we do not know that they *are* evil (or, until we know this), we may expect good of the wicked'.

of [their] virtue and knowledge, but we may also have the same opinion about the vicious, not knowing that they *are* vicious. The first proposition [of the syllogism immediately above] is false, for we honour people because of our opinion not only of [their] virtue and knowledge but also of their other powers, such as civil power, riches, and high birth.

19. That the virtuous have more friends than the wicked have he reasons thus: We make friends through [mutual] advice, help, consolation, and agreement over many things and disagreement over few; but this cannot be expected of bad people; therefore it is not the evil but the good that have friends. [His] second proposition is patently false, for what about Achitophel? Was he a good man? Was he particularly backward in giving advice? And depraved though they are, do not conspirators aid and comfort one another, and share common designs?

FOL. 442

20. Lastly, in order to demonstrate that even that pleasure seen [to derive] from banqueting and amours[34] is greater in good people than in evil, he uses an argument like this: 'If you derived pleasure from feasting, from love-affairs, or from any other of the senses (he says), wouldn't you take steps beforehand, by means of a previous strong desire, or fasting, to make your body best prepared?'[35] And a little later, "'Say some prudent and wise person exercised his judgment and the whole force of his mind [to creating for himself a pleasure or a power that was firm, enduring and prosperous].[36] Then imagine that someone else tried to do the same hastily, with precipitation, and by fits and starts. Which of the two men, possessed of the same faculty in other respects and of the same means of indulgence, would be the more likely to achieve his desires?" "Only too clearly he who used method and judgment; otherwise we should have no confidence in reasoning."'

---

[34] The pejorative sense in thus translating *amores* is perhaps justified here by the preceding word 'even', though in the context 'mutual affection' may seem preferable.

[35] White means either that previous continence will make the vice more pleasurable, or that previous abstemiousness may make the body very fit and able to resist the excesses that are to come.

[36] These words, not included by Hobbes (or omitted by the copyist), are inserted by Mersenne from the printed text of White; see the opening sentence of the next paragraph.

FOL. 442v   Then he says: '"Again, the one who used reasoning in the way I have suggested, do you think he would be given over to the passions or not?" "Quite clearly he cannot be a prey to the passions."'

These words, here abridged,[37] have the [same] force as if White had reasoned thus: 'He who, through eager anticipation, prepares himself for pleasure–the unpleasant one we are discussing–does he derive from it a pleasure greater than that of a person who does not so prepare himself? But the same person is prudent and wise, and he uses reason and method; but he who uses reason and method is not given over to his passions; in sum, he who is not given over to the passions is a good man; therefore a good man takes greater delight in banquetings, etc., than a bad'.

(i) I deny that such an art or method of preparing the body for such pleasures is any part of that [power to converse] by which men are marked out from the brute animals; it is a

FOL. 443   wickedness stemming from the recollection of past wantonness.

(ii) By declaring that those who prepare themselves in this way are not given over to the passions, he contradicts himself. Someone who employs artifice in order to become more aroused is more exposed to the passions than he who is the victim of a longing that arises only from nature.

In conclusion, he who is prudent and wise as regards preparations of this kind is not a good man; he merely shows ingenuity in his wickedness.[38]

---

[37] Probably a reference to Hobbes's omitting the clause later inserted by Mersenne (see last note); but Hobbes may mean: 'White is speaking briefly; he could have said far more if he had wished'.

[38] Fol. 443v is blank.

CHAPTER THIRTY-NINE

❧

On Problem 17 of the Third Dialogue:
Why most men fall from happiness

1. Causes impairing happiness. 2. That the reason why people are turned aside from [the pursuit of] happiness is not 'that events *may* occur' (i.e. contingencies). 3. That non-existence is not preferable to being in eternal torment.[1] An examination (4) of an argument to the contrary, based on the nature of goodness, and (5) of an argument taken from our evaluation of miseries.[2] 6. In which is sifted the reply to a quotation from Holy Scripture. 7. What kind of philosophy our author believes is to be made subject to theology, and what kind not. 8. A scrutiny of the reasons why he thinks that true philosophy is not to be thus subjected.

1. If by 'happiness', as in the Problem immediately above, we understand 'worldly happiness'; and if this consists, as our author has wished it to, in riches, titles and physical pleasure, then the many reasons why most men do not attain happiness, and why these are different for different people, come to light immediately. Most of these reasons do not depend on choice: as some persons are born to riches and great place (namely those descended, as to their families, from the affluent and the noble), and are thus fully possessed of the understanding and effort required if they are to become fortunate, so others are of poor parentage and obscure origins. Hence the latter must toil for the greatest part of their life to acquire possessions; and in the remainder of it, oppressed by the inconveni-

---

[1] See the note at fol. 448v.
[2] Perhaps, '...our evaluation of the wretched'.

ences of old age, they cannot use what they have secured. Again, when they give themselves up to vices and lust, those who even at the threshold of life have great riches sometimes lose their physical health and at other times experience a surfeit. Whichever of these two forms it takes, every bodily pleasure decays. Further, such riches are sometimes lost through imprudence, i.e. through unforeseen troubles, which closely attend joys. Lastly, even if one has been very prudent in choosing what is for one's good, nevertheless upon any one choice [made] there depends a chain of good and bad things so lengthy that one cannot foresee all the evils it contains. These unforeseen evils, which our author calls contingent, sometimes obstruct the path to well-being. Such are the reasons why men do not win or retain that happiness which he says consists in riches, honours, and physical pleasure.

Now if happiness is based on a steady progression from the acquiring of one good thing to the acquiring of another, then—to leave aside unexpected evils—in general the following impair[3] happiness: (i) someone's entering a path which there is no means of leaving, as he who pursues unlawful practices and sustains punishment, or who (at the least) dissipates his effort and slides back; and this is to feel remorse; (ii) someone's wanting [something] but lacking the strength to get it. This gives rise to intellectual sniping at, and to envy of, others; and (iii) his setting his course, vain-glory pressing him, towards the contemplation of the things already secured [by him]. This last is but a brief joy and ends in a brutish dullness of sense. For instance, when Nebuchadnezzar congratulated himself on the dimensions of his own power, his joy was short-lived, for he had only the mind of a beast and was later deprived even of a human body. In sum, whatever reason precludes a runner from performing well on the track, [this] same reason may prevent a person from achieving prosperity in this life.

For the rest, if we look at true happiness, namely that which has been laid up for the faithful in a future age, we see that the sole cause of enmities is lack of faith. All men without exception fall into three categories: (i) the main one, consisting of those who themselves know what it is best to do; (ii) those who, knowing nothing themselves, admit that they must be ruled by those more wise; and (iii) those who neither know [anything]

---

[3] Or, '...to leave aside the unexpected, in general the following ills impair...'

themselves nor will trust others. In the first category are those who know the path to happiness, namely those to whom the Holy Ghost has pointed the way through supernatural revelation; such were the Apostles, and such now alone is, and ever will be, the Church, which coalesces into one Person.[4] In the second category are those who, having no supernatural revelation accorded them, admit that they must be ruled by one who *has* [had this revelation], i.e. the Church. In the third are those who themselves neither know what the path to a future life is, nor believe in the Church, nor perform what the Church teaches as being necessary to the winning of felicity. So the main reason why men fall out of that everlasting happiness is a lack of belief in the Church, i.e. simply a lack of Christian faith.

2. The reasons our author assigns [for this failure] are different [from mine],[5] but he seems to be collecting them all under one head at the opening of the present problem. 'I suppose', he says, 'that the genus man occupies the most inferior station of [those held by] the creatures endowed with understanding. For what can be a worse consequence than that [men] are so many instruments towards an end determined by nature, even though [their] motions [towards it] are so difficult?'[6] Indeed, [I observe,] of the creatures we call rational the lowest is, without dispute, humankind, yet this is not the humblest of the creatures endowed with understanding, because sense, imagination and memory we share with the beasts as modes-of-perception even though the latter are not forms of reasoning such as syllogisms. White, however, divides this cause[7] into three parts. These he calls 'the three operative grades of contingency' that bring men to the attainment of their end as

FOL. 446v

---

[4] So the literal rendering of the last clause. It may be of no significance that the MS. spells 'Persona' with a small letter; but possibly *cuius regio, eius religio* is intended.
[5] Or, 'different [from one another]'.
[6] A near-literal rendering of an obscure and doubtful passage, the last sentence having intermediary punctuation neither in the MS. nor (except for one obvious error corrected by a reader in the British Library copy) in *DM*. The context does not make clear whether or not Hobbes himself understands this extract fully. It is perhaps more satisfactory to insert a question-mark, producing: 'For what is worse? May we expect, in consequence, that...' Things would be rather easier if White had said: 'Let us suppose' instead of 'I suppose'.
[7] Presumably the first of the 'reasons' named at the start of the article.

FOL. 447

expeditiously as possible. 'From this (he continues) I see that there must necessarily occur three grades in man, all these operating by contingency, of which grades individuals take the earlier ones as prerequisites to achieving the end we seek', etc. (From the words 'I see that there must necessarily occur', we should note, incidentally, that here we are not to take 'contingency' as opposed to 'necessity'.)

The first grade of contingency he places in the physical constitution necessary if the body is to thrive, the second in the discipline of conduct, and the third in an adherence to what is taught us about a future age[8] which we do not [yet] know. He wishes the fabric of the human body to be the more subject to contingency because it is set apart from the bodies of other [creatures] by reason of the greater sophistication of its workmanship. Now, the discipline of conduct is not complete; this, he says, gives rise of another contingency: that one is properly educated.[9]

He concludes by saying: 'In order to be blessed in the future state we must unlearn and shun everything we have grown used to in this life and must cultivate what we do not know'. Hence he postulates the third [grade of] contingency: '...that one be blessed in it'.[10] Any one of these grades, he says, must depend upon the one preceding, i.e. if we are to achieve discipline of

FOL. 447v

conduct, everything dependent on the fabric of the body must occur 'correctly'.[11] For a man to unlearn what he is familiar with, [White concludes] and to hold fast to the [yet] unknown things concerning a future state, i.e. to believe in the Church, i.e. to be inspired by the Christian Faith, everything to do with the discipline of conduct must befall him 'rightly'.

As for (i), it does not seem necessary that, if discipline of conduct is to be attained, everything requisite should arise from one's bodily make-up. For geometry, physics, rhetoric and similar branches of knowledge a strong bodily constitution

---

[8] Perhaps, 'world, state'; cf. fol. 424v.
[9] 'Complete' (*perfectus*): 'perfect, absolute'. This 'other' contingency seems to be a 'part two' of the second contingency.
[10] Or, 'Who may be blessed therein?' Hobbes seems to be taking White's third grade-distinction as meaning: 'Earthly knowledge has no reference to the future life'.
[11] Again obscure: see also the sentence following, where Hobbes seems to take White as saying: 'If a man is to profit from religion, the rules of personal conduct (over which, ultimately, he has little personal control) must help him to do so (because they are divinely determined)'.

is perhaps necessary; hence a person skilled in [one of] these subjects was born to it. As for the discipline of conduct, however, it is enough if one has the will to obey the laws of Church and state, i.e. those of the Church *simpliciter*; for conformity to the civil laws is included in conformity to those of the Church. Now, I do not see why anyone, whatever his physical constitution, should not wish to obey the laws of the Church: discipline of conduct requires no knowledge other than that of the [civil] laws as published, which is gained from public proclamations. Indeed, someone who with great lucidity disputes about the virtues is not instantly a person best instructed in the rules of conduct but is one who very diligently seeks to learn what the laws of state and Church are, in order to comply with them.[12]

FOL. 448

But that second point–that discipline of behaviour is needed if one is to unlearn what one has learnt and to believe in what one does not know–is in no way true.[13] It is the holding to that discipline of obedience, but not a going beyond it,[14] that causes a person to obey the Church, i.e. by faith to cling and grow accustomed to the unknown things concerning a future state. So it seems we must not concede that future happiness is either lost or gained through contingency; it is won through someone's seeing the path in advance himself, or his believing those who do so, but it is lost when someone neither knows the path himself nor believes those who do.

3. From this, White slips into the question: 'Is it better not to exist at all than to suffer eternal torment, when happiness is lost?' In my opinion, not to exist is preferable to existence in such a state. The nature of goodness consists in its pleasing [us] or in its being eagerly desired, but of evil in its being repellent or despicable. We may say, then: 'Nothing is worse or more to be shunned than the greatest evil, from which we can never extricate ourselves'. But if 'not to exist' were *worse* than to

FOL. 448v

---

[12] Is not the converse expected?: 'A person best instructed in the rules of conduct is not he who disputes about the virtues, but rather he who seeks to learn the laws of Church and state'.

[13] The text, here ambiguous, may mean: 'But according to that point, i.e. "discipline...does not know", this [i.e. the closing words of the preceding paragraph] is in no way true'.

[14] *Adjiciendo*; i.e. not adding to that discipline through unreasonable self-denial and the like?

suffer everlasting affliction, 'not to be' would be more hateful than torment, because clearly non-being endures no pain; therefore [non-being] is not evil. That we prefer a torment *which lasts [only] for a fixed, finite, and short period* to perishing or to death[15] is due not to the loathsomeness associated with death or with our ceasing-to-be, but either to the hope of receiving, with life itself, the joys of life, or to the fear of [suffering] pain as we die.

FOL. 449

4. Our author holds the opposite view, and does so in two arguments, the one *ex natura rei*, the other from the opinion held by those who endure evil.

(i) His argument *ex natura rei* runs thus: 'Come now, if we consider nature, what in nature does something better than someone else does it? Or, under what heading is improvement to be examined? Improvement comes under the heading *ens* [and is better placed under the heading *ens*][16] because it embraces more *ens*'. In consequence, because he who is tortured eternally has a greater share of *ens* than he who does not exist at all, it is better to be in eternal torment than never to exist.

In this reasoning it ought to have been proved that one entity has more *ens* than another. For the philosophers deny that *ens* is found in substances to a greater or less degree; indeed, one substance can be larger or smaller than another, but it cannot be to a greater or lesser extent *ens*. This conclusion is not, as he undertook at the start of his book [to adopt as his general policy], confirmed by reasons 'taken wholly from nature', nor does religion suggest we must believe that one entity has more

FOL. 449v

ens than another. In no way, therefore, is the above proposition to be allowed.

(ii) It is untrue that a good and a bad thing are [respectively] called more good or less good, or more bad or less bad, according to the extent to which they are entities; they are so called according to the degree to which they please. As far as we have

---

[15] Translator's italics, here and in the previous sentence. Without them, the narrative would seem to stray beyond the wording of the article title given at the head of the chapter, though the title represents White's proposition. For 'because clearly' the context seems to require 'even though'.

[16] The words within square brackets do not occur in White. They are of less doubtful meaning once it is conceded that 'the more *ens* there is, the better'.

seen up to this point, non-existence is preferable to a great and everlasting sorrow, unless, on the evidence of the judgment and the experience of those who endure such torments, our author has shown that things are otherwise.

5. He now comes to this. 'The torments that nature shows us to be found among the wretched', he says, 'are the states of mind of these very people. Of envy the poet [Horace] writes: "The Sicilian tyrants have not discovered a worse torment". All men know pain and how heavy an evil it is–what we call "sadness" I call "pain". Moreover, the fable of Narcissus demonstrates yearning, and many experiences testify how evil it is. It is not uncommon to call to mind men who have died of fear. But I should like to learn this of you: You've never heard, have you, of an envious person who did not make the most of envy itself? The poets sing of yearnings and jealousy, as though lovers took no pleasure in anything but these, etc. If all this be true, do you think these wretched people would want to be without such hardships?'[17]

FOL. 450

Suppose it is true that the greatest and everlasting torments that await the wicked in a future age are none other than envy, covetousness, sadness, fear, and similar mental stresses. Without doubt there are a great many people who would rather endure them than endure non-existence, because associated with them are the good things that outweigh them: with envy, a pleasant expectation of the evil that can befall him whom we envy; with yearning, the pleasant hope of getting what we want; with sadness, the hope of a change for the better; and with fear, the hope of running away and saving oneself for better things. This must be allowed as true from [the case of] someone who has said: 'Such pitiable people *want*–even love–their own wretchedness'. Again, the only reason why something appeals to, or is craved by, a person is that it benefits him, self-advantage being the proper and sufficient object of the will. If, then, someone in such a state desired not to exist, he would certainly prefer non-existence to living in ease. The argument based on such wretched people's preferring their condition to non-existence is therefore invalid: on the contrary, we should

FOL. 450v

---

[17] Where Hobbes has 'etc.', White has (p. 431): 'Daily we strive to seek out dread darkness (*moestos tenebras*–sic) and everything that makes evil greater; the timid eschew evil, but cherish this fear of theirs'.

refer the question to those who are tormented not only in mind but also in body, and this with very great and everlasting tortures beyond hope of relief. If the last-named persons, as well as [the others], say they would rather be thus than not exist, then we may believe that to suffer eternal torment *is* better [than non-being]—otherwise not.

Now, the question posed concerns evils of the same kind [as those just referred to], for, as White himself says a little earlier, 'Philosophy acknowledges [the existence of] torments even in a future state; and she herself has not hesitated to swear that, if you compare with them those which our present life endures, the latter are, in comparison with these, nothing but sports, flowers and roses, and delights'. In my opinion, however, philosophers can maintain this truthfully, not indeed as knowledge gained through philosophy, but as a result of our belief in the Holy Scriptures and in the authority of the Church.

FOL. 451

6. It seems that an objection [raised by White himself] hinders acceptance of the above doctrine.[18] This is the place in Scripture where Our Lord tells Judas the betrayer it were better he had never been born. White answers: Those words mean that it were better for Judas to have been born prematurely and carried from the womb [directly] to the grave, but not that he should never have existed, 'if the souls of the prematurely born indeed survive, as those of the rest of mankind do'.[19] This suggests that Our Saviour avoided saying that it were better for Judas if he had not been conceived (for those who have never been begotten can have no souls). Whether such was Our Lord's intention, however, I am not examining.[20]

7. Finally, White saw that an opportunity to complain about the way in which the world is run arises from dogmas of the following kind: 'What men do of their own free-will depends on causes linked together from all time', and 'By the law of nature most men lack caution', and 'It is through imprudence that they are evil', and hence wretched. Conse-

---

[18] Hobbes seems to refer to the doctrine that non-existence is not preferable to eternal torment, i.e. the topic of Article 3.
[19] This quotation from White seems to attach to 'it were better...to the grave'. There is good cause to alter the order (followed here) in which Hobbes places the members of his ill-constructed sentence.
[20] Malebranche (*Recherche*, IV.v.11) uses the same allusion in a similar context.

CHAPTER XXXIX: 7

quently such philosophy must have been most ill-conceived, because it is offensive to theology. So White stumbles on the question: 'Ought theology tell the philosophers what opinions they should hold?' He distinguishes philosophers into 'those who truly philosophise', i.e. 'those who proceed by a sure route and by the settled path of demonstration', and 'those who make a display of philosophy but in fact use only logic (i.e. the ability to argue in either direction) when handling the *materia philosophica*.[21] As the mind of the latter people is of itself wavering in its opinion, it rightly submits to control by theology, and it defers to this one control alone; true philosophy never so bends'.

FOL. 451v

In this distinction the several parts are not sufficiently differentiated. 'To proceed by a sure route and by the settled path of demonstration' applies only to logic; 'to be able to argue in both directions' stems from the subject of rhetoric. Now, there are many who think they are making a [philosophical] demonstration, though actually nothing is less like it than what they are doing–indeed, the very fact that they are claiming to make one shows that they do not know what a demonstration is. There are those [too] who assert that they are making demonstrations in physics, theology, geometry and metaphysics; but no-one claims to do so for logic, because no demonstration exists except that based on logic. To whatever material it is applied, therefore, a demonstration is nothing but the advance from the definitions or explanations of names to the conclusion to be proved by means of syllogisms shown to be true. Which of the two roles[22] our author has taken upon himself is fit subject for consideration by someone who would wish his own philosophy to be demonstrable, incapable of distortion, and not answerable to theology. As for those who use logic, let [our author] require that [their philosophy] be governed by theology.[23] Certainly, for my part, everything I have said seems to me to have been demonstrated; nevertheless, because many things seem [to be demonstrated] which are not,[24] I am willing to submit all my claims to the religious authorities.

FOL. 452

[21] '*When handling...philosophica*' may go with the bracketed words; the original (both in the MS. and in the *De mundo*) lacks punctuation.
[22] I.e. of those who only imagine themselves to be disputing correctly; and of those who extend the realm of logic to extraneous disciplines?
[23] Dubious. It is just possible that Hobbes means: '...let [their] theology insist that [logic] be subservient [to it]'.
[24] Or, 'Many things seem to be what they are not (...to exist that do not)'.

8. There remains the argument by which he proves that true philosophy is not subject to theology. These are his words: 'The highest and best part of philosophy is none other than the way, and the preparations for the journey, to the truths of theology. But how you can want theology to rule over philosophy I do not know, for theology claims neither to know nor to teach philosophical matters'.

FOL. 452v

The first part of this proof I can use to show the contrary, as follows: If the dogmas of philosophy are indeed a way and a journey to those of theology, the truths of theology ought to be understood and adhered to before we make up our minds about the way, i.e. about the dogmas of philosophy; and the path can be known only by comparison with the goal we are making for. Hence philosophy, which is the path, is ruled by theology, which is the goal. To the latter part [of his statement just quoted] I say: 'Either it is untrue that theology does not teach things philosophical, or else the things which our author says are concluded in accordance with "reasons taken wholly from nature" are not philosophical'. To settle whether the world is infinite, whether unique, whether existing from all time, whether eternal; likewise whether the human soul is immortal, whether free-will exists, and whether or not there are incorporeal substances lacking quantity;[25] likewise what titles we must set upon God, and whether He could have founded a better world; likewise concerning the blessedness and unhappiness of a future age–all these [five] questions, which are discussed in the present Dialogues as philosophical matters, the theologians claim to determine. Moreover, what White says he has attempted in [using these] five metaphysical elements[26]–that he has penetrated both to God and to the angels–all this is avowedly the function of theology. Either, then, these are not philosophical matters; or philosophy, even the true, must be subject to theology. However, when I say 'theology' I am saying 'the leaders of the Church', whose task it is to regulate all dogmas; for these will be seen to be able either to strengthen or to overthrow the fixed tenets of faith.

---

[25] Or, 'whether or not incorporeal substances lack quantity'.
[26] I.e. 'metaphysical in their nature' rather than 'included in the subject of metaphysics'? 'Elements': basic materials?

CHAPTER FORTY

❦

On Problems 18 and 19 of the Third Dialogue:

[Must the world come to an end?]

1. The state of the question: 'Must the world come to an end?' 2. An explication of the definition of 'motion' in Aristotle. 3. Our author's explanation of this. 4. The term 'end' ambiguous. 5. An inspection of some parts of an argument based on the nature of motion. 6. The structure of the whole argument set out. 7. That an argument from the nature of the infinite is invalid, as is (8) that from the nature of things.[1] 9. The definition of 'world' according to our author.

1. In this Problem, Number 18, it seems to me that one thing is put forward for disputation, but that another is disputed. Now, by 'world' we understand 'the aggregate of all created bodies, or at least everything included within the ambit of the empyrean heaven'; so the world will be said 'to come to an end' or 'to be no more' only when all bodies that comprise it have been brought to nothing. If the earth is reduced to ashes; if the sun and stars disintegrate into atoms, or the face of the present world becomes different, this is not called 'perishing'. The reason is that all these things make up the aggregate of bodies that we shall always understand by the name 'world'. Now, undoubtedly God, who created all things from nothing, can reduce every one of them to nothing. He will be thought unable to perform this, however, if we are to believe what our author lays down in Problem 3 of the First Dialogue. Here he claims [at folio 23] that a vacuum cannot be contained within a

---

[1] Cf. fol. 449 above.

body because the sides of the surrounding body will close up in order to force it out.[2] Hence, if the things that exist on all sides from here [earth] to the empyrean heaven were to be destroyed, the entire concave sphere of such a firmament would necessarily collapse into one point, i.e. its own centre. Whichever [account of 'perishing'] is the true one, what he argues in the present context is not whether the world itself, but whether its motion must come to an end. He declares that it must, basing his arguments on the nature of motion and of infinity, and on the nature of things.

FOL. 454

2. He therefore takes the following definition of motion from Aristotle: 'Motion is the act of *ens* in potential, so long as it is in potential'.[3] In order to understand Aristotle's sense correctly in these words of his, we must look at the example Aristotle uses to explain the definition. It runs thus: 'Let us imagine some stones being delivered at a definite site where a house is to be built, and being collected and made ready in some part of the house under construction. This delivering and this preparation clearly constitute a movement of the stone, i.e. the very motion by which one stone unites with all the others in the form of a house. Now, while being transported, all the stones were not yet a house, but were of a future house; at this stage, therefore, they were said to be "a house in potential" or to have the potential to become a house'. So Aristotle noticed that this potential of the stones consisted in their becoming a house if any one of them were moved to a definite position and were put in its correct place. He therefore declared that such potential is nothing but the actual movement of a stone itself. Further, [he says,] not only with respect to

FOL. 454v

the motion of the stones that were carried but also in general, every motion is the act of that body which is in potential, inasmuch as it is in potential.

This pronouncement seems to be not so much a definition as [the description of a physical] property, for the words carry just as much meaning as if he had said: 'Any body may be understood as being or existing, though unmoved'. That a body without motion can perform, or be acted on by, anything, i.e. that it can possess any potential, is inconceivable. Indeed, it

[2] Because nature does not tolerate a vacuum?
[3] 'So long as...' (or, 'inasfar as'): When the potential of a body to move dies, that body is no longer in motion.

is unbelievable that anything acts or is acted on without motion, for action and being acted on are undoubtedly motion,[4] either of the entire body of the agent and the thing acted on, or of some of its parts. This property can be inferred from the true definition of motion by means of a chain of many syllogisms. A commendable and lucid definition [of motion] is: the continuous quitting of one place and the acquiring of another.

3. In order to make the Aristotelian definition fit his present purpose better, our author in his interpretation departs from Aristotle's sense as much as one possibly could. White says that motion is 'the act of *ens* in potential, inasmuch as it is in potential; i.e. (in other words) motion is either [i] the path of a thing that strives to perfection, or [ii] its advance towards perfection, as long as the thing does not possess perfection and yet its good genius urges it in that direction'. Aristotle, however, did not think that someone hurled into a pit tends to perfection or presses on towards it; yet such a fall very obviously constitutes motion. Again, those who kill and thieve, they are all moved. If Aristotle interpreted 'motion' in this way, then the same Aristotle must have believed that murderers and robbers are driven to homicide and stealing by a good genius, which is inconceivable.

FOL. 455

4. If we do allow the above interpretation [of Aristotle's definition] we may reasonably infer that that good genius which moves the earth has had either [i] a goal or objective towards which it moves the world or [ii] motivation towards, or a purpose in, the moving. We may not infer, however, that the world was endowed with motion with the express purpose that the motion so imparted, i.e. with the intention of [its reaching] an end, should at some time or other cease and be terminated. (Here 'end' is taken as 'limit', as is done in the present question.) Motion was imparted to the world because of a goal [to be reached]–whence he infers that the world is moved because it has an end. So the ambiguity creates a paralogism: for 'goal' taken as 'objective' and 'intention' does not mean the same as when it is taken as 'final point' or 'the *terminus ad quem* of the motion'.[5]

FOL. 455v

---

[4] Or, 'motions' (see in relation to the words following).
[5] Hobbes means that *finis* (here rendered 'goal') differs from *terminus* ('end').

5. Given also that motion exists because of the end [made for], he goes on to say that the end is superior to the motion itself; that a motion is good or desirable only with reference to the end; and that those motions in which the *terminus ad quem* is not better than, but is either equal to or worse than, the *terminus a quo*, cannot be keenly desired or sought after for their own sakes.

FOL. 456

This is true enough if, as regards motion, the *terminus ad quem* is the same as the goal of the mover generating that particular motion in things, or is [the motion's] target and objective. Granting, however, that God has imparted every motion of things because of some 'end' that the motion itself has, White infers that if humankind were removed [from the earth] the motion of the other creatures would have no 'end'. Hence, 'Only imagine that humankind were absent from our earth. Wouldn't the successive comings- and ceasings-to-be of the rest of the animals and plants, and of the two combined, be merely a kind of continuous motion that lacked an end?'

Isn't this the self-contradiction of someone who has previously undertaken to prove[6] that every motion will cease at some time or other? For how, if man were taken away, could the movements of everything else lack an end, since every motion were eventually to cease? Unless, perhaps, we replace 'the target of the mover' by 'the end of the motion', I do not believe that the motions of animals or of inanimates lack an objective. But even if we concede that, what follows? This: Centuries cannot succeed one another indefinitely.[7] Let us now see the whole connection.

FOL. 456v

6. 'Motion is the act of *ens* in potential, inasfar as it is in potential', i.e. [motion is] a progression to the perfection of him whom his good genius acts on. The nature of motion is therefore such as to give it some goal or objective; so no motion is good except as regards its relationship with its goal.[8] But the goal is the end; therefore there is no good or desirable motion, nor can motion take place, save with respect to the end. Remove man, and the motion of the rest of things lacks an end;[9] hence

---

[6] Cf. fols 40v, 305f.
[7] The text lacks adequate punctuation and other interpretations are possible. 'Centuries' should perhaps be 'generations'.
[8] Or, 'except because it has an end [to strive for]'. Cf. fol. 14v.

CHAPTER XL: 8

if you do take away man, motion cannot be activated or desired. That which cannot [thus] be set in train would, if it were done, be evil and inordinate. *Entia* will not accept bad rule;[10] hence if man be taken away no motion is possible; therefore the centuries (i.e. the motions of things) cannot be prolonged beyond the duration of the human species, i.e. after Resurrection Day, as was to be shown.

Such then is the argument (based on the nature of motion) by which he explains not only that the world must come to an end but also when it will.

FOL. 457

7. The second argument, one based on the nature of infinity, is as follows: If the world is to last for ever, souls must eventually be infinite (in number). There will come a time when someone will see included [in the period] between himself and Adam souls infinite in number. This, [I say,] is most manifestly false, because no sum or number is, or ever will be, infinite.[11] In this reasoning the consequent, on which hangs the validity of the whole argument, is untrue, namely that if the world were to last for ever souls would eventually be infinite in number.[12] Nay, even though a thousand myriad myriads of souls were brought into being daily, and this bringing-into-being never ceased, the number of souls conceived would never be infinite, but the number of those *to be* conceived would. These latter will therefore be said to be infinite in number, not because the number of them will ever become infinite but because they can never be counted.

FOL. 457v

8. The argument from 'the nature of things' is like this: 'Anyone who does not see that our globe depends on the sun is blind. We have explained above that the sun is nothing but a huge fire, but that a continuous fire is a flow nourished by, and dispersed from, the body which is consumed in feeding it. Hence this body must continually diminish and grow drier, as it cannot sustain so great a destructive force for ever. So because there will no longer be forces in nature to provide the nourishment required [to keep the sun aflame], this earth of ours must perish with it'.

Many men of science, and famous ones, have believed that

---

[10] Possibly: 'Inanimates must be properly controlled–by man'.
[11] Cf. fol. 10.
[12] Cf. fol. 330.

FOL. 458

in respect of its annual and daily motion the world depends on the sun. As no-one has yet furnished any proof, however, other scientists are not to be called blind if they do not share the belief. Undoubtedly the vicissitudes of the ages, the growth of crops, and many other things depend on the sun; but the certainty of the earth's being body, and of its existing and not being reduced to nothingness is dependent not on the sun but on the Author of all things, equally as the sun itself is.[13] Next, it is perhaps true that the sun is fire, but I do not think we must believe it to be a flaming mass, i.e. the material of fire such as kindling-wood, or that, in consequence, it needs to draw its nourishment from somewhere [other than itself]. Nor should anyone take this as a reason for thinking that the sun will shrink through lack of fuel, though there may be some people who already see that the sun, or the fuel it replenishes itself with, has diminished. Furthermore, if what White says be true–that the world depends on the sun–then the earth's orb will not be said 'to nourish' the sun: if it did, the earth would not depend on the sun, but the sun would depend on the earth. So if the sun receives no nourishment from the earth, then if the sun is extinguished the world will not necessarily perish with it. Therefore up to now there appears to be no reason why the world should seem about to perish or motion to disappear universally even though it were true that the generation of men and many other particular motions might cease.

FOL. 458v

9. Last of all, he sets down what 'the world' is. Why he does so I do not fully see, because, so far, the general view seems to be that we do not know. So whatever has been debated in these Dialogues about the world appears to have been argued incautiously and about a thing not yet understood, whereas every demonstration should begin with definitions. The definition [of 'world'] that he at last puts forward we deduce from, rather than read in, the following statement of opinion: 'The world is an instrument made for man, and in its parts, not illuminating of themselves but illuminated, are the habitations of different races of men'.[14] In other words, there are as many illuminated bodies as there are putative separate races of man; their size matches the number of men to be begotten,

---

[13] Cf. fol. 453v.
[14] Only up to 'for man' are White's words.

and their duration is limited. From the above definition [of 'world'] we may infer a sequel of sorts: that bodies which shine of themselves, such as the sun, are less noble than the rest of the illuminated bodies because the former are the more 'instrumental'.[15] Such questions, however, are too slight to be argued against; and whether they are true we shall not be in a position to say before the human race (which has happened to secure its dwelling on earth) has found some way of conversing with the rest of the species of living beings: the inhabitants of the moon and of the other planets.

FOL. 459

---

[15] Either: 'they serve a purpose' or 'they are more useful or necessary'. See fol. 446v and Chap. IX.

APPENDIX I:
Geometrical figures in the text

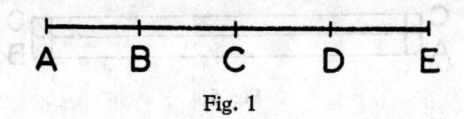

Fig. 1

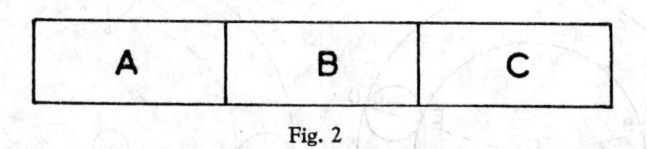

Fig. 2

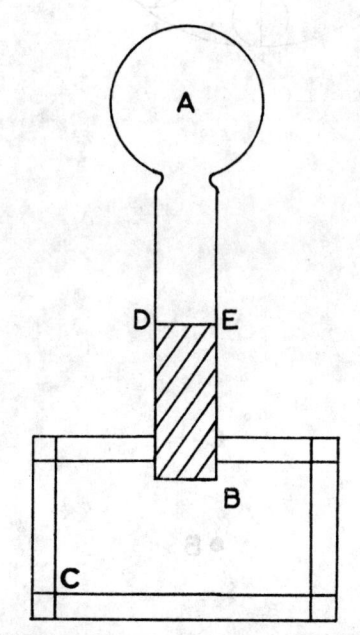

Fig. 3 Reproducing the proportions of the original.

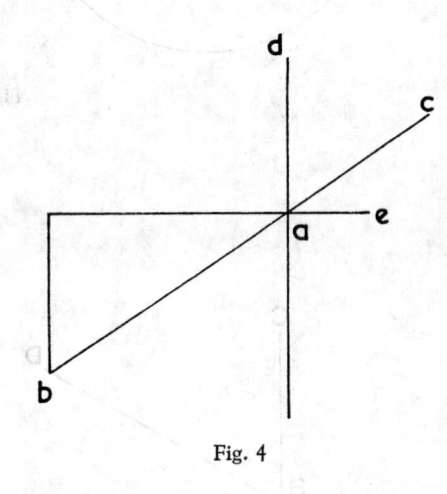

Fig. 4

Fig. 5

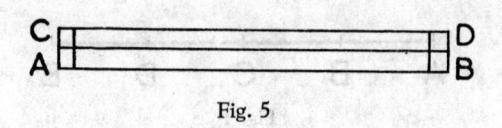

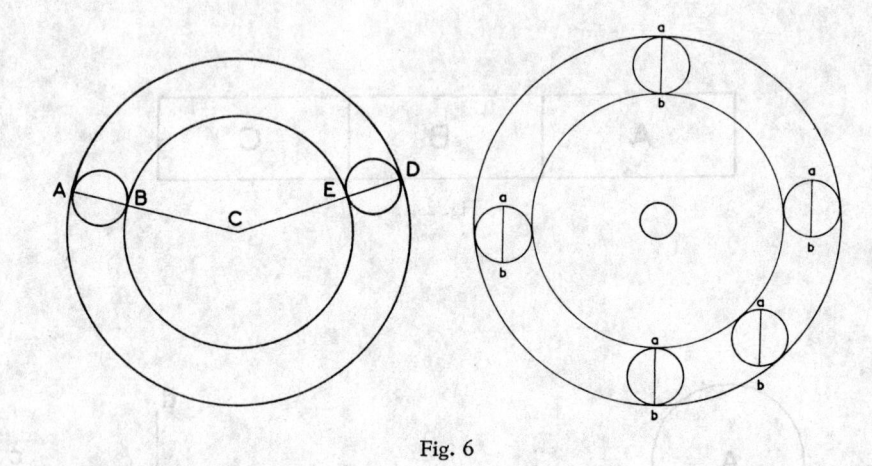

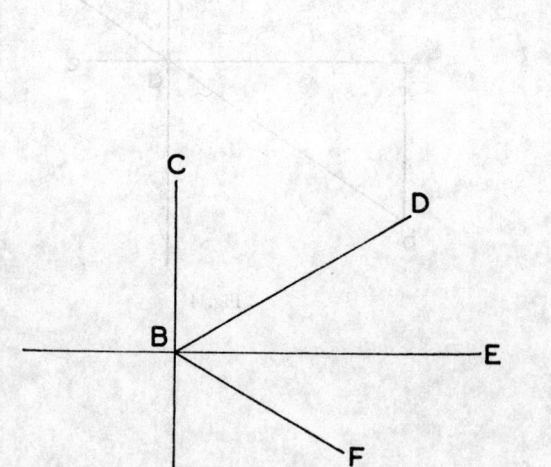

Fig. 6

Fig. 7

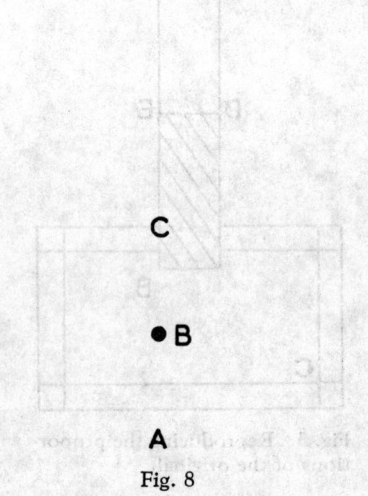

Fig. 8

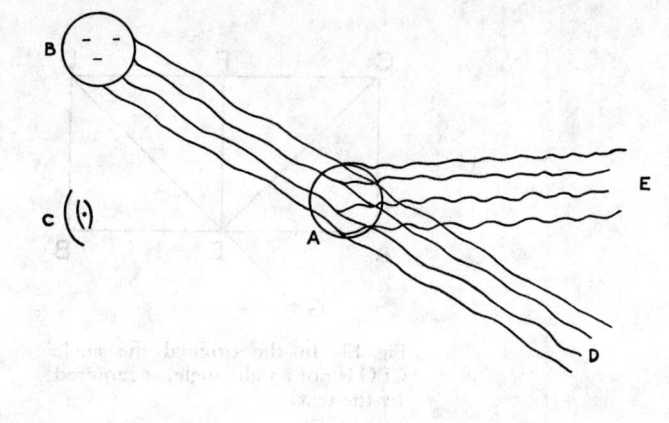

Fig. 9 In the MS. BAD was originally an obtuse angle, D being further from E than now. The copyist himself makes the correction.

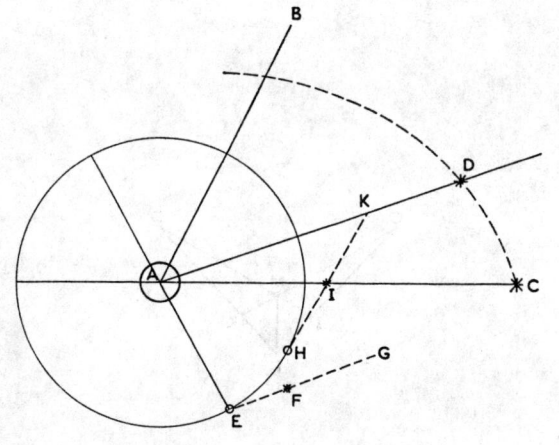

Fig. 10 A tentative reconstruction of a diagram heavily erased and altered.

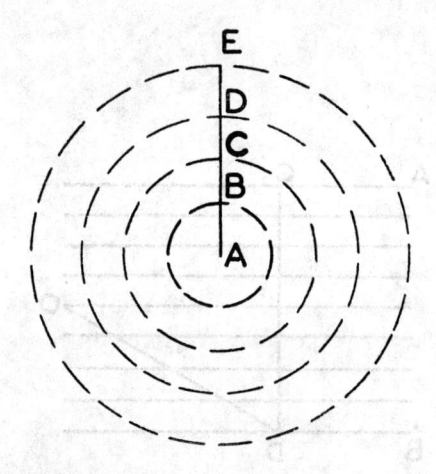

Fig. 11

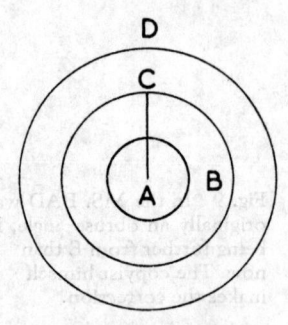

Fig. 12 A circle of radius AE, mentioned in the text, does not appear in the MS.'s diagram.

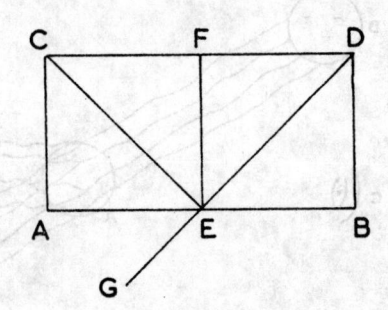

Fig. 13 In the original the angle CEG is not a right angle, as required by the text.

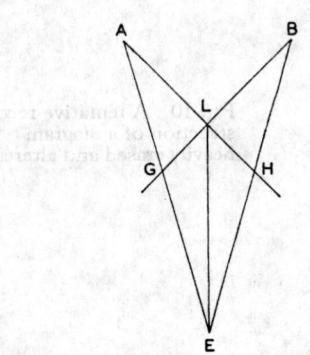

Fig. 14

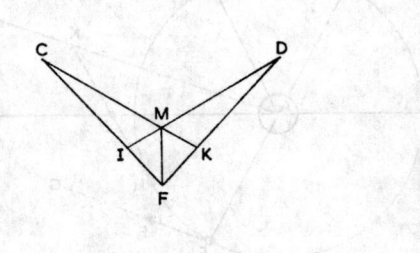

Fig. 15 Modified to fit the text: DC is made perpendicular to the rays A, B, etc., and equal to DO. The figure is identical with that in Galileo (Drake, pp. 80, 338). Both he and Hobbes envisage a plane figure.

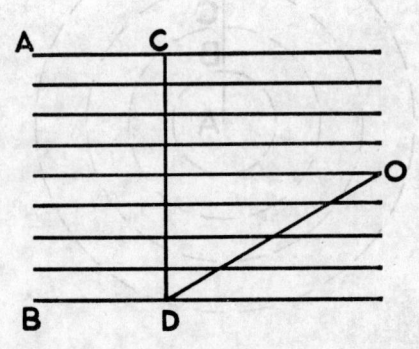

Fig. 16

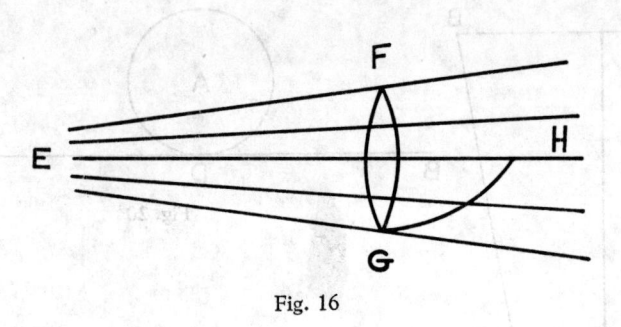

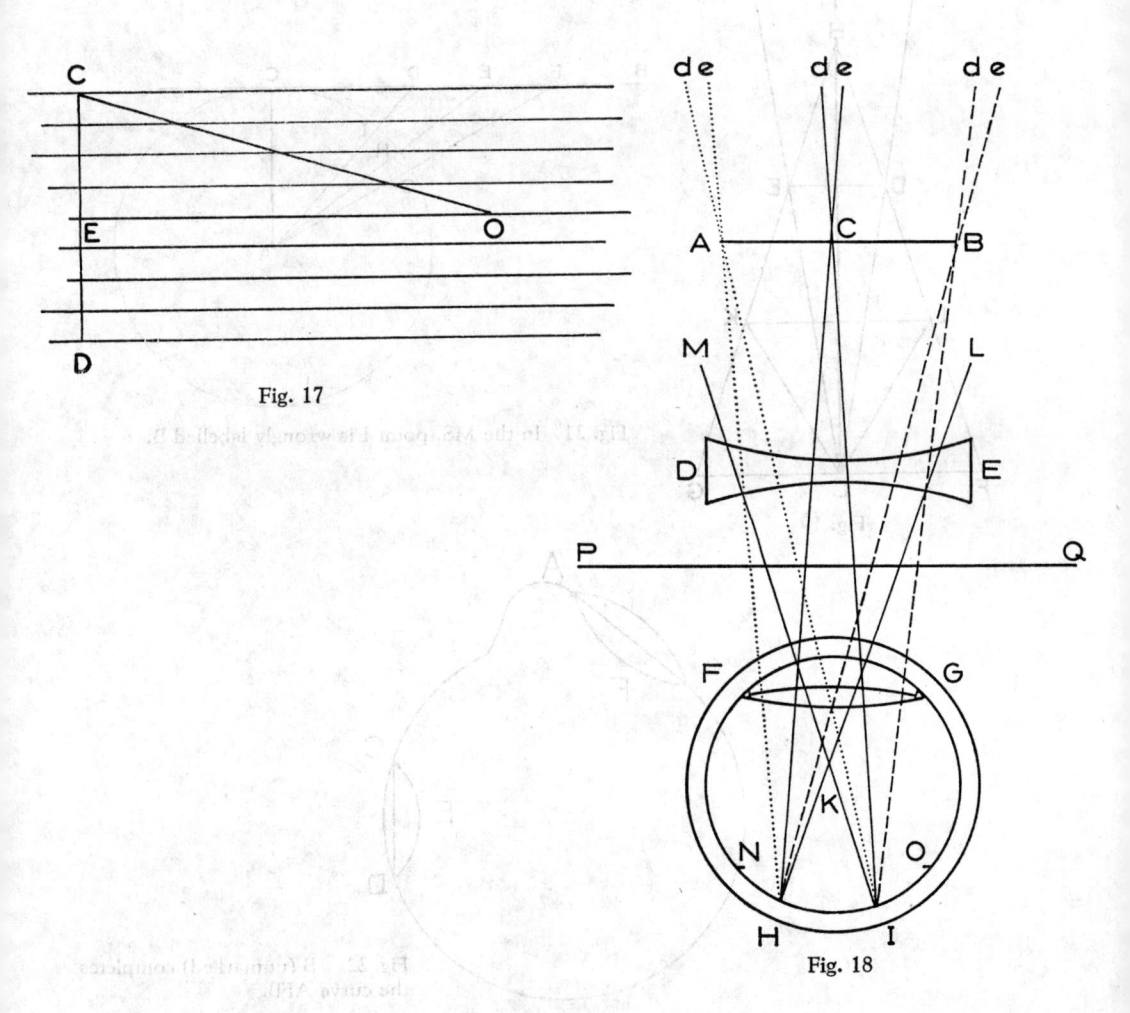

Fig. 17

Fig. 18

505

Fig. 20

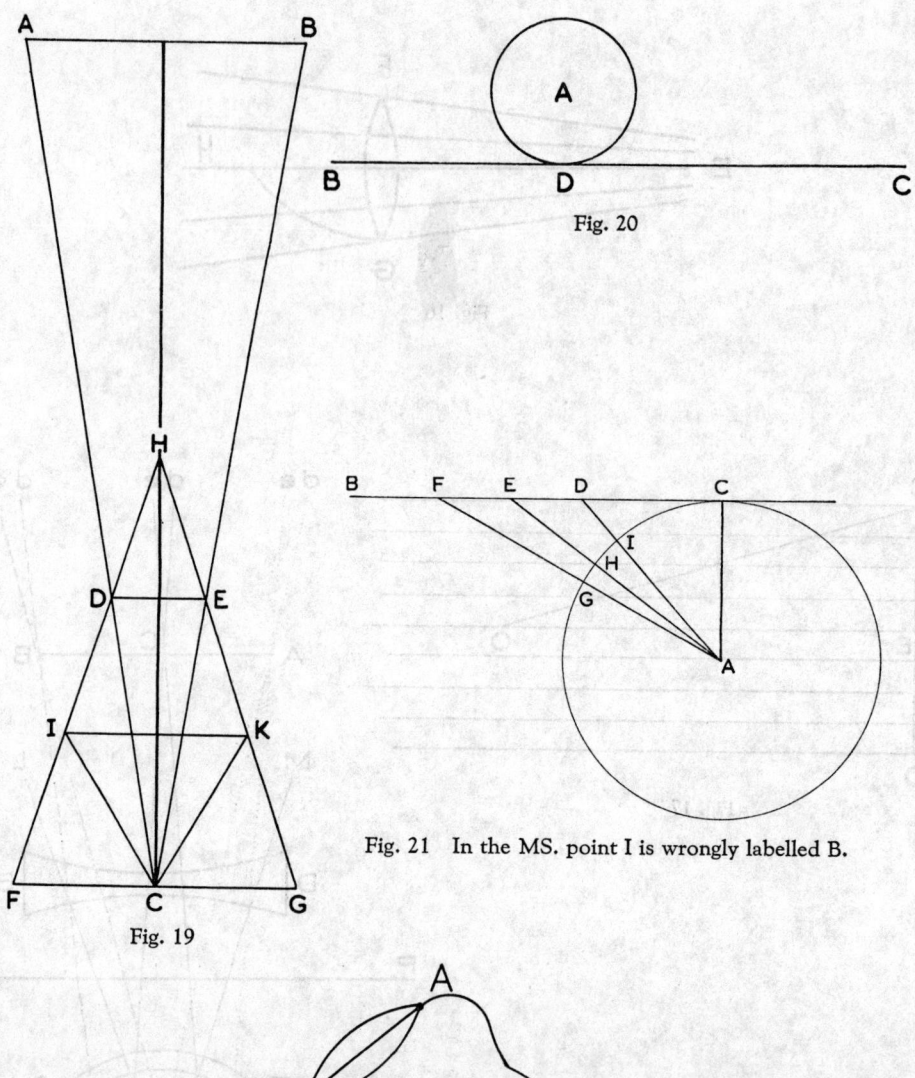

Fig. 19

Fig. 21 In the MS. point I is wrongly labelled B.

Fig. 22 B (unmarked) completes the curve AFB.

506

Fig. 23

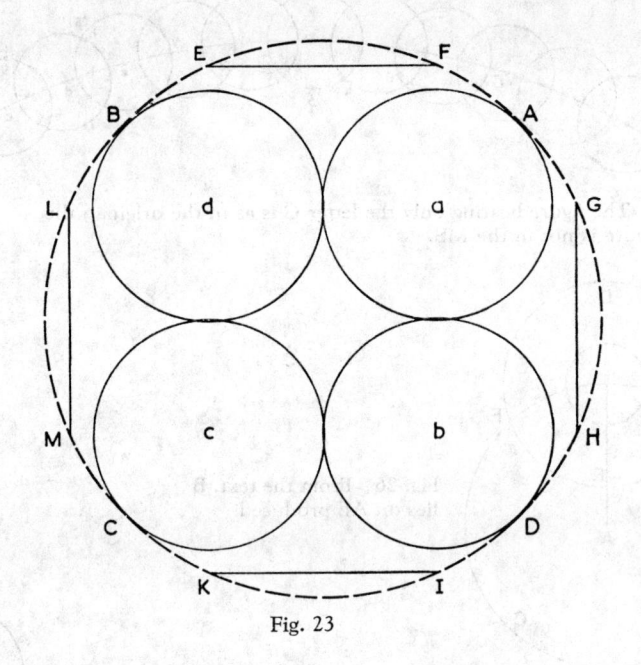

Fig. 24

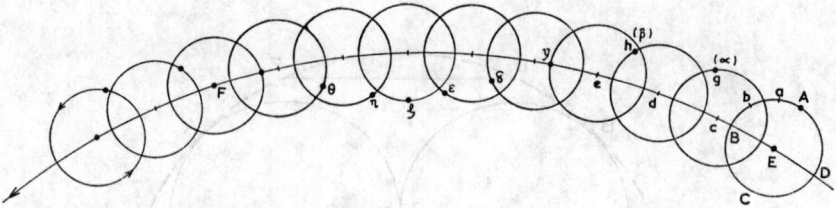

Fig. 25 The figure bearing only the letter C is as in the original; the other figure is not in the MS.

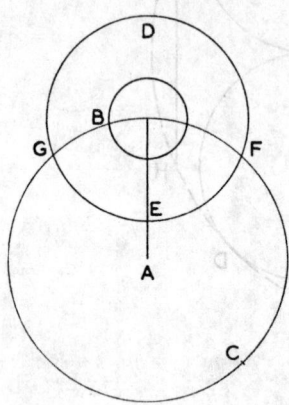

Fig. 26 From the text, B lies on AE produced.

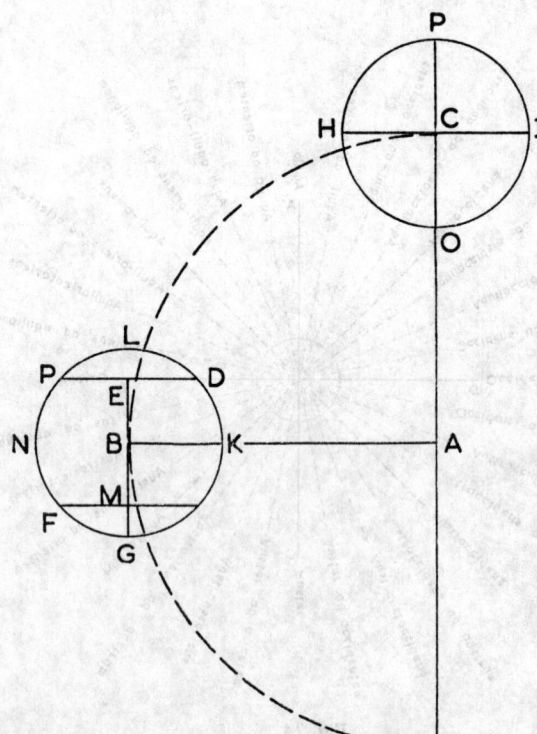

Fig. 27 The text requires L to lie on GE produced.

Fig. 28

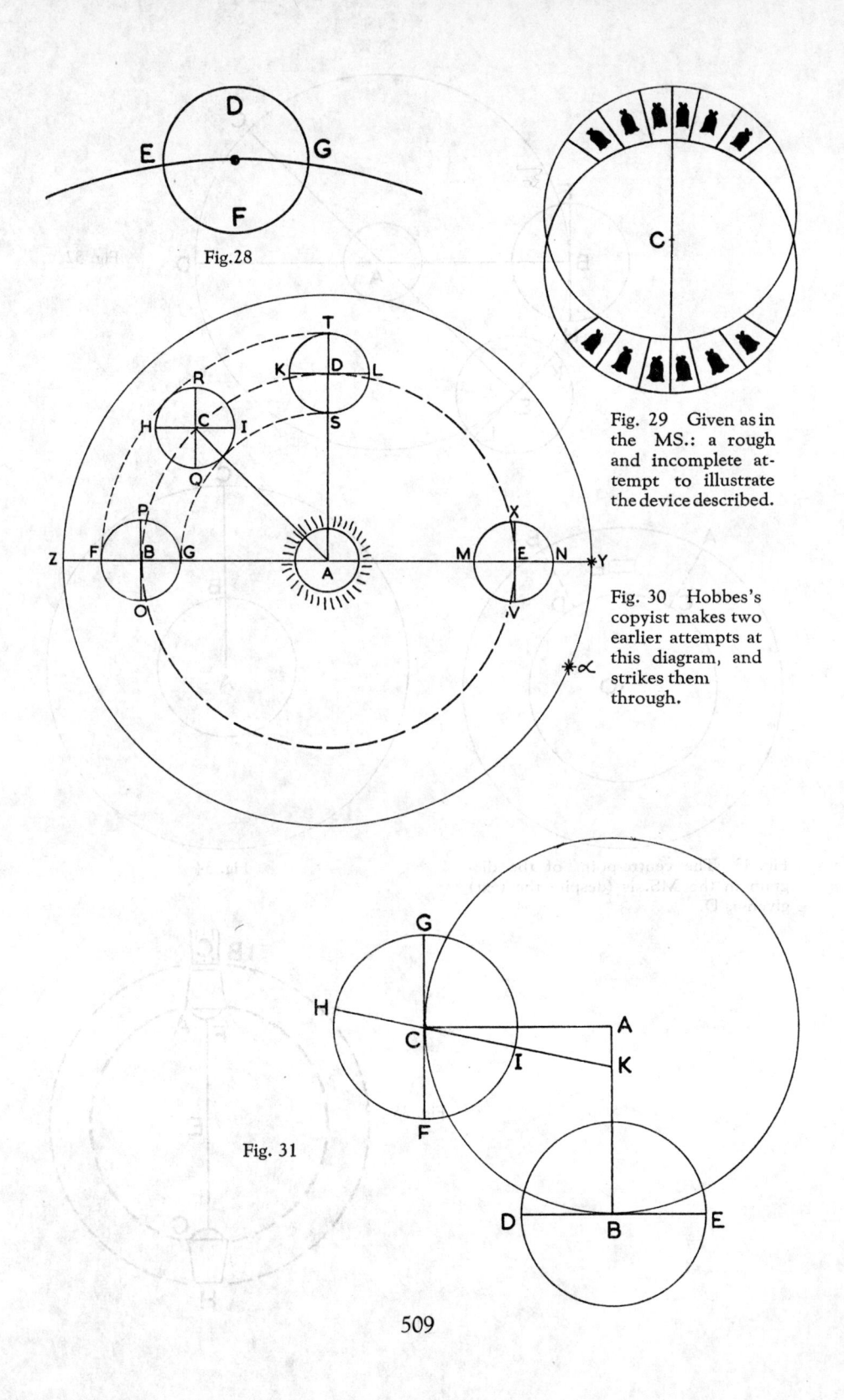

Fig. 29 Given as in the MS.: a rough and incomplete attempt to illustrate the device described.

Fig. 30 Hobbes's copyist makes two earlier attempts at this diagram, and strikes them through.

Fig. 31

509

Fig. 32

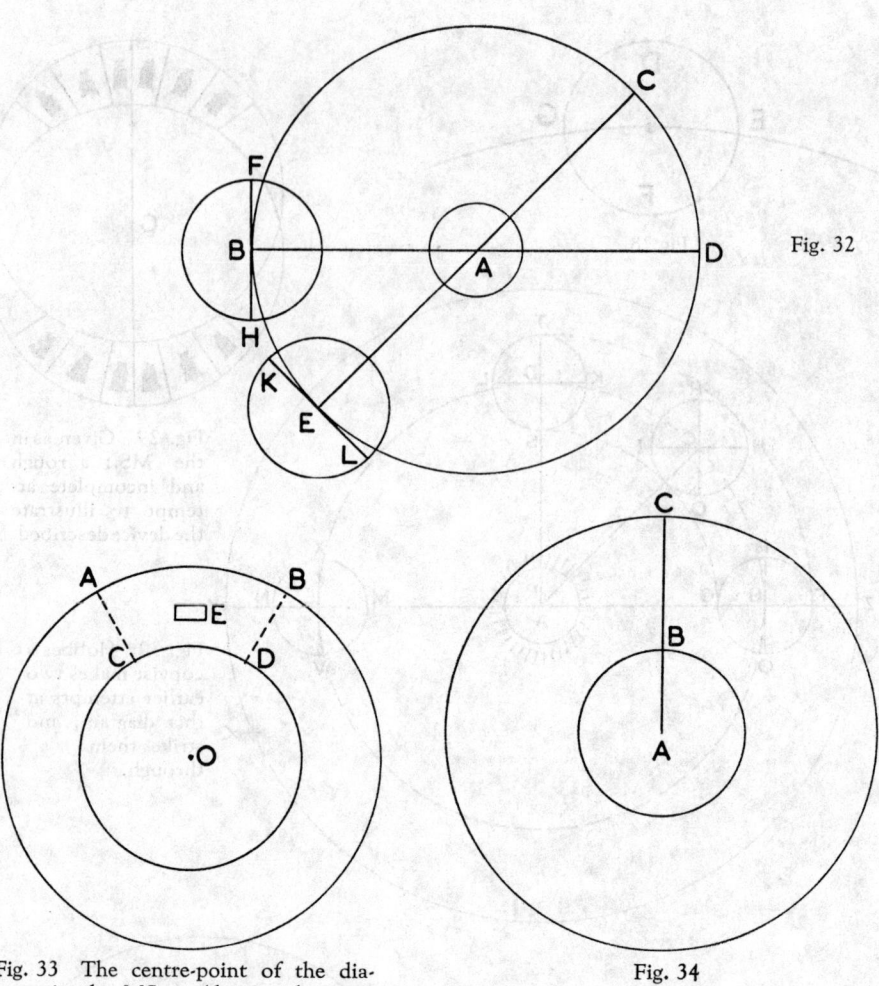

Fig. 33 The centre-point of the diagram in the MS. is (despite the text) given as D.

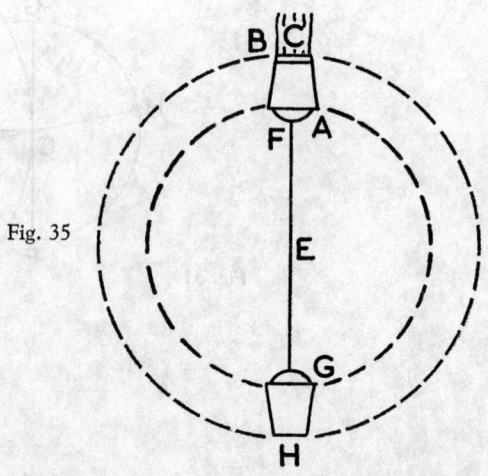

Fig. 34

Fig. 35

510

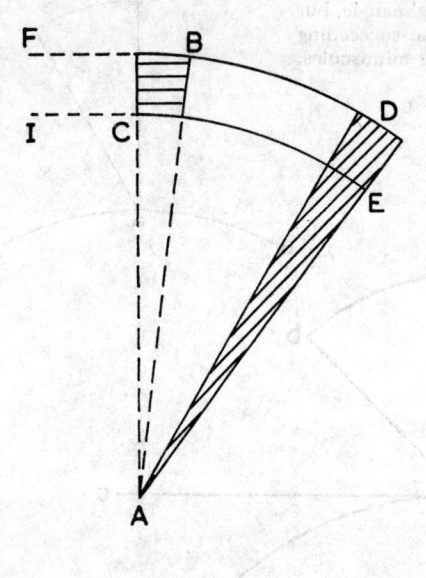

Fig. 36 In the original the tangents to the arcs are omitted.

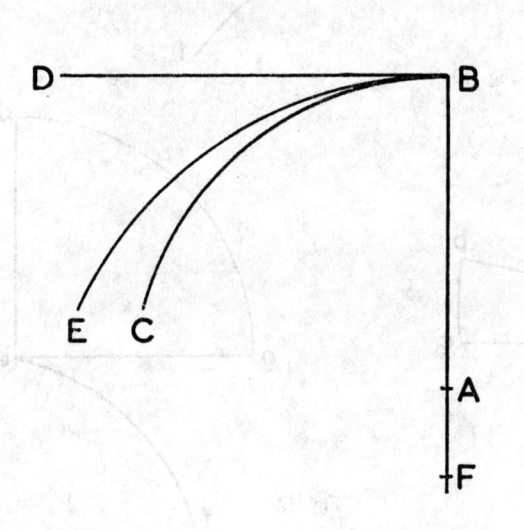

Fig. 37

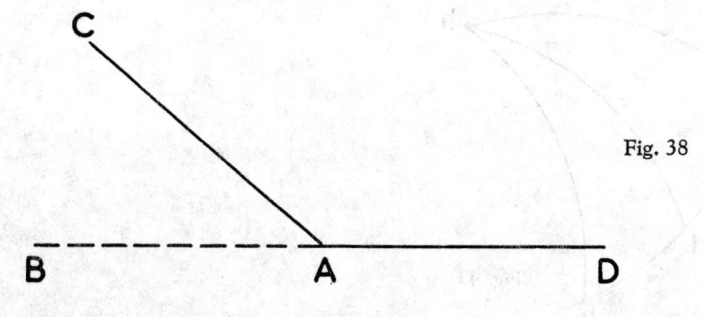

Fig. 38

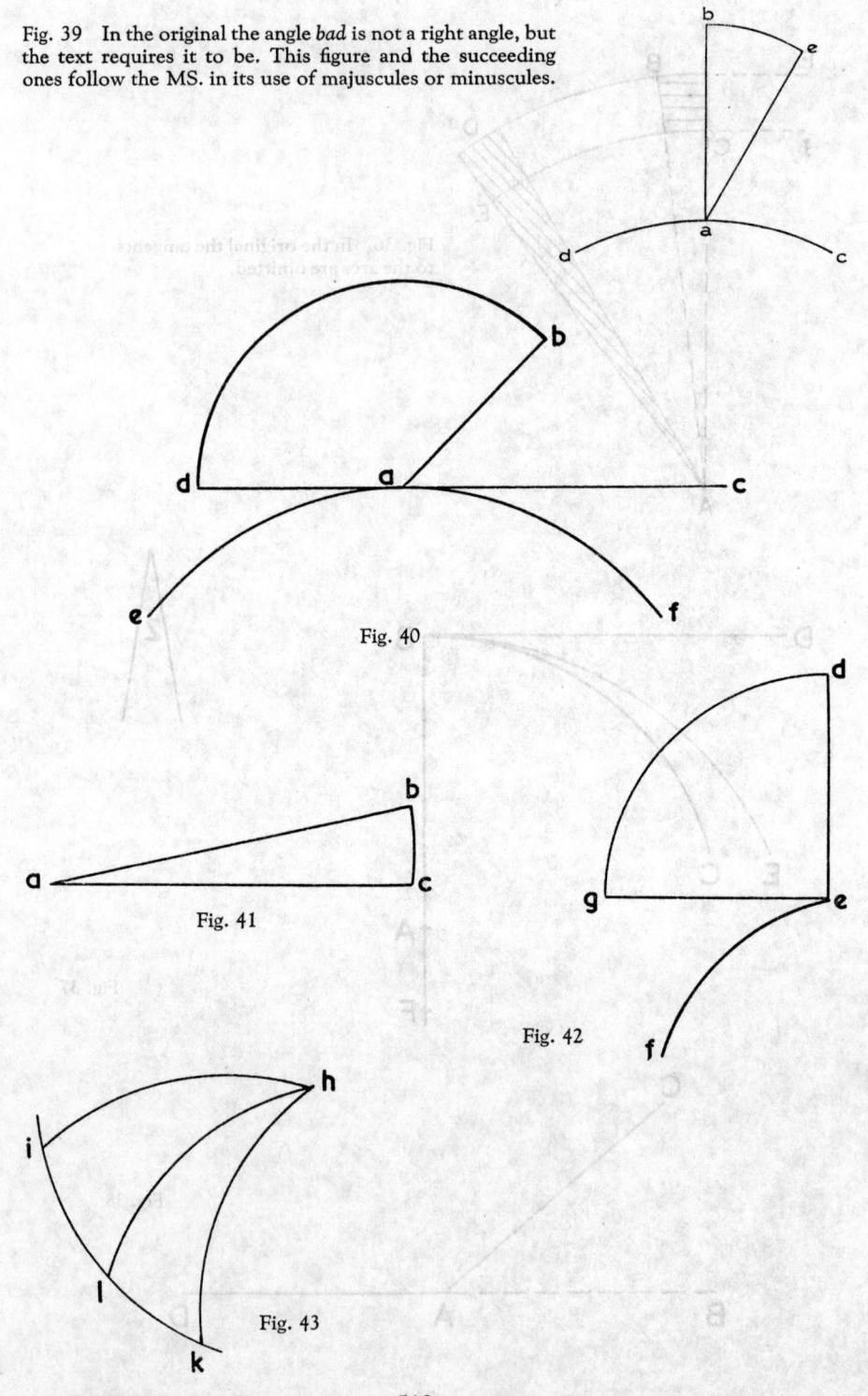

Fig. 39 In the original the angle *bad* is not a right angle, but the text requires it to be. This figure and the succeeding ones follow the MS. in its use of majuscules or minuscules.

Fig. 40

Fig. 41

Fig. 42

Fig. 43

512

Fig. 44 In the MS. and in Clavius.

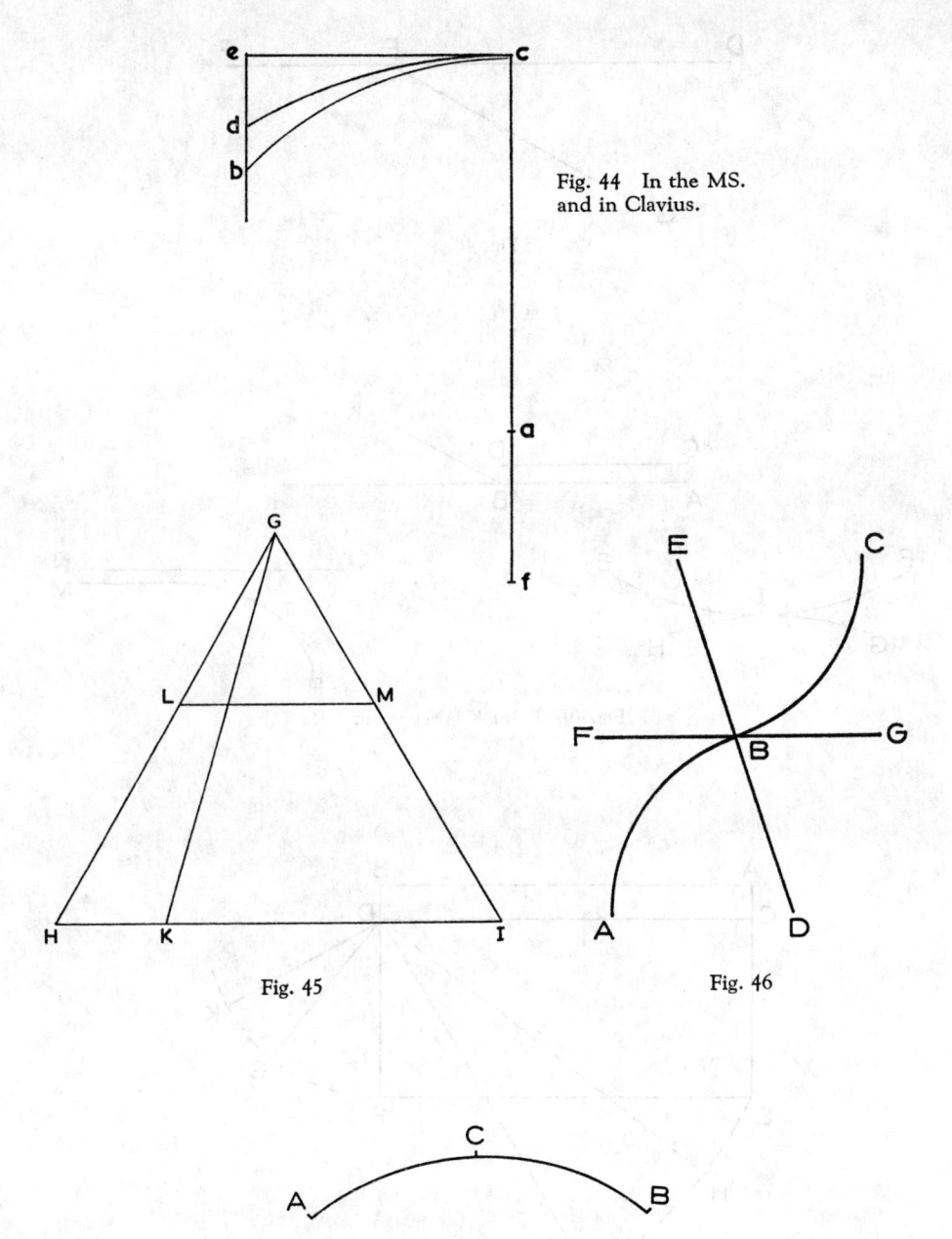

Fig. 45

Fig. 46

Fig. 47

513

Fig. 48

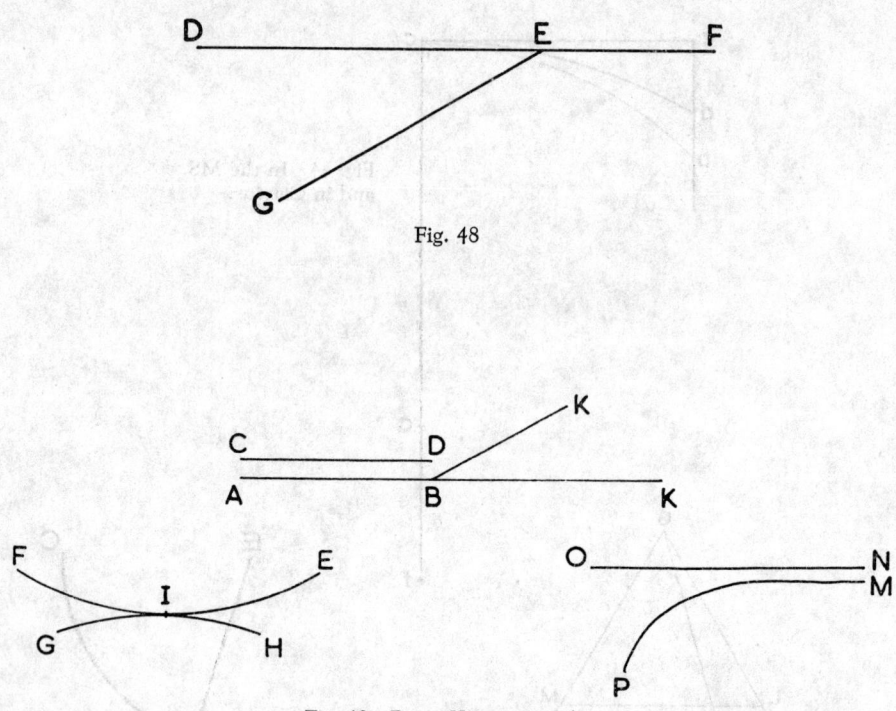

Fig. 49 Point K is repeated.

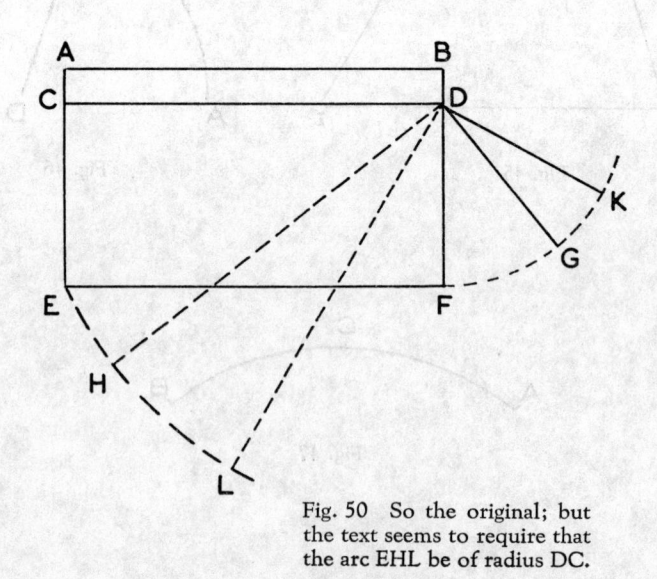

Fig. 50 So the original; but the text seems to require that the arc EHL be of radius DC.

514

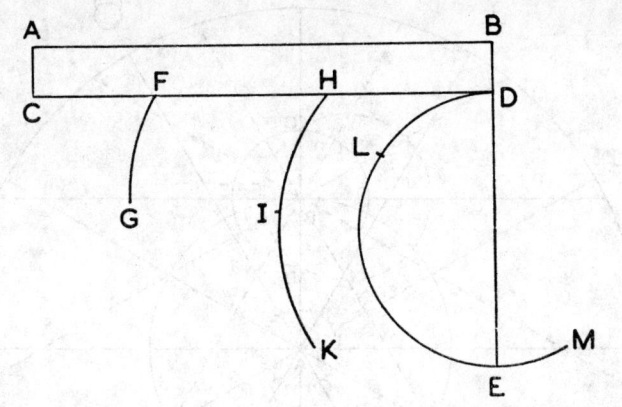

Fig. 51

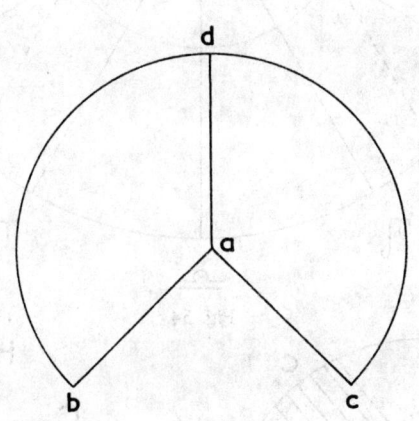

Fig. 52

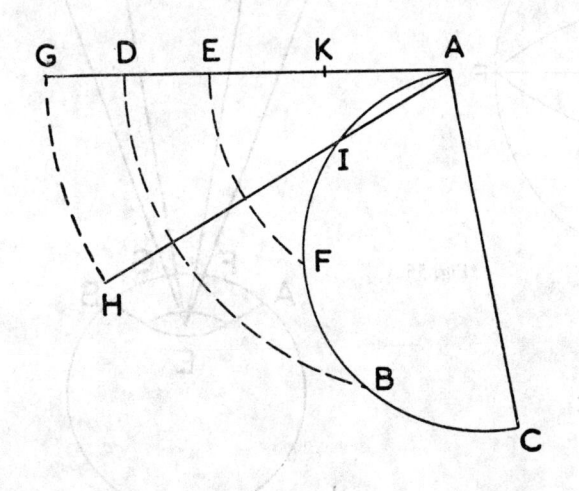

Fig. 53 So the original; but in the light of the text GH should continue as far as C, and DAC should be a right angle.

515

Fig. 54

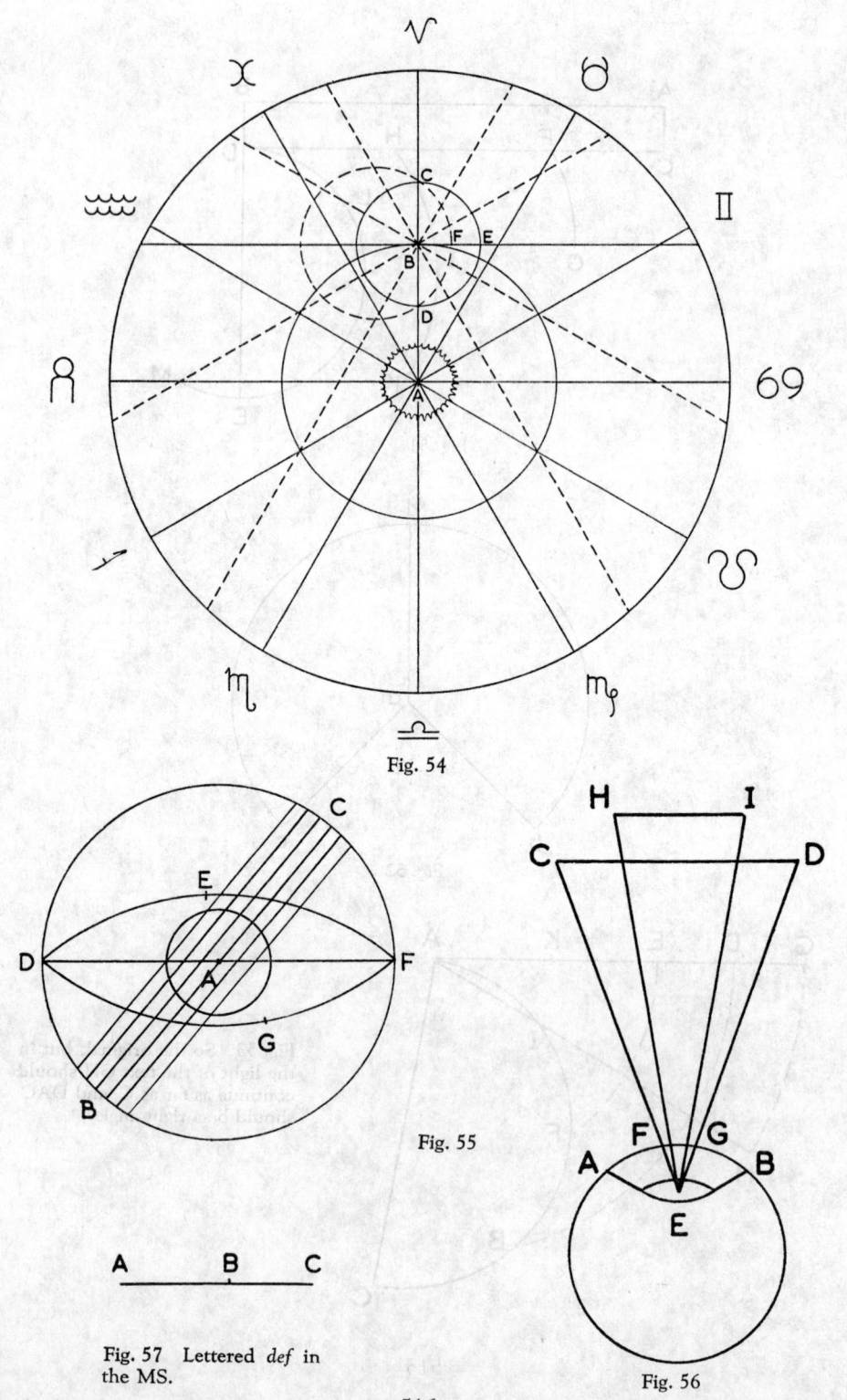

Fig. 55

Fig. 56

Fig. 57 Lettered *def* in the MS.

APPENDIX II:

Translation of an extract from Kepler
(see folio 266, note)

Epitome astronomiae Copernicanae, usitatâ forma quaestionum & responsionum...Authore Ioanne Kepplero...1618, p. 562.

Is the disparity in the effect, i.e. that the greatest acceleration in the moon's motion takes place in the syzygies DH [in the figure below], and no acceleration in the quadratures FK, due to this reason [i.e. that the sun is a contributory factor in the earth's motion, and even in the lunar movement about earth]?

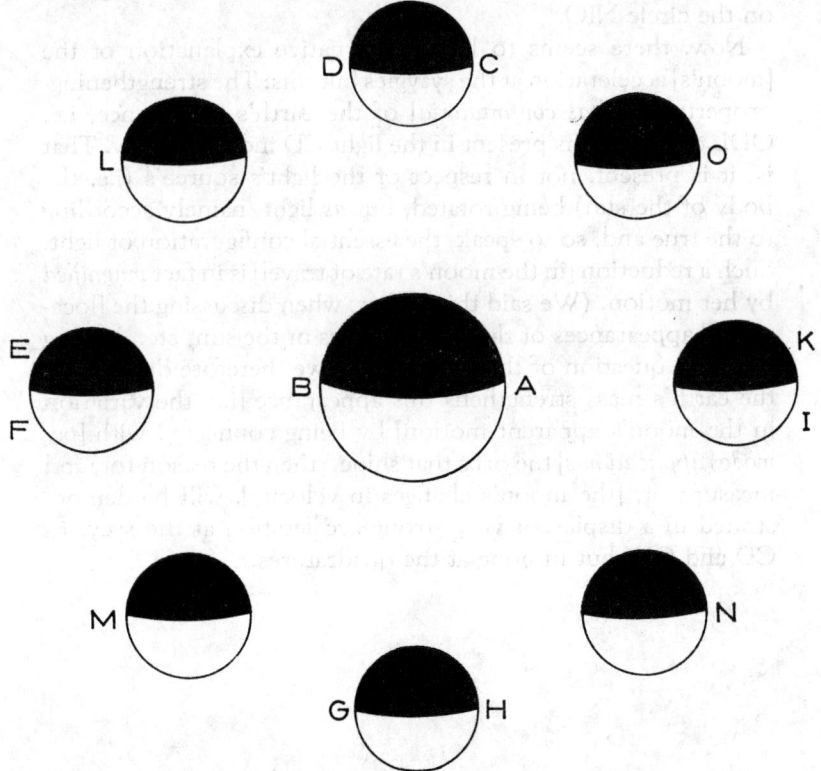

No part of the physics of the firmament has been harder to explain than this. In order to deal with it as effectively as possible we shall have to use the figure on [my] page 560.

APPENDIX II

You will recall that all the circles that represent the limits of the illumination from the moon's sphere, e.g. CD and GH and all the others, are equally parts of the spherical surfaces over which the illumination, issuing from the sun as its centre, is dispersed. The circle DFHK represents the appearance of AB, the earth's mass (for the earth moves the moon), the earth being located in the centre of the circle. You see that at the syzygies, D [and] H, there are drawn together [*applicari*], and into contact, an image of light, CD, and an image of the earth's mass, OCDL. These cut one another in LMNO at oblique angles, so that their connection [*applicatio*] is somewhat imperfect; but at EF and IK, the quadratures, the section is at right angles. Here, clearly, there is no juncture [*applicatio*, between CD and OCDL], because the section of the moon tends towards the centre of the earth and is represented by a mere point on the circle NIO.

Now, there seems to be no alternative explanation of the [moon's] acceleration at the syzygies but this: The strengthening-property [*facultas confortatoria*] of the earth's appearance, i.e. ODL the mover, is present in the light CD independently. That is, it is present, not in respect of the light's source's (i.e. the body of the sun) being rotated, but *as* light, namely according to the true and, so to speak, the essential configuration of light. Such a reduction [in the moon's rate of travel] is in fact *magnified* by her motion. (We said this earlier, when discussing the [locational] appearances of the actual bodies of the sun, etc., leaving aside the question of their shining.) If we therefore decide that the earth's mass strengthens this appearance [i.e. the variation in the moon's apparent motion] by being connected with [*per modos applicationis*] the orbs that shine,[1] then the reason for, and measure of, [the moon's changes in velocity], will be demonstrated in a display of very strong acceleration at the syzygies CD and GH, but in none at the quadratures.

---

[1] By 'the orbs that shine' Kepler may mean 'the heavenly bodies that emit light (by reflection)', i.e. 'circles of light'.